TECHNOLOGY AND VALUES

TECHNOLOGY AND VALUES

Edited by
KRISTIN SHRADER-FRECHETTE AND
LAURA WESTRA

ROWMAN & LITTLEFIELD PUBLISHERS, INC.
Lanham • Boulder • New York • Oxford

ROWMAN & LITTLEFIELD PUBLISHERS, INC.

Published in the United States of America
by Rowman & Littlefield Publishers, Inc.
4720 Boston Way, Lanham, Maryland 20706

12 Hid's Copse Road
Cummor Hill, Oxford OX2 9JJ, England

British Library Cataloguing in Publication Information Available

Library of Congress Cataloging-in-Publication Data

Technology and values / edited by Kristin Shrader-Frechette, Laura
    Westra.
        p.   cm.
    Includes bibliographical references and index.
    ISBN 0-8476-8630-2 (cloth : alk. paper). — ISBN 0-8476-8631-0
(pbk. : alk. paper)
    1. Technology—Philosophy. 2. Technology—Moral and ethical
aspects. I. Shrader-Frechette, K. S. (Kristin Sharon) II. Westra,
Laura.
    T14.T389   1997
    601—dc21   97-13505
                                                            CIP

ISBN 0-8476-8630-2 (cloth : alk. paper)
ISBN 0-8476-8631-0 (pbk. : alk. paper)

Printed in the United States of America

♾ ™ The paper used in this publication meets the minimum requirements of
American National Standard for Information Sciences—Permanence of Paper
for Printed Library Materials, ANSI Z39.48–1984.

For Marie Rush, and Mildred House Shrader

For Sari Kaluzny, William Vajda, and Sari Halász in memoriam

Contents

Acknowledgments

"What Is Moral Philosophy?" by Louis Pojman is reprinted here by permission of the author and publisher. Copyright © 1995 by Louis Pojman.

"Technology and Ethical Issues" by Kristin Shrader-Frechette appeared as "Technology" in *Encyclopedia of Ethics*, by Lawrence C. Becker and Charlotte B. Becker (New York: Garland Publishing, 1992), pp. 1231–1234. Reprinted by permission of Garland Publishing. Copyright © 1992 by Garland Publishing.

"Heidegger on Gaining a Free Relation to Technology" by Hubert Dreyfus appeared in *Technology and the Politics of Knowledge*, edited by Andrew Feenberg and Alastair Hannay (Bloomington: Indiana University Press, 1995), pp. 97–107. Reprinted by permission of Indiana University Press. Copyright © 1995 by Indiana University Press.

"Technologies as Forms of Life" by Langdon Winner is excerpted from chapter 1 of his book, *The Whale and the Reactor* (Chicago: University of Chicago Press, 1986). Reprinted by permission of University of Chicago Press. Copyright © 1986 by University of Chicago Press.

"The Role of Technology in Society" by Emmanuel Mesthene originally appeared in *Technology and Culture* 10 (1969), pp. 489–536. The essay also appeared as chapter 6 in *Technology and the Future*, edited by A. Teich (New York: St. Martin's Press, 1997). Reprinted by permission of the author and the University of Chicago Press. Copyright © 1997 by University of Chicago Press.

"Technology: The Opiate of Intellectuals" by John McDermott originally appeared in *The New York Review of Books*, July 31, 1969. The essay also appeared as chapter 7 in *Technology and the Future*, edited by A. Teich (New York: St. Martin's Press, 1997). Reprinted by permission of the author. Copyright © 1969 by John McDermott.

"The Social Construction of Safety" by Rachelle Hollander appeared in *The International Journal of Applied Ethics* 8 (1994), pp. 15–18.

"The Political Construction of Technology" by Jesse Tatum appeared in *Social and Philosophical Constructions of Technology* 15, edited by Carl Mitcham, of annual *Research in Philosophy and Technology* (Greenwich, Conn.: JAI Press, 1995). Reprinted by permission of JAI Press Inc. Copyright © 1995 by JAI Press Inc.

"Frankenstein's Problem: Autonomous Technology" by Langdon Winner is excerpted from chapter 8 of his book, *Autonomous Technology* (Cambridge: MIT Press, 1977). Reprinted by permission of MIT Press. Copyright © 1977 by the Massachusetts Institute of Technology.

"Technology, Demography, and the Anachronism of Traditional Rights" by Robert McGinn appeared in *The Journal of Applied Philosophy* 11 (1994). Reprinted by permission of the Society for Applied Philosophy and Blackwell Publishers. Copyright © 1994 by the Society for Applied Philosophy, Blackwell Publishers.

"Economic Evaluations of Technology" by Kristin Shrader-Frechette appeared as "Technology Assessment as Applied Philosophy of Science" in *Science, Technology, and Human Values* 6 (1980), pp. 33–50. Reprinted by permission of the Massachusetts Institute of Technology and the President and Fellows of Harvard College. Copyright © 1980 by the Massachusetts Institute of Technology.

"Sociological versus Metascientific Views of Technological Risk Assessment" by Deborah Mayo appeared as "Sociological versus Metascientific Views of Risk Assessment" in her book, coauthored with Rachelle Hollander, *Acceptable Evidence* (New York: Oxford University Press, 1991), pp. 249–268, 276–279. Reprinted by permission of Oxford University Press. Copyright © 1991 by Oxford University Press.

"Engineering Design Research and Social Responsibility" by Carl Mitcham appeared as "Engineering Design and Social Responsibility" in Kristin Shrader-Frechette's book, *Ethics of Scientific Research* (Lanham, Md.: Rowman and Littlefield, 1994), pp. 153–168. Reprinted by permission of Rowman and Littlefield. Copyright © 1994 by Rowman and Littlefield.

"Ethics and Automobile Technology: The Pinto Case" by Richard De George appeared as "Ethical Responsibilities of Engineering in Large Organizations: The Pinto Case" in *Business and Professional Ethics Journal* 1 (1981), pp. 1–14. Reprinted by permission of the Human Dimensions Center, Rensselaer Polytechnic Institute. Copyright © 1981 by Rensselaer Polytechnic Institute Human Dimensions Center.

"Computers and Privacy" by Stacey Edgar appeared as "Privacy," chapter 7 of her book, *Morality and Machine: Perspectives in Computer Ethics* (Boston: Jones and Bartlett Publishers, 1997). Reprinted by permission of Jones and Bartlett. Copyright © 1997 by Jones and Bartlett Publishers, Inc.

Part One

Introduction

1.1

Overview: Ethical Studies about Technology

Kristin Shrader-Frechette and Laura Westra

Why is it that humans can put people in space but cannot clean up the slums of Harlem or Calcutta, that they can discover the secrets of subatomic particles but cannot stop the loss of 17,000 plant and animal species each year?[1] One reason is that people seem more competent at developing science or technology than at correcting the social and environmental problems associated with it. Humans' intellectual progress often outstrips their moral and ethical development.

Technology Takes a Toll

Perhaps nowhere is the gap between intellectual and moral development more evident than in the way society uses various technologies. Society employs nuclear technology to build bombs and generate electricity, chemical technology to manufacture products and arrest cancer, and biotechnology to help infertile couples and to improve the genetics of food crops. Despite the many benefits of these technologies, they all have associated ethical, social, public health, and environmental consequences. Often society focuses on the economic benefits associated with technology but pays less attention to its ethical costs.

Of all the ethical costs associated with technology, perhaps the most serious are the threats to life. In part because of technologically induced threats (such as asbestos, benzene, chlorine, and radiation), the U.S. Office of Technology Assessment (OTA) claimed that up to 90 percent of all cancer is "environmentally induced and theoretically preventable."[2] If the OTA numbers are correct, then apparently people

3

accept the benefits associated with various cancer-causing technologies, but they fail to assess the ethical, social, public health, and environmental consequences arising because of these technologies. In fact, more North Americans die annually from environmentally induced cancer than from what society calls "murder." If current trends are not reversed, then shortly after the year 2000, cancer will be the leading cause of mortality in the United States and in other developed nations.[3]

Each year roughly a million new cases of cancer appear in the United States, and approximately 500,000 Americans die of cancer. Breast cancer has been increasing by about 1 percent per year since 1973; colorectal cancer has increased 19 percent since 1950, prostate cancer has increased 69 percent since 1950; Hodgkin's disease has increased 24 percent during this same time period; non-Hodgkin's lymphoma by 123 percent; cancer of the larynx by 70 percent; stomach cancer by 42 percent; bladder cancer by more than 50 percent; kidney and renal pelvis cancer by 82 percent; malignant melanoma of the skin by more than 200 percent. For all cancer sites combined, there has been a 36 percent increase since 1950. Most disturbing, the incidence of cancers among children under age 15 has increased by 32 percent since 1950. Although medical progress has been significant, cancer incidence is increasing roughly six times faster than cancer mortality is decreasing. Moreover, cancer is no longer mainly a disease of old people. The average cancer victim dies fifteen years earlier than other people.[4]

People in developed nations, however, do not face the most serious environmental and technological threats. Countries stressed by population and poor economies are typically in worse shape. The World Health Organization estimates, for instance, that 40,000 people are killed each year by pesticides (chemical technologies) used in developing nations; developed nations export most of these products. In the United States, for example, nearly one-third of the pesticides produced are banned for local use.[5] Such export raises the ethical question whether standards for protection against technology ought to be the same in developing, as in developed, countries.[6] Is it better to use technology to produce a tainted loaf of bread than to have no bread at all?

Technology and the Primacy of Risk Issues

As these data on technology-induced cancers and deaths indicate, perhaps the most important set of ethical issues raised by technology concerns the risks it poses to life, welfare, and values. Because of technology, we humans are now killing ourselves in massive numbers. For centuries we have feared the military, industrial, and agricultural technologies of our enemies, but in this century, our own military, industrial, and agricultural technologies have been killing our own people. For the past fifty years the United States, for example, has developed technologies associated with nuclear weapons. The justification for this military technological development is the protection of our own citizens. Yet despite all the fears that the enemies of the United States might use atomic bombs, only U.S. weapons have ever hurt U.S. citizens. More than 500,000 U.S. soldiers were victims of fallout from weapons testing in the western United States and the Pacific, and many of them have died from radiation-induced cancer. "Downwinders," living in Nevada and Utah, have endured birth defects, mental retardation, cancer, and even death because of the fallout from U.S. weapons tested by the United States. Oddly, the only U.S. casualties of nuclear military technology have been the victims of the United States, not its enemies. As Pogo says, we have met the enemy and he is us.[7]

Because our chemical, nuclear, biological, industrial, and agricultural technologies now cause more premature cancers and deaths than any other source, including crime and natural causes, we humans are literally killing ourselves. We are increasing the probability that we shall die at our own hands. We say that we have rights to life, but our technologies pollute the water that is necessary for life. We say that we have rights to liberty, but our technologies take away the liberty to breathe clean air. Risk—the probability of harm—created by technology is perhaps the crucial ethical locus for questions about technology because without our lives and our health, it is not possible to raise any other questions—metaphysical, political, or aesthetic—about technology. Risk is also the focus of questions about technology and ethics because, of all rights, those to life and to bodily security are the most absolute and the most inalienable. Hence whatever threatens our lives and bodily security—as technological risks do—threatens our most basic rights and our most cherished ethical notions. Risk issues are thus important because resolving them is necessary for resolving all other questions about technology and because risk threatens our most cherished ethical values and rights. Risk issues also provide a barometer that reveals how drastically we humans have created the instruments of our destruction. The primacy of these technological instruments suggests that we may have lost our bearings, our sanity, and our values in the quest for greater technogical efficiency. Indeed we may have lost ourselves.

Technology puts at risk not only present humans but also life itself and the gene pool. A perennial philosophical question is whether humans have duties to future generations, to protect the gene pool against the damaging effects of chemical- and nuclear-related technologies. British and Russian scientists recently reported, for example, that radiation (from the Chernobyl nuclear reactor accident in 1986) has altered the genetic legacy that radiation-exposed Russian parents have passed on to their children. Not only are nine million people still living in heavily contaminated areas because of this accident, but future people will be living with the consequences of the accident as well. Scientists have found twice the number of mutations in Russian children born eight years after the Chernobyl accident as in English children of the same age.[8]

Still other ethical questions raised by technological development concern risks not merely to present and future humans but to other inhabitants of the planet. How should one balance the social and economic benefits from technology with its environmental costs? E. O. Wilson estimates that the current rate of plant and animal extinction is about a thousand times greater than before human intervention in the biosphere. To prevent such massive destruction, Wilson claims that stewardship of the planet's species will require at least an $8 billion investment by the U.S. government, apart from the efforts of other nations.[9]

Evaluating the environmental effects of technology, some moral philosophers argue that humans have duties not to allow their technologies to interfere with the basic structures and functions of the natural world; they say humans have duties to preserve the biotic integrity of the holistic natural world, and not merely to protect individual species.[10] Other moral philosophers argue that nature has no intrinsic value and that the justification for protecting nature against technological threats relies on human "feelings."[11] Still other moral philosophers believe that it is unrealistic to attempt to preserve the biotic integrity of the holistic natural world, both because society must use technology to meet some human needs and because holistic environmental views do not provide adequately for the welfare of individual human beings. These philosophers argue for prima facie principles such as "respect nature"; prima facie principles create a strong presumption in their favor, not an absolute presumption against violating them, and they place the burden of proof on those who wish to violate them.[12]

Organization of the Volume

Before one can evaluate the ethical, social, public health, and environmental consequences of various technologies—as well as the value questions they raise—it is important to understand both moral philosophy or ethics and technology. The purpose of this first part of the book is to introduce some of the background ethical concepts necessary for assessing questions of technology and values. The general plan of the volume is to move from theoretical to applied issues—from the ethical theory necessary to assess technology to more and more concrete applications of this ethical theory to technological cases such as computers, nuclear power plants, and pesticides.

After outlining the basic ethical theories necessary to evaluate technology and its applications, the second part of the book surveys different ways to look at technology. Some people view technology as a savior that prolongs our lives and lightens our workloads. Others look at technology as something that harms us more than it helps us. Some scholars view technology in metaphysical ways, as raising fundamental questions about the meaning and value of life. Still others view technology in primarily social and political ways. Part Two of the book examines several of the prominent alternative views of technology. These alternatives are important because they color the kinds of ethical questions we ask about technology. If we view technology as primarily beneficial, then our ethical questions are likely to focus on how to make technology better. If we view technology as primarily harmful, then our ethical questions are likely to focus on how to avoid these harms and prevent their recurrence. Fundamental to examining technology is the notion that we must also examine the alternative ways that we view technology, since these views may be as biased as the technologies we favor.

Part Three of the volume begins the task of examining not our ethics about technology or our views of technology, but technology itself. This part discusses the various ways that humans have evaluated technology, and it provides some insight into how our assessments may have failed us. It suggests how and why our evaluations of technology have led to some disasters, from the Challenger space shuttle explosion, to the Chernobyl nuclear core melt, to the Bhopal chemical accident. One essay in Part Three suggests that technology has escaped appropriate evaluation in part because it is dragging humanity behind it; it is autonomous. Other essays in this part show that we have evaluated technology by the norms of speed, economic efficiency, and money—always maximizing these three items but forgetting that they are means to meaningful lives rather than the ends of life.

After examining various evaluations of technology, the reader will be ready in Part Four to evaluate specific technologies—chemical, electronic, biological, nuclear, and so on. In this fourth and most practical part of the book, the reader should be able to apply the various tools of ethical theory (from Part One), of alternative technological viewpoints (from Part Two), and of alternative ways to evaluate technology (from

Part Three) to analyze the assets and liabilities of specific technologies. The reader, by this time, should be able to understand why philosophical values, viewpoints, and evaluations are as essential to sorting out what is good from what is bad about a particular technology as are its particular consequences and applications.

Methodology of the Volume

Because this volume emphasizes alternative ethical theories for understanding technology (Part One), alternative views of technology (Part Two), alternative means of evaluating technology (Part Three), and alternative technologies (Part Four), its method is analytical. The purpose of the volume is to provide the student with the opportunity to evaluate alternative stances on technology, rather than to present a particular ideology that the student must adopt. Most of the book, especially Parts Two and Three, attempts to provide different stances on questions about technology, so that students can analyze various points of view and decide for themselves which is right. Is Emmanuel Mesthene correct in believing that technology itself is self-correcting, or is John McDermott right to say that it is not self-correcting but merely a powerful tool of powerful people?

As John Stuart Mill once observed, the best way to discover the truth about any particular position is to hear all relevant opinions about it. Hoping to encourage analysis rather than indoctrination, criticism rather than ideology, this volume has no single theoretical framework for evaluating technology, such as that of Aristotle, Karl Marx, or Martin Heidegger. To provide any single such framework is to beg the crucial question of how best to evaluate technology. The editors believe that textbooks do better to avoid begging such questions and to offer the student a buffet line of quite different intellectual treats rather than a single course permeated with the same taste and smell, the same allegedly correct underlying theory. There are correct and incorrect theories, but the best way to learn to distinguish them from each other is not through propaganda or prescription but through sorting among their differences, one by one.

True to this fundamentally analytical rather than theoretical method, the volume emphasizes understanding each essay on its own terms, through its strengths and its weaknesses. This understanding develops by attempting to discover the assumptions of the author and the desirable or undesirable consequences to which a particular point of view or a particular essay might lead. Indeed, the questions at the end of each selection emphasize first understanding the selection and then evaluating it with analytical tools (assumptions, consequences) that are as free of question-begging prejudices as possible.

Part One

In chapter 1.2, "What Is Moral Philosophy?," Louis Pojman compares moral philosophy with other normative subjects such as religion and law, and he shows how moral philosophy differs from these other areas. He also explains that in ethical assessment of some subject or issue (such as a technological problem), philosophers are concerned with evaluating the act, its consequences, the character of the moral agent, and her motive. Pojman also discusses some of the purposes of morality or ethics.

In chapter 1.3, "Technology and Ethical Issues," Kristin Shrader-Frechette explains what technology is and why it raises new ethical questions. She also argues that those who do ethical assessments of technological issues must possess technical expertise, especially in economics, as well as ethical skills. Shrader-Frechette explains the five types of philosophical questions about technology and gives examples of them. Next she discusses the five main philosophical issues concerning technology and ethics. These are (1) how to define technological risk; (2) how to evaluate technologies in the face of uncertainty about their consequences and potential benefits and threats; (3) how to assess technological threats to due process rights, that is, rights to protection from harm by means of the procedures of law; (4) how to determine how safe is safe enough; and (5) whether, when, and how to achieve free informed consent to the risks and benefits brought by technology. Together, these introductory articles should help to provide some of the background information and philosophical tools necessary to assess the various values issues that technological issues raise.

Notes

1. E. O. Wilson, "The Current State of Biological Diversity," in E. O. Wilson and F. M. Peters (eds.), *Biodiversity* (Washington, D.C.: National Academy Press, 1988), 13; see also pp. 3–18.

2. For the 90 percent figure, see J. C. Lashof et al., Health and Life Sciences Division of the U.S. Office of Technology Assessment, *Assessment of Technologies for Determining Cancer Risks from the Environment* (Washington, D.C.: Office of Technology Assessment, 1981), 3, 6ff. Some scientists, however, disagree with the OTA; see J. R. Trotter, "Spontaneous Cancer," *Proceedings of the National*

Academy of Sciences 77, 4 (April 1980): 1763–67. See also E. Efron, *The Apocalyptics* (New York: Simon and Schuster, 1984), especially section 8 of chapter 6.

3. National Research Council, *Carcinogens and Anticarcinogens in the Human Diet* (Washington, D.C.: National Academy Press, 1996), 335. See also Bureau of the Census, *Statistical Abstract of the United States* (Washington, D.C.: Department of Commerce, 1994), 99. See also Abraham Lilienfeld, Morton L. Levin, and Irving I. Kessler, *Cancer in the United States* (Cambridge: Harvard University Press, 1992).

4. National Institutes of Health, *1987 Annual Cancer Statistics Review* (Bethesda, Md.: National Cancer Institute, 1987), I.4–I.8.

5. J. T. Mathews et al., *World Resources 1986* (New York: Basic Books, 1986), 48–49. See also R. Repetto, *Paying the Price: Pesticides in Developing Countries*, Research Report No. 2 (Washington, D.C.: World Resources Institute, 1985), 3.

6. For discussion of this point, see K. S. Shrader-Frechette, *Risk and Rationality* (Berkeley: University of California Press, 1991), ch. 10.

7. See Carole Gallagher, *American Ground Zero: The Secret Nuclear War* (Cambridge: MIT Press, 1993).

8. "Radiation Damages Chernobyl Children," *Science News* 149, 17 (April 27, 1996).

9. Wilson and Peters, *Biodiversity*, 16.

10. See Laura Westra, *An Environmental Proposal for Ethics: The Principle of Integrity* (Lanham, Md.: Rowman & Littlefield, 1994).

11. Bryan Norton, "Why I Am Not a Nonanthropocentrist," *Environmental Ethics* 17, 4 (Winter 1995): 357–58.

12. Paul Taylor, *Respect for Nature* (Princeton: Princeton University Press, 1986), 118, 243.

1.2

What Is Moral Philosophy?

Louis P. Pojman

We are discussing no small matter, but how we ought to live.

(Socrates in Plato's *Republic*)

What is it to be a moral person? What is the nature of morality and why do we need it? What is the good and how shall I know it? Are moral principles absolute or simply relative to social groups or individual decision? Is it in my interest to be moral? Is it sometimes in my best interest to act immorally? What is the relationship between morality and religion? What is the relationship between morality and law? What is the relationship between morality and etiquette?

These are some of the questions we shall examine in this chapter. We want to understand the foundation and structure of morality. We want to know how we should live.

The terms "moral" and "ethics" come from Latin and Greek respectively (mores and ethos), deriving their meaning from the idea of custom. Although philosophers sometimes use these terms interchangeably, it is useful to have a clearer conceptual scheme. In this essay I shall use "morality" to refer to certain customs, precepts, and practices of people and cultures. This is sometimes referred to as "positive morality." I shall use "moral philosophy" to refer to philosophical or theoretical reflection on morality. Specific moral theories issuing from such philosophical reflection I shall call "ethical theories," in line with a common practice. "Ethics" I shall use to refer to the whole domain of morality and moral philosophy, since they have many features

11

in common. For example, they both have to do with values, virtues, practices, and principles, though in different ways. I shall refer to specific moral theories as "ethical theories," in line with a common practice.

Moral philosophy refers to the systematic endeavor to understand moral concepts and justify moral principles and theories. It undertakes to analyze such concepts as right, wrong, permissible, ought, good, and evil in their moral contexts. Moral philosophy seeks to establish principles of right behavior that may serve as action guides for individuals and groups. It investigates which values and virtues are paramount to the worthwhile life or society. It builds and scrutinizes arguments in ethical theories, and it seeks to discover valid principles (e.g., Never kill innocent human beings) and the relationship between those principles (e.g., Does saving a life in some situations constitute a valid reason for breaking a promise?).

Morality as Compared with Other Normative Subjects

Moral precepts are concerned with norms; roughly speaking they are concerned not with what is, but with what *ought* to be. How should I live my life? What is the right thing to do in this situation? Should one always tell the truth? Do I have a duty to report a student I have seen cheating in class? Should I tell my friend that his wife is having an affair? Is premarital sex morally permissible? Ought a woman ever to have an abortion? Morality has a distinct action guiding or *normative* aspect,[1] an aspect it shares with other practical institutions, such as religion, law, and etiquette.

Moral behavior, as defined by a given religion, is often held to be essential to the practice of that religion. But neither the practices nor the precepts of morality should be identified with religion. The practice of morality need not be motivated by religious considerations. And moral precepts need not be grounded in revelation or divine authority—as religious teachings invariably are. The most salient characteristic of ethics—by which I mean both philosophical morality (or morality, as I will simply refer to it) and moral philosophy—is that it is grounded in reason and human experience.

To use a spatial metaphor, secular ethics is horizontal, omitting a vertical or transcendental dimension. Religious ethics has a vertical dimension, being grounded in revelation or divine authority, though generally using reason to supplement or complement revelation. These two differing orientations will often generate different moral principles and standards of evaluation, but they need not. Some versions of religious ethics, which posit God's revelation of the moral law in nature or conscience, hold that reason can discover what is right or wrong even apart from divine revelation.

Morality is also closely related to law, and some people equate the two practices. Many laws are instituted in order to promote well-being and resolve conflicts of interest and/or social harmony, just as morality does, but ethics may judge that some laws are immoral without denying that they are valid laws. For example, laws may permit slavery or irrelevant discrimination against people on the basis of race or sex. A Catholic or antiabortion advocate may believe that the laws permitting abortion are immoral.

In a 1989 television series, *Ethics in America,* James Neal, a trial lawyer, was asked what he would do if he discovered that his client had committed a murder some years back for which another man had been convicted and would soon be executed. Neal said that he had a legal obligation to keep this information confidential and that if he divulged it, he would be disbarred. It is arguable that he has a moral obligation that overrides his legal obligation and that demands that he take action to protect the innocent man from being executed.

Furthermore, there are some aspects of morality that are not covered by law. For example, while it is generally agreed that lying is usually immoral, there is no general law against it (except under special conditions, such as in cases of perjury or falsifying income tax returns). Sometimes college newspapers publish advertisements for "research assistance," where it is known in advance that the companies will aid and abet plagiarism. The publishing of such research paper ads is legal, but it is doubtful whether it is morally correct. In 1963, thirty-nine people in Queens, New York, watched from their apartments for some forty-five minutes as a man beat up a woman, Kitty Genovese, and did nothing to intervene, not even call the police. These people broke no law, but they were very likely morally culpable for not calling the police or shouting at the assailant.

There is one other major difference between law and morality. In 1351 King Edward III of England promulgated a law against treason that made it a crime merely to think homicidal thoughts about the king. But, alas, the law could not be enforced, for no tribunal can search the heart and fathom the intentions of the mind. It is true that *intention,* such as malice aforethought, plays a role in the legal process in determining the legal character of the act, once the act has been committed. But preemptive punishment for people presumed to have bad intentions is illegal. If malicious intentions (called in law mens rea) were criminally illegal, would we not all deserve imprisonment? Even if it were possible to detect intentions, when should the punishment be administered? As soon as the subject has the intention? But how do we know that he will not change his mind? Furthermore, is there not a continuum between imagining some harm to X, wishing a harm to X, desiring a harm to X, and intending a harm to X?

While it is impractical to have laws against bad intentions, these intentions are still bad, still morally wrong. Suppose I buy a gun with the intention of killing Uncle Charlie in order to inherit his wealth, but never get a chance to fire it (for example, Uncle Charlie moves to Australia). While I have not committed a crime, I have committed a moral wrong.

Finally, law differs from morality in that there are physical and financial sanctions[2] (e.g., imprisonment and fines) enforcing the law but only the sanction of conscience and reputation enforcing morality.

Morality also differs from etiquette, which concerns form and style rather than the essence of social existence. Etiquette determines what is polite behavior rather than what is *right* behavior in a deeper sense. It represents society's decision as to how we are to dress, greet one another, eat, celebrate festivals, dispose of the dead, express gratitude and appreciation, and in general, carry out social transactions. Whether we greet each other with a handshake, a bow, a hug, or a kiss on the cheek will differ in different social systems, whether we uncover our heads in holy places (as males do in Christian churches) or cover them (as females [used to] do in Catholic churches and males do in synagogues), none of these rituals has any moral superiority.

People in Europe wear their wedding rings on the third finger of their right hand, whereas in the United States people wear them on the left hand. People in England hold their fork in their left hand when they eat, whereas people in other countries hold it in their right hand or in whichever hand a person feels like holding it, while people in India typically eat without a fork at all, using the forefingers of their right hand for conveying food from their plate to their mouth.

Polite manners grace our social existence but they are not what social existence is about. They help social transactions to flow smoothly, but are not the substance of those transactions.

At the same time, it can be immoral to disregard or flaunt etiquette. The decision whether to shake hands when greeting a person for the first time or putting one's hands together and forward as one bows, as people in India do, is a matter of cultural decision, but once the custom is adopted, the practice takes on the importance of a moral rule, subsumed under the wider principle of Show Respect to People. Similarly, there is no moral necessity of wearing clothes, but we have adopted the custom partly to keep us warm in colder climates and partly out of modesty. But there is nothing wrong with people who decide to live together naked in nudist colonies. But it may well be the case that people running nude outside nudist colonies, say in classrooms and stores and along the road, would constitute such offensive behavior as to count as morally insensitive. Recently there was a scandal on the beaches of southern India where American tourists wore bikinis, shocking the more modest Indians. There was nothing immoral in itself about wearing bikinis, but given the cultural context, the Americans, in willfully violating etiquette, were guilty of moral impropriety.

Although Americans pride themselves on tolerance, pluralism, and awareness of other cultures, custom and etiquette can be—even among people from similar backgrounds—a bone of contention. A friend of mine, John, tells of an experience early in his marriage. He and his wife, Gwen, were hosting their first Thanksgiving meal. He had been used to small celebrations with his immediate family, whereas his wife had been used to grand celebrations. He writes, "I had been asked to carve, something I had never done before, but I was willing. I put on an apron, entered the kitchen, and attacked the bird with as much artistry as I could muster. And what reward did I get? [My wife] burst into tears. In *her* family the turkey is brought to the *table*, laid before the [father], grace is said, and *then* he carves! 'So I fail patriarchy.' I hollered later. 'What do you expect?' "[3]

Law, etiquette, and religion are all important institutions, but each has limitations. The limitation of the law is that you can't have a law against every social malady, nor can you enforce every desirable rule. The limitation of etiquette is that is doesn't get to the heart of what is of vital importance for personal and social existence. Whether or not one eats with one's fingers pales in significance compared with the importance of being honest or trustworthy or just. Etiquette is a cultural invention, but morality claims to be a discovery.

The limitation of the religious injunction is that it rests on authority, and we are not always sure of or in agreement about the credentials of the authority, nor on how the authority would rule in ambiguous or new cases. Since religion is not founded on reason but on revelation, you cannot use reason to convince someone who does not share your religious views that your view is the right one. I hasten to add that when moral differences are caused by fundamental moral principles, it is unlikely that philosophical reasoning will settle the matter. Often, however, our moral differences turn out to be rooted in worldviews, not moral principles. For example, pro-life and pro-choice advocates often agree that it is wrong to kill innocent people, but differ on the facts. The pro-life advocate may hold a religious view that states that the fetus has an eternal soul and thus possesses a right to life, while the pro-choice advocate may deny that anyone has a soul and believes that only self-conscious, rational beings have a right to life.

The chart characterizes the relationship between ethics, religion, etiquette, and law.

Subject	Normative Disjuncts	Sanctions
Ethics	Right—Wrong—Permissible as defined by conscience or reason	Conscience—Praise and Blame—Reputation
Religion	Right—Wrong (Sin)—Permissible as defined by religious authority	Conscience—Eternal Reward and Punishment caused by a supernatural agent or force
Law	Legal and Illegal as defined by a judicial body	Punishments determined by the legislative body
Etiquette	Proper and Improper as defined by culture	Social Disapprobation and Approbation

In summary, morality distinguishes itself from law and etiquette by going deeper into the essence of rational existence. It distinguishes itself from religion in that it seeks reasons, rather than authority, to justify its principles. The central purpose of moral philosophy is to secure valid principles of conduct and values that can be instrumental in guiding human actions and producing good character. As such, it is the most important activity known to humans, for it has to do with how we are to live.

Domains of Ethical Assessment

It might seem at this point that ethics concerns itself entirely with rules of conduct based solely on an evaluation of acts. However, the situation is more complicated than this. There are four domains of ethical assessment:

Domain	Evaluative terms
1. Action, the act	Right, wrong, obligatory, permissible
2. Consequences	Good, bad, indifferent
3. Character	Virtuous, vicious, neutral
4. Motive	Good will, evil will, neutral

Let us examine each of these domains.

Types of Action

The most common distinction may be the classification of actions as right and wrong, but the term *right* is ambiguous. Sometimes it means "obligatory" (as in *"the right* act"), but sometimes it means permissible (as in "a right act"). Usually, philosophers define *right* as permissible, including under that category what is obligatory.

(1) A "right act" is an act that it is permissible for you to do. It may be either (a) optional or (b) obligatory.

(a) An *optional* act is an act that is neither obligatory nor wrong to do. It is not your duty to do it, nor is it your duty not to do it. Neither doing it nor not doing it would be wrong.

(b) An *obligatory* act is one that morality requires you to do; it is not permissible for you to refrain from doing it. It would be wrong not to do it.

(2) A "wrong act" is an act that you have an obligation, or a duty, to refrain from doing. It is an act you ought not to do. It is not permissible to do it.

Let me briefly illustrate these concepts. The act of lying is generally seen as wrong (prohibited), whereas telling the truth is generally seen as obligatory. But some acts do not seem to be either obligatory or wrong. Whether you take a course in art history or English literature, or whether you write your friend a letter with a pencil or a pen, seems morally neutral. Either is permissible. Whether you listen to pop music or classical music is not usually considered morally significant. Listening to both is allowed and neither is obligatory. A decision to marry or remain single is of great moral significance. (It is, after all, an important decision about how to live one's life.) The decision reached, however, is usually considered to be morally neutral or optional. Under most circumstances, to marry (or not to marry) is thought to be neither obligatory nor wrong, but permissible. Within the range of permissible acts is the notion of "supererogatory" or highly altruistic acts. These acts are not required or obligatory, but are acts that exceed what morality requires, going beyond the call of duty. You may have an obligation to give a donation to help people in dire need, but you are probably not obliged to sell your car, let alone become destitute, in order to help them.

Theories that place the emphasis on the nature of the act are called "deontological" (from the Greek word for duty). These theories hold that there is something inherently right or good about such acts as truth-telling and promise-keeping and something inherently wrong or bad about such acts as lying and promise-breaking. Illustrations of deontological ethics include the Ten Commandments in Exodus 20, natural law ethics, such as is found in the Roman Catholic church, and Immanuel Kant's theory of the Categorical Imperative.

Kant (1724–1804) argued that two kinds of commands or imperatives existed: *hypothetical* and *categorical*. Hypothetical imperatives are conditional, having the form "If you want X, do act A!" For example, if you want to pass this course, do your homework and study this book! Categorical imperatives, on the other hand, are not conditional, but universal and rationally necessary. Kant's primary version of the Categorical Imperative (he actually offered three versions that he thought were equivalent) states: "Act only on that maxim whereby you can at the same time will that it would become a universal law." Examples were "Never break your promise" and "Never commit suicide." Contemporary Kantians often interpret the Categorical Imperative as yielding objective, though not absolute principles. That is, while in general it is wrong to break promises or commit suicide, sometimes other moral principles may override them. Here is where Consequences enter the picture.

Kant gave a second formulation of the Categorical Imperative, referred to as the *principle of ends:* "So act as to treat humanity, whether in your own person or in that of any other, in every case as an end and never as merely a means." Each person as a rational agent has dignity and profound worth that entails that he or she must never be exploited or manipulated or merely used as a means to our idea of what is for the general good—or to any other end. The individual is sacred and our acts must reflect as much.

Consequences

We said above that lying is generally seen as wrong and telling the truth is generally seen as right. But consider this situation. You are hiding in your home an innocent woman named Laura, who is fleeing gangsters. Gangland Gus knocks on your door and when you open it, he asks if Laura is in your house. What should you do? Should you tell the truth or lie? Those who say that morality has something to do with consequences of actions, would prescribe lying as the morally right thing to do. Those who deny that we should look at the consequences when considering what to do when there is a clear and absolute rule of action, will say that we should either keep silent or tell the truth. When no other rule is at stake, of course, the rule-oriented ethicist will allow the foreseeable consequences to determine a course of action. Theories that focus primarily on consequences in determining moral rightness and wrongness are called "teleological" ethical theories (from the Greek *telos*, meaning goal directed). The most famous of these theories is Utilitarianism, set forth by Jeremy Bentham (1748–1832) and John Stuart Mill (1806–1873), which enjoins us to do the act that is most likely to have the best consequences: do that act which will produce the greatest happiness for the greatest number.

Character

While some ethical theories emphasize principles of action in themselves and some emphasize principles involving consequences of action, other theories, such as Aristotle's ethics, emphasize character or virtue. According to Aristotle, it is most important to develop virtuous character, for if and only if we have good people can we ensure habitual right action. Although it may be helpful to have action-guiding rules, what is vital is the empowerment of character to do good. Many people know that cheating or gossiping or overindulging in eating or imbibing alcohol is wrong, but are incapable of doing what is right. The virtuous person may not be consciously following the moral law when he or she does what is right and good. While the virtues are not central to other types of moral theories, most moral theories include the virtues as important. Most reasonable people, whatever their notions about ethics, would judge that the people who watched the assault on Kitty Genovese lacked good character. Different moral systems emphasize different virtues and emphasize them to different degrees.

Motive

Finally, virtually all ethical systems, but especially Kant's system, accept the relevance of *motive*. It is important to the full assessment of any action that the intention of the agent be taken into account. Two acts may be identical, but one may be judged morally culpable and the other excusable. Consider John's pushing Joan off a ledge, causing her to break her leg. In situation A he is angry and intends to harm her, but in situation B he sees a knife flying in her direction and intends to save her life. In A what he did was clearly wrong, whereas in B he did the right thing. On the other hand, two acts may have opposite results but the action be equally good judged on the basis of intention. For example, two soldiers may try to cross the enemy lines to communicate with an allied force, but one gets captured through no fault of his own and the other succeeds. In a full moral description of any act, motive will be taken into consideration as a relevant factor.

The Purposes of Morality

What is the role of morality in human existence? I believe that morality is necessary to stave off social chaos, what Thomas Hobbes called a "state of nature" wherein life becomes "solitary, poor, nasty, brutish and short." It is a set of rules that if followed by nearly everyone will promote the flourishing of nearly everyone. These rules restrict our freedom but only in order to promote greater freedom and well-being. More specifically morality seems to have these five purposes:

1. To keep society from falling apart.
2. To ameliorate human suffering.
3. To promote human flourishing.
4. To resolve conflicts of interest in just and orderly ways.
5. To assign praise and blame, reward, and punishment and guilt.

Let me elaborate these purposes. Imagine what society would be like if everyone or nearly everyone did whatever he or she pleased without obeying moral rules. I would promise to help you with your philosophy homework tomorrow if you fix my car today. You believe me. So you fix my car, but you are deeply angry when I laugh at you on the morrow when I drive away to the beach instead of helping you with your homework. Or you loan me money but I run off with it. Or I lie to you or harm you when it is in my interest or even kill you when I feel the urge.

Parents would abandon children and spouses betray each other whenever it was convenient. Under such circumstances society would break down. No one would have an incentive to help anyone else because reciprocity (a moral principle) was not recognized. Great suffering would go largely unameliorated and, certainly, people would not be very happy. We would not flourish or reach our highest potential.

I have just returned from Kazakhstan and Russia, which is undergoing a difficult transition from communism to democracy. In this transition (one hopes it will be resolved favorably), with the state's power considerably withdrawn, crime is on the increase and distrust is prevalent. At night when trying to navigate my way up the staircases to apartments throughout one city, I had to do so in complete darkness. I inquired as to why there were no lightbulbs in the stairwells, only to be told that the residents stole them, believing that if they did not take them, their neighbors would. Absent a dominant authority, the social contract has been eroded and everyone must struggle alone in the darkness.

We need moral rules to guide our actions in ways that light up our paths and prevent and reduce suffering, that enhance human (and animal, for that matter) well-being, that allow us to resolve our conflicts of interest according to recognizably fair rules, and to assign responsibility for actions, so that we can praise and blame, reward and punish people according to how their actions reflect moral principles.

Even though these five purposes are related, they are not identical, and different moral theories emphasize different purposes and in different ways. Utilitarianism fastens on human flourishing and the amelioration of suffering, whereas contractual systems rooted in rational self-interest accent the role of resolving conflicts of interest. A complete moral theory would include a place for each of these purposes. Such a system has the goal of internalizing the rules that promote these principles in each moral person's life, producing the virtuous person, someone who is "a jewel that shines in [morality's] own light," to paraphrase Kant. The goal of morality is to create happy and virtuous people, the kind that create flourishing communities. That's why it is the most important subject on earth.

Let us return to the questions asked at the beginning of this chapter. You should be able to answer each of them.

(1) What is the nature of morality and why do we need it? It has to do with discovering the rules that will promote the human good, elaborated in the five purposes discussed above. Without morality we cannot promote that good.

(2) What is the good and how shall I know it? The good in question is the human good, specified by happiness, reaching one's potential, and so forth. Whatever we decide that meets human needs and helps us develop our deepest potential is the good that morality promotes.

(3) Are moral principles absolute or simply relative to social groups or individual decision? It would seem that moral principles have universal and objective validity, since similar rules are needed in all cultures to promote human flourishing. So moral rules are not justified by cultural acceptance and are not relative. But neither are they absolute, if absolute means that they can never be broken or overridden. Most moral rules can be overridden by other moral rules in different contexts. For example, it is sometimes justified to lie in order to save an innocent life.

(4) Is it in my interest to be moral? In general and in the long run it is in my interest to be moral, for morality is exactly the set of rules that are most likely to help (nearly) all of us if nearly all of us follow them nearly all the time. The good is good for you—at least most of the time. Furthermore, if we believe in the superior importance of morality we will bring up children so that they will not be happy when they break the moral code. Instead they will feel guilt. In this sense the commit-

ment to morality and its internalization nearly guarantee that if you break the moral rules, you will suffer—both because of external sanctions and internal sanctions—moral guilt.

(5) What is the relationship between morality and religion? We have seen that while religion relies more on revelation, morality relies more on reason, on rational reflection. But religion can provide added incentive for the moral life, offering the individual a relationship with God, who sees and will judge all our actions.

(6) What is the relationship between morality and law? Morality and law should be very close, and morality should be the basis of the law, but there can be both unjust laws and immoral acts that cannot be legally enforced. The law is not as deep as morality and has a harder time judging human motives and intentions. You can be morally evil, intending to do evil things, but so long as you don't get the opportunity to do them, you are legally innocent.

(7) What is the relationship between morality and etiquette? Etiquette consists in the customs of a culture, but they are typically morally neutral in that a culture could flourish with a different set of etiquette. In our culture we eat with knives and forks, but a culture that cannot afford silverware and eats with chopsticks or its fingers is no less moral because of that fact.[4]

Notes

1. The term *normative* means seeking to make certain types of behavior a norm or standard in a society. Webster's Dictionary defines it as "of, or relating or conforming to or prescribing norms or standards."

2. A sanction is a mechanism for social control, used to enforce society's standards. It may consist in rewards or punishment, praise or blame, approbation or disapprobation.

3. John Buehrens in his book *Our Chosen Faith* (Boston: Beacon Press, 1989), 140.

4. I am grateful to Laura Westra and Kristin Shrader-Frechette for criticisms of an earlier version of this essay.

Study Questions

1. What does Pojman say are the differences among ethics, religion, law, and etiquette?
2. What are the four domains of ethical assessment? Show how a person could perform an act that is morally right according to three of

the domains of assessment but wrong according to the other domain.

3. Do you think Pojman is right to say that it can be immoral to disregard etiquette? Does he have too strict a view of etiquette?

4. Distinguish among deontological, teleological, and utilitarian ethics. Do you think that proponents of one of these ethical theories would be likely to have different views of the acceptability of technological risks than proponents of one of the other ethical theories? Why or why not?

5. What are the five purposes of morality, according to Pojman?

1.3

Technology and Ethical Issues

Kristin Shrader-Frechette

Aristotle (384–322 B.C.) pointed out in Book III of the *Nicomachean Ethics* that one can deliberate only about what is within one's power to do. Technologies such as gene splicing and nuclear fission were not within the power of the Greeks, so there was no ethical deliberation about them until centuries later. Throughout history, technology (knowledge associated with the industrial arts, applied sciences, and various forms of engineering) has opened new possibilities for actions. As a result, it has also raised new ethical questions.

Most of these questions have not generated new ethical concepts; instead they have expanded the scope of existing ones. For example, because hazardous technologies threaten those who live nearby, ethicists have expanded the notion of "equal treatment" to include "geographical equity," equal treatment of persons located different distances from dangerous facilities.

Because new developments force the expansion of ethical concepts, those who investigate technology and ethics need both technical and philosophical skills. To assess the ethical desirability of using biological (versus chemical) pest control, for example, one must know the relevant biology and chemistry, as well as the economic constraints on the choice. Although such factual knowledge does not determine the ethical decision, it constrains it in important and unavoidable ways.

Since policymakers evaluate virtually all technologies, at least in part, by methods such as benefit-cost or benefit-risk analysis, knowledge of economics is essential for informed discussions of technology

Originally published as "Technology," in *Encyclopedia of Ethics* 2, pp. 1231–34. Copyright © 1992 by Garland Publishing. Published by Garland Publishing, 1992.

and ethics. Philosophers investigate both the ethical constraints on developing or implementing particular technologies and the ethical acceptability of various economic and policy methods used to evaluate technology.

Philosophical questions about technology and ethics generally fall into one of at least five categories. These are (1) conceptual or *metaethical* questions; (2) *general normative* questions; (3) *particular normative* questions about specific technologies; (4) questions about the ethical *consequences* of technological developments; and (5) questions about the ethical justifiability of various *methods* of technology assessment.

Examples of (1) are: "how ought one to define 'free, informed consent' to risks imposed by sophisticated technologies?" or "how ought one to define 'equal protection' from such risks?" Examples of (2) are: "does one have a right, as Alan Gewirth argues, not to be caused to contract cancer?" or "are there duties to future generations potentially harmed by various technologies?" Examples of (3) are: "should commercial nuclear power licensees, contrary to the Price-Anderson Act, be subject to strict and full liability?" or "should the US continue to export banned pesticides to developing nations?" Examples of (4) are: "would development of a plutonium-based energy technology threaten civil liberties?" or "would deregulation of the airline industry result in less safe air travel?" Examples of (5) are: "does benefit-cost analysis ignore noneconomic components of human welfare?" or "do Bayesian methods of technology assessment ignore the well-being of minorities likely to be harmed by a technological development?"

The leading philosophical issues concerning technology and ethics are the following:

How to Define Technological Risk

Engineers and technical experts tend to define "technological risk" as a probability of physical harm, usually as "average annual probability of fatality." Philosophers and other humanistic critics claim both that technological risk cannot be defined purely quantitatively, and that it includes more than physical harm. Instead they argue that technology often threatens other goods, such as civil liberties, personal autonomy, or rights such as due process.

The technologists argue for a quantitative definition of risk, claiming that we need a common denominator for evaluating diverse technological hazards. They also claim that it is impossible to evaluate nonquantitative notions, such as the technological threat to democracy. Those who oppose the quantitative definition argue not only that it excludes qualitative factors (like equity of risk distribution) affecting welfare, but also that the nonquantitative factors are sometimes more important than the quantitative ones. Hence they argue, for example, that an equitably distributed technological risk could be more desirable than a quantitatively smaller one (in terms of probability of fatality) that is inequitably distributed.

How to Evaluate Technologies in the Face of Uncertainty

Whether one technological risk is quantitatively greater than another, in terms of average annual probability of fatality, however, is often difficult to determine. Most evaluations of technology are conducted in the face of probabilistic uncertainty about the magnitude of potential hazards. Typically this uncertainty ranges from two to four orders of magnitude. It arises because the developments most needing evaluation, e.g., biotechnology, are new. We have limited experience with them and hence limited data about their accident frequency.

How should one evaluate technologies whose level of risk is uncertain? According to John Harsanyi, the majority position is that, in such situations, one should either use subjective probabilities of experts, or assume that all uncertain events are equally probable. The desirable technological choice is then the one having the highest "average expected utility," as measured by the probability and utility of the outcomes associated with each choice. Critics of the majority position, like John Rawls, maintain that it has all the flaws of utilitarianism. It fails, they say, to take adequate account of minorities likely to be harmed by high-consequence, low-probability risks. Rawls argues instead that we should use a maximin rule in situations of probabilistic uncertainty. Such a rule, like the difference principle, would direct us to avoid the outcome having the worst possible consequences, regardless of its alleged probability.

Critics of the Rawlsian position claim that it is irrational to choose so as to avoid worst-case technological accidents. They claim that taking small chances with technology often brings great economic benefits for everyone. Opponents of the majority Bayesian position respond, however, that such benefits are neither assured nor worth the risk, and that the subjective probabilities of experts often exhibit an "overconfidence bias" that there will be no serious accidents or negative health effects from a given technology.

Technological Threats to Due Process

Ethicists also charge that technology threatens due-process rights. To the extent that hazardous technologies cause (what Judith Thomson calls) "incompensable risks," like death, due process is impossible because the victim cannot be compensated.

One of the most controversial due-process debates concerns commercial nuclear fission for generating electricity. Current United States law limits liability of the nuclear licensees to less than three percent of total possible losses in a catastrophic accident. Critics maintain that this law (the Price-Anderson Act) violates citizens' due-process rights. Defenders argue that it is needed to protect the industry from possible bankruptcy, and that a catastrophic nuclear accident is unlikely. Critics respond that if a catastrophic nuclear accident is unlikely, then industry needs no protection from bankruptcy caused by such an event.

How Safe Is Safe Enough?

Because a zero-risk society is impossible, philosophers and policymakers debate both how much risk is acceptable and how it ought to be distributed. The distribution controversies raise all the classical problems associated with utilitarian *versus* egalitarian ethical schemes. Conflicts over how much technological risk is acceptable typically raise issues of whether the public has certain welfare rights, like the right to breathe clean air. The controversies also focus on how much economic progress can be traded for the negative health consequences of technology-induced risks.

Philosophers are particularly divided about how to evaluate numerous negligible risks, from a variety of technologies, that together pose a serious hazard. Small cancer risks that are singly harmless, but cumulatively and synergistically harmful, provide a good example of such cases. They raise the classical ethical problem of the contributor's dilemma. This dilemma occurs because the benefit of avoiding imposing a single small technological risk is imperceptible, although the cumulative benefit of everyone's doing so is great. Some philosophers view such small risks as ethically insignificant, while others claim they are important. Those in the latter group argue that agents are responsible for the effects of *sets* of acts (that together cause harm) of which their individual act is only one member.

Consent to Risk

The sophistication of many technologies, from genetically engineered organisms to the latest nuclear weapons, makes it questionable whether many individuals understand them. If they do not, then it is likewise questionable whether persons are able to give free, informed consent to the risks that they impose. Critics of some contemporary technologies point out that those persons most likely to take technological risks (e.g., blue-collar workers in chemical or radiation-related industries) are precisely those who are least able to give free, informed consent to them. This is because they are often persons with limited education and no alternative job skills.

Those who claim that both workers and the public have given consent to technological risks use notions like "compensating wage differential" to defend their position. They say that, since workers in hazardous technologies receive correspondingly higher pay because of the greater risks that they face, they are compensated. Likewise they maintain that accepting a risky job constitutes a form of consent. They also claim that society's acceptance of the economic benefits created by hazardous technologies constitutes implicit acceptance of the technologies.

In response, more conservative ethicists argue both that economic analysis does not show the existence of a compensating wage differential in all cases, and that mere acceptance of a job in a risky technology does not constitute consent to the hazard, especially if the worker has no other realistic employment alternatives. They also argue that acceptance of the benefits of hazardous technologies does not constitute acceptance of the technologies themselves since many people are inadequately informed about such risks.

Bibliography

Baier, Kurt, and Nicholas Rescher, eds. *Values and the Future: The Impact of Technological Change on American Values.* New York: Free Press, 1969. Philosophical analysis of problems associated with technological development.

Durbin, Paul T., ed. *Philosophy and Technology.* Boston: Kluwer, 1985–. Annual publication of the Society for Philosophy and Technology. Essays, many of which deal with ethics and technology, written from a variety of philosophical perspectives.

Ellul, Jacques. *The Technological Society.* Translated by John Wilkinson. New York: Knopf, 1964. General overview of problems associated with technology; widely viewed as anti-technology.

———. *The Technological System.* Translated by Joachim Neugroschel. New York: Continuum, 1980. General overview of problems associated with technology; widely viewed as anti-technology.

Goodpaster, Kenneth, and Kenneth Sayre, eds. *Ethics and the Problems of the 21st Century.* South Bend, Ind.: University of Notre Dame Press, 1979. Original philosophical essays on various problems associated with technological development.

Harsanyi, John C. "Can the Maximin Principle Serve as a Basis for Morality? A Critique of John Rawls' Theory." *American Political Science Review* 59, no. 2 (1975): 594–605.

Jonas, Hans. *The Imperative of Responsibility.* Chicago: University of Chicago Press, 1985. Criticism of contemporary technological society and proposals for reform.

Kasperson, Roger, and Mimi Berberian, eds. *Equity Issues in Radioactive Waste Management.* Boston: Oelgeschlager, 1983. Descriptive analysis of equity issues related to radwaste management.

MacLean, Douglas. *Values at Risk.* Totowa, N.J.: Rowman and Allanheld, 1984. Consideration of the ethical problems posed by issues of nuclear security and deterrence.

MacLean, Douglas, and Peter Brown, eds. *Energy and the Future.* Totowa, N.J.: Rowman and Allanheld, 1983. Essays on ethical issues associated with energy technologies.

Rescher, Nicholas. *Unpopular Essays on Technological Progress.* Pittsburgh, Pa.: University of Pittsburgh Press, 1980. Widely ranging essays on a variety of topics related to technology.

Shrader-Frechette, Kristin. *Nuclear Power and Public Policy.* Boston: Reidel, 1983. Criticism of nuclear technology on grounds that it violates various ethical principles.

———. *Science Policy, Ethics, and Economic Methodology.* Boston: Reidel, 1984. Analysis of problems with benefit-cost methods and proposals for amending them.

———. *Risk Analysis and Scientific Method.* Boston: Reidel, 1984. Analysis of problems associated with assessment of technological risk assessment and proposals for amending them.

Thomson, Judith Jarvis. *Rights, Restitution, and Risk.* Cambridge: Harvard University Press, 1986. Treats a variety of ethical issues related to technological risk.

Study Questions

1. What is technology? What are some examples of technologies?
2. What are the five leading philosophical issues concerning technology?
3. Do you think that technological risk should be defined quantitatively or qualitatively? Why?
4. Explain and defend Rawls's way of dealing with technological risks in a situation of uncertainty. Explain and defend Harsanyi's way of dealing with technological risks in a situation of uncertainty.
5. What is the Price-Anderson Act? Do you think it is ethically defensible?

6. What is the compensating wage differential? Do you think that it is ethically defensible?
7. What is the contributor's dilemma? Do you think that the government should ignore very small risks or impose them on citizens without their free informed consent?

1.4

Further Reading

Agassi, Joseph. 1985. *Technology*. Dordrecht, Netherlands: Reidel.

Aronowitz, Stanley. 1984. *Science as Power*. Minneapolis: University of Minnesota Press.

Bayertz, Kurt (ed.). 1994. *The Concept of Moral Consensus*. Dordrecht: Kluwer.

Cooper, Barry. 1991. *Action into Nature: An Essay on the Meaning of Technology*. Notre Dame: University of Notre Dame Press.

De George, Richard. 1995. *Business Ethics*, 4th ed. Englewood Cliffs, N.J.: Prentice-Hall.

Donagan, Alan. 1977. *The Theory of Morality*. Chicago: University of Chicago Press.

Drengson, Alan R. 1986. "Applied Philosophy of Technology: Reflections on Forms of Life and the Practice of Technology," *International Journal of Applied Philosophy* 3 (Spring): 1–13.

Durbin, Paul T. 1992. *Social Responsibility in Science, Technology, and Medicine*. Bethlehem, Pa.: Lehigh University Press.

——— (ed.). 1984. *A Guide to the Culture of Science, Technology and Medicine*. New York: Free Press.

Ehrenfeld, David. 1978. *The Arrogance of Humanism*. New York: Oxford University Press.

Ellul, Jacques. 1989. "The Search for Ethics in a Technicist Society," *Research in Philosophy and Technology* 9: 23–36.

Ferre, Frederick. 1988. *Philosophy of Technology*. Englewood Cliffs, N.J.: Prentice-Hall.

Garrett, Thomas. 1966. *Business Ethics*. Englewood Cliffs, N.J.: Prentice-Hall.

Hastedt, Heiner. 1994. "Enlightenment and Technology: Outline for a General Ethics of Technology," *Research in Philosophy and Technology* 14: 205–17.

Heidegger, Martin. 1977. *The Question Concerning Technology and Other Essays*, Wm. Lovitt (trans.). New York: Harper Colophon Books.

Ihde, Don. 1993. *Philosophy of Technology: An Introduction*. New York: Paragon House.

—— 1991. *Instrumental Realism: The Interface Between Philosophy of Science and Philosophy of Technology*. Bloomington: Indiana University Press.

Jonas, Hans. 1984. *The Imperative of Responsibility*. Chicago: University of Chicago Press.

Lowrance, William W. 1985. *Modern Science and Human Values*. New York: Oxford University Press.

Mitcham, Carl. 1994. *Thinking Through Technology: The Path Between Engineering and Philosophy*. Chicago: University of Chicago Press.

National Research Council. National Academy of Sciences. 1993. *National Issues in Science and Technology, 1993*. Washington, D.C.: National Academy Press.

Rawls, John. 1975. *A Theory of Justice*. Cambridge: Cambridge University Press.

Sagoff, Mark. 1989. *The Economy of the Earth*. Cambridge: Cambridge University Press.

Scherer, D. 1990. *Upstream/Downstream*. Philadelphia: Temple University Press.

Shrader-Frechette, Kristin. 1994. "Technology Assessment." In W. Reich (ed.), *Encyclopedia of Bioethics*, vol. 5. Englewood Cliffs, N.J.: Prentice-Hall.

——. 1992. "Technology and Ethics." In L. C. Becker (ed.), *Encyclopedia of Ethics*, vol. 2. New York: Garland.

——. 1991. *Risk and Rationality*. Berkeley: University of California Press.

Strong, David. 1994. "Disclosive Discourse, Ecology, and Technology," *Environmental Ethics* 16, 1 (Spring): 89–102.

Taitte, W. Lawson. 1987. *Traditional Moral Values in the Age of Technology*. Dallas: University of Texas Press.

Velasquez, Manuel. 1991. *Business Ethics Concepts and Cases*. Englewood Cliffs, N.J.: Prentice-Hall.

Part Two

Alternative Views of Technology

2.1

Introduction and Overview

Kristin Shrader-Frechette and Laura Westra

There is a famous story of three people, all in darkness, who feel parts of an elephant. Each person draws different conclusions about the animal being discovered. The first person feels the elephant's trunk and concludes that the animal is long and thin with rough skin. The second person feels the elephant's tusk and concludes that the animal is smooth and hard. The third person feels the elephant's belly and concludes that the animal is large and hairy. Although each person is correct in describing the elephant, each describes it from a particular point of view. So it is with technology. Businesspeople may see the labor-saving effects of technologies such as commercial nuclear power (which requires less fuel and lower fuel-transport costs than coal-generated electricity), and they may conclude that technology is a money saver. Scientists and engineers may see the implementation of their technical designs and inventions, and they may view technologies such as commercial nuclear power as opportunities to learn more about how to apply their inventions and theories. Physicians may see the health effects of nuclear accidents such as Chernobyl, and they may view technologies as threats to public health and safety. Environmentalists may worry about the longevity of radioactive wastes from nuclear reactors, and they may fear that, over the next million years, these wastes eventually will contaminate groundwater. They may see technologies as threats to a clean and sustainable planet.

Each of the essays in this second part of the book presents an alternative view of technology. These alternative views of technology are important because different views lead to different ethical conclusions. Using the example of commercial nuclear fission, for example, businesspeople, scientists, and engineers are likely to find reactors more acceptable, from an ethical point of view, because they view nuclear technology as an opportunity to reach certain goals, such as saving money or learning about application of scientific theories. Medical doctors and environmentalists, however, are less likely to believe that nuclear reactors are acceptable, from an ethical point of view, because they view nuclear technology as a threat to important societal and environmental goals such as health and safety.

Because there often is a grain of truth in each of the different points of view about technology, it is important to study alternative views of technology so that the insights of people with different backgrounds, experiences, and beliefs can be shared by everyone. Studying alternative views of technology also enables people to understand why others arrive at particular ethical conclusions about technology. On the one hand, if a person believes that nuclear power is very safe and efficient,[1] for example, then she is likely to claim that the risks associated with reactors are ethically acceptable. On the other hand, if a person believes that nuclear power is risky and expensive,[2] then she is likely to claim that the risks associated with reactors are not ethically acceptable. Ethical and values conclusions about technology obviously depend, in part, on the factual aspects of the technology being evaluated and on the worldview that a person uses to interpret technology. For both of these reasons, studying alternative views of technology is likely to provide insights into the ethical and values issues associated with technology.

Just because different views of technology may each have a grain of truth in them does not mean that all views of technology are equally correct. One view may be more factually correct than another, or one view may lead to more correct predictions about a technology than another view. A good approach is to take the insights of each view of technology, whatever they are, or however slight they may be, and combine them into a reasonable position of one's own.

The authors in this part have diverse views of technology. In chapter 2.2, "Heidegger on Gaining a Free Relation to Technology," Hubert Dreyfus presents and accepts much of Martin Heidegger's position on technology. Heidegger believes that technologies such as chemical pesticides present both fundamental dangers and saving powers for society. They are saving powers because they open communities to new, future possibilities for their existence. Technologies are fundamental dangers because they encourage merely scientific or calculative thinking and not deeper, meditative thinking that opens us to the truth of being, the truth of what exists. To save ourselves and the world from the dangers of technology, Heidegger and Dreyfus believe that we must have new ways of thinking and living, ways that do not rely on efficiency, control, and power over other beings. To save ourselves, they believe that we must change our cultural paradigm and its heavy reliance on technology.

In chapter 2.3, Langdon Winner continues to develop some insights similar to those of Heidegger and Dreyfus. Winner argues that technology, such as television, creates so much "virtual reality" that it misleads us as to its moral significance. Television is not real, yet it has many effects on people that they may miss because of "technological somnambulism," lack of awareness of the dangers of technology. One of the effects of technological somnambulism is that people emphasize how to do things—technologically speaking—but they rarely question why they should do them. As a result, Winner says that technology often appears to be morally neutral when it is not. Winner argues that society needs to thoroughly examine all technologies before it implements them. Otherwise, technology may control society and its values in ways that no one realizes.

Disagreeing with Dreyfus and Winner, Emmanuel Mesthene in chapter 2.4 has a more traditional view of technology. He believes that technology is neither a blessing nor a curse nor indifferent. Rather, he maintains that technology creates both opportunities and problems for people. He claims that the ends of technology are not inherent in it but created by the ways humans use technology. Unlike Winner, Mesthene believes that technology is open-ended in its potential effects.

John McDermott, in chapter 2.5, criticizes Mesthene's view that technology has no inherent ends and is open-ended in its effects. Instead, McDermott argues that technology is not open-ended because it benefits first and foremost the "techno-elites" and promotes centralized and nondemocratic concerns. Using the Vietnam War as an example, McDermott claims that advanced technologies enabled soldiers to eliminate "targets" and to fail to recognize the fact that these targets included human beings and those who opposed the war.

In chapter 2.6, "The Social Construction of Safety," Rachelle Hollander argues that scholars of technology ought to avoid emphasis on risk assessment and management and instead ought to stress collective moral responsibility for developing a culture of safety. She shows that in order to emphasize safety regarding technology, people must evaluate the expectations they place on humans who often cannot guarantee safety. Theoretically, something may be safe, but not if humans do not perform according to ideal expectations. In other words, by exercising due care and feasible control, people construct safety. Hollander notes that most scientific and engineering discussions of technology wrongly assume that risk is impervious to human influence. Instead, she argues that human error and ignorance may play a key role in creating or constructing safety.

Jesse Tatum, in chapter 2.7, "The Political Construction of Technology," pursues themes similar to those of Hollander. He argues that technology is not a naturally unfolding process but rather is determined by human choices or indifferences. He claims that the unmistakable stamp of political and economic power is present in various new technologies, such as genetic and chemical technologies for creating "hard tomatoes" that can be picked by machine. To counteract the political and economic interests that often produce technologies fitting their interests rather than the public interest, Tatum argues that society ought to perform constructive technology assessments that evaluate all possible technological options rather than merely the particular consequences of one technological option. Tatum argues that by becoming aware of the political and economic power that promotes individual technologies, people can help to create more technological options, options that take account of the knowledge, participation, and permission of citizens.

Notes

1. See, for example, H. W. Lewis, *Technological Risk* (New York: Norton, 1990).

2. See, for example, Kristin Shrader-Frechette, *Burying Uncertainty: Risk and the Case Against Geological Disposal of Nuclear Waste* (Berkeley: University of California Press, 1993).

2.2

Heidegger on Gaining a Free Relation to Technology

Hubert L. Dreyfus

Introduction: What Heidegger Is Not Saying

In *The Question Concerning Technology* Heidegger describes his aim:

> We shall be questioning concerning technology, and in so doing we should like to prepare a free relationship to it.

He wants to reveal the essence of technology in such a way that "in no way confines us to a stultified compulsion to push on blindly with technology or, what comes to the same thing, to rebel helplessly against it."[1] Indeed, he claims that "When we once open ourselves expressly to the *essence* of technology, we find ourselves unexpectedly taken into a freeing claim."[2]

We will need to explain essence, opening, and freeing before we can understand Heidegger here. But already Heidegger's project should alert us to the fact that he is not announcing one more reactionary rebellion against technology, although many respectable philosophers, including Jürgen Habermas, take him to be doing just that; nor is he doing what progressive thinkers such as Habermas want him to do, proposing a way to get technology under control so that it can serve our rationally chosen ends.

The difficulty in locating just where Heidegger stands on technology is no accident. Heidegger has not always been clear about what distin-

---

Originally published in *Technology and the Politics of Knowledge*. Copyright © 1995 by Indiana University Press. Published by Indiana University Press, 1995.

guishes his approach from a romantic reaction to the domination of nature, and when he does finally arrive at a clear formulation of his own original view, it is so radical that everyone is tempted to translate it into conventional platitudes about the evils of technology. Thus Heidegger's ontological concerns are mistakenly assimilated to humanistic worries about the devastation of nature.

Those who want to make Heidegger intelligible in terms of current anti-technological banalities can find support in his texts. During the war he attacks consumerism:

> The circularity of consumption for the sake of consumption is the sole procedure which distinctively characterizes the history of a world which has become an unworld.[3]

And as late as 1955 he holds that:

> The world now appears as an object open to the attacks of calculative thought. . . . Nature becomes a gigantic gasoline station, an energy source for modern technology and industry.[4]

In this address to the Schwartzwald peasants he also laments the appearance of television antennae on their dwellings.

> Hourly and daily they are chained to radio and television. . . . All that with which modern techniques of communication stimulate, assail, and drive man—all that is already much closer to man today than his fields around his farmstead, closer than the sky over the earth, closer than the change from night to day, closer than the conventions and customs of his village, than the tradition of his native world.[5]

Such statements suggest that Heidegger is a Luddite who would like to return from the exploitation of the earth, consumerism, and mass media to the world of the pre-Socratic Greeks or the good old Schwartzwald peasants.

Heidegger's Ontological Approach to Technology

As his thinking develops, however, Heidegger does not deny these are serious problems, but he comes to the surprising and provocative conclusion that focusing on loss and destruction is still technological.

> All attempts to reckon existing reality . . . in terms of decline and loss, in terms of fate, catastrophe, and destruction, are merely technological behavior.[6]

Seeing our situation as posing a problem that must be solved by appropriate action turns out to be technological too:

> [T]he instrumental conception of technology conditions every attempt to bring man into the right relation to technology. . . . The will to mastery becomes all the more urgent the more technology threatens to slip from human control.[7]

Heidegger is clear this approach cannot work.

> No single man, no group of men, no commission of prominent statesmen, scientists, and technicians, no conference of leaders of commerce and industry, can brake or direct the progress of history in the atomic age.[8]

His view is both darker and more hopeful. He thinks there is a more dangerous situation facing modern man than the technological destruction of nature and civilization, yet a situation about which something *can* be done—at least indirectly. The threat is not a *problem* for which there can be a *solution* but an ontological *condition* from which we can be *saved*.

Heidegger's concern is the human distress caused by the *technological understanding of being*, rather than the destruction caused by specific technologies. Consequently, Heidegger distinguishes the current problems caused by technology—ecological destruction, nuclear danger, consumerism, etc.—from the devastation that would result if technology solved all our problems.

> What threatens man in his very nature is the . . . view that man, by the peaceful release, transformation, storage, and channeling of the energies of physical nature, could render the human condition . . . tolerable for everybody and happy in all respects.[9]

The "greatest danger" is that

> the approaching tide of technological revolution in the atomic age could so captivate, bewitch, dazzle, and beguile man that calculative thinking may someday come to be accepted and practiced as *the only* way of thinking.[10]

The danger, then, is not the destruction of nature or culture but a restriction in our way of thinking—a leveling of our understanding of being.

To evaluate this claim we must give content to what Heidegger means by an understanding of being. Let us take an example. Normally we deal with things, and even sometimes people, as resources to be used until no longer needed and then put aside. A styrofoam cup is a perfect example. When we want a hot or cold drink it does its job, and when we are through with it we throw it away. How different this understanding of an object is from what we can suppose to be the everyday Japanese understanding of a delicate teacup. The teacup does not preserve temperature as well as its plastic replacement, and it has to be washed and protected, but it is preserved from generation to generation for its beauty and its social meaning. It is hard to picture a tea ceremony around a styrofoam cup.

Note that the traditional Japanese understanding of what it is to be human (passive, contented, gentle, social, etc.) fits with their understanding of what it is to be a thing (delicate, beautiful, traditional, etc.). It would make no sense for us, who are active, independent, and aggressive—constantly striving to cultivate and satisfy our desires—to relate to things the way the Japanese do; or for the Japanese (before their understanding of being was interfered with by ours) to invent and prefer styrofoam teacups. In the same vein *we* tend to think of politics as the negotiation of individual desires while the Japanese seek consensus. In sum the social practices containing an understanding of what it is to be a human self, those containing an interpretation of what it is to be a thing, and those defining society fit together. They add up to an understanding of being.

The shared practices into which we are socialized, then, provide a background understanding of what counts as things, what counts as human beings, and ultimately what counts as real, on the basis of which we can direct our actions toward particular things and people. Thus the understanding of being creates what Heidegger calls a *clearing* in which things and people can show up for us. We do not produce the clearing. It produces us as the kind of human beings that we are. Heidegger describes the clearing as follows:

> [B]eyond what is, not away from it but before it, there is still something else that happens. In the midst of beings as a whole an open place occurs. There is a clearing, a lighting. . . . This open center is . . . not surrounded by what is; rather, the lighting center itself encircles all that is. . . . Only this clearing grants and guarantees to human beings a passage to those entities that we ourselves are not, and access to the being that we ourselves are.[11]

What, then, is the essence of technology, i.e., the technological understanding of being, i.e., the technological clearing, and how does opening ourselves to it give us a free relation to technological devices?

To begin with, when we ask about the essence of technology we are able to see that Heidegger's question cannot be answered by defining technology. Technology is as old as civilization. Heidegger notes that it can be correctly defined as "a means and a human activity." He calls this "the instrumental and anthropological definition of technology."[12] But if we ask about the *essence* of technology (the technological understanding of being) we find that modern technology is "something completely different and . . . new."[13] Even different from using styrofoam cups to serve our desires. The essence of modern technology, Heidegger tells us, is to seek more and more flexibility and efficiency *simply for its own sake*. "[E]xpediting is always itself directed from the beginning . . . towards driving on to the maximum yield at the minimum expense."[14] That is, our only goal is optimization:

> Everywhere everything is ordered to stand by, to be immediately at hand, indeed to stand there just so that it may be on call for a further ordering. Whatever is ordered about in this way has its own standing. We call it standing-reserve. . . .[15]

No longer are we subjects turning nature into an object of exploitation:

> The subject-object relation thus reaches, for the first time, its pure "relational," i.e., ordering, character in which both the subject and the object are sucked up as standing-reserves.[16]

A modern airliner is not an object at all, but just a flexible and efficient cog in the transportation system.[17] (And passengers are presumably not subjects but merely resources to fill the planes.) Heidegger concludes: "Whatever stands by in the sense of standing-reserve no longer stands over against us as object."[18]

All ideas of serving God, society, our fellow men, or even our own calling disappear. Human beings, on this view, become a resource to be used, but more important to be enhanced—like any other.

> Man, who no longer conceals his character of being the most important raw material, is also drawn into this process.[19]

In the film *2001*, the robot HAL, when asked if he is happy on the mission, answers: "I'm using all my capacities to the maximum. What more could a rational entity desire?" This is a brilliant expression of what anyone would say who is in touch with our current understanding of being. We pursue the growth or development of our potential simply for its own sake—it is our only goal. The human potential movement perfectly expresses this technological understanding of being, as does the attempt to better organize the future use of our natural resources. We thus become part of a system which no one directs but which moves toward the total mobilization of all beings, even us. This is why Heidegger thinks the perfectly ordered society dedicated to the welfare of all is not the solution of our problems but the distressing culmination of the technological understanding of being.

What Then Can We Do?

But, of course, Heidegger uses and depends upon modern technological devices. He is no Luddite and he does not advocate a return to the pre-technological world.

> It would be foolish to attack technology blindly. It would be shortsighted to condemn it as the work of the devil. We depend on technical devices; they even challenge us to ever greater advances.[20]

Instead, Heidegger suggests that there is a way we can keep our technological devices and yet remain true to ourselves:

> We can affirm the unavoidable use of technical devices, and also deny them the right to dominate us, and so to warp, confuse, and lay waste our nature.[21]

To understand how this might be possible we need an illustration of Heidegger's important distinction between technology and the technological understanding of being. Again we can turn to Japan. In contemporary Japan a traditional, non-technological understanding of being still exists alongside the most advanced high-tech production and consumption. The TV set and the household gods share the same shelf— the styrofoam cup co-exists with the porcelain one. We can thus see that one can have technology without the technological understanding of being, so it becomes clear that the technological understanding of being can be dissociated from technological devices.

To make this dissociation, Heidegger holds, one must rethink the history of being in the West. Then one will see that although a technological understanding of being is our destiny, it is not our fate. That is, although our understanding of things and ourselves as resources to be ordered, enhanced, and used efficiently has been building up since Plato and dominates our practices, we are not stuck with it. It is not the way things have to be, but nothing more or less than our current cultural clearing.

Only those who think of Heidegger as opposing technology will be surprised at his next point. Once we see that technology is our latest understanding of being, we will be grateful for it. We did not make this clearing nor do we control it, but if it were not given to us to encounter things and ourselves as resources, nothing would show up *as* anything at all and no possibilities for action would make sense. And once we realize—in our practices, of course, not just in our heads—that we *receive* our technological understanding of being, we have stepped out of the technological understanding of being, for we then see that what is most important in our lives is not subject to efficient enhancement. This transformation in our sense of reality—this overcoming of calculative thinking—is precisely what Heideggerian thinking seeks to bring about. Heidegger seeks to show how we can recognize and thereby overcome our restricted, willful modern clearing precisely by recognizing our essential receptivity to it.

> [M]odern man must first and above all find his way back into the full breadth of the space proper to his essence. That essential space of man's essential being receives the dimension that unites it to something beyond itself . . . that is the way in which the safekeeping of being itself is given to belong to the essence of man as the one who is needed and used by being.[22]

But precisely how can we experience the technological understanding of being as a gift to which we are receptive? What is the phenomenon Heidegger is getting at? We can break out of the technological understanding of being whenever we find ourselves gathered by things rather than controlling them. When a thing like a celebratory meal, to take Heidegger's example, pulls our practices together and draws us in, we experience a focusing and a nearness that resists technological ordering. Even a technological object like a highway bridge, when experienced as a gathering and focusing of our practices, can help us resist the very technological ordering it furthers. Heidegger describes the bridge so as to bring out both its technological ordering function and its continuity with pre-technological things.

> The old stone bridge's humble brook crossing gives to the harvest wagon its passage from the fields into the village and carries the lumber cart from the field path to the road. The highway bridge is tied into the network of long-distance traffic, paced as calculated for maximum yield. Always and ever differently the bridge escorts the lingering and hastening ways of men to and fro. . . . The bridge *gathers* to itself in *its own way* earth and sky, divinities and mortals.[23]

Getting in sync with the highway bridge in its technological functioning can make us sensitive to the technological understanding of being
as the way our current clearing works, so that we experience our role
as receivers, and the importance of receptivity, thereby freeing us from
our compulsion to force all things into one efficient order.

This transformation in our understanding of being, unlike the slow
process of cleaning up the environment which is, of course, also necessary, would take place in a sudden Gestalt switch.

> The turning of the danger comes to pass suddenly. In this turning, the
> clearing belonging to the essence of being suddenly clears itself and lights
> up.[24]

The danger, when grasped as the danger, becomes that which saves us.
"The self-same danger is, when it is *as* the danger, the saving power."[25]

This remarkable claim gives rise to two opposed ways of understanding Heidegger's response to technology. Both interpretations
agree that once one recognizes the technological understanding of
being for what it is—a historical understanding—one gains a free relation to it. We neither push forward technological efficiency as our only
goal nor always resist it. If we are free of the technological imperative
we can, in each case, discuss the pros and cons. As Heidegger puts it:

> We let technical devices enter our daily life, and at the same time leave
> them outside . . . as things which are nothing absolute but remain depen-
> dent upon something higher [the clearing]. I would call this comportment
> toward technology which expresses "yes" and at the same time "no", by
> an old word, *releasement towards things.*[26]

One way of understanding this proposal—represented here by Richard Rorty—holds that once we get in the right relation to technology,
viz. recognize it as a clearing, it is revealed as just as good as any other
clearing. Efficiency—getting the most out of ourselves and everything
else—is fine, so long as we do not think that efficiency for its own sake
is the *only* end for man, dictated by reality itself, to which all others
must be subordinated. Heidegger seems to support this acceptance of
the technological understanding of being when he says:

> That which shows itself and at the same time withdraws [i.e., the clearing]
> is the essential trait of what we call the mystery. I call the comportment
> which enables us to keep open to the meaning hidden in technology,
> *openness to the mystery.* Releasement toward things and openness to the
> mystery belong together. They grant us the possibility of dwelling in the
> world in a totally different way. They promise us a new ground and foun-

dation upon which we can stand and endure in the world of technology without being imperiled by it.[27]

But acceptance of the mystery of the gift of understandings of being cannot be Heidegger's whole story, for he immediately adds:

> Releasement toward things and openness to the mystery give us a vision of a new rootedness which *someday* might even be fit to recapture the old and now rapidly disappearing rootedness in a changed form.[28]

We then look back at the preceding remark and realize *releasement* gives only a "possibility" and a "promise" of "dwelling in the world in a totally different way."

Mere openness to technology, it seems, leaves out much that Heidegger finds essential to human being: embeddedness in nature, nearness or localness, shared meaningful differences such as noble and ignoble, justice and injustice, salvation and damnation, mature and immature—to name those that have played important roles in our history. *Releasement*, while giving us a free relation to technology and protecting our nature from being distorted and distressed, cannot give us any of these.

For Heidegger, there are, then, two issues. One issue is clear:

> The issue is the saving of man's essential nature. Therefore, the issue is keeping meditative thinking alive.[29]

But that is not enough:

> If releasement toward things and openness to the mystery awaken within us, then we should arrive at a path that will lead to a new ground and foundation.[30]

Releasement, it turns out, is only a stage, a kind of holding pattern, awaiting a new understanding of being, which would give some content to our openness—what Heidegger calls a new rootedness. That is why each time Heidegger talks of *releasement* and the saving power of understanding technology as a gift he then goes on to talk of the divine.

> Only when man, in the disclosing coming-to-pass of the insight by which he himself is beheld . . . renounces human self-will . . . does he correspond in his essence to the claim of that insight. In thus corresponding man is gathered into his own, that he . . . may, as the mortal, look out toward the divine.[31]

The need for a new centeredness is reflected in Heidegger's famous remark in his last interview: "Only a god can save us now."[32] But what does this mean?

The Need for a God

Just preserving pre-technical practices, even if we could do it, would not give us what we need. The pre-technological practices no longer add up to a shared sense of reality and one cannot legislate a new understanding of being. For such practices to give meaning to our lives, and unite us in a community, they would have to be focused and held up to the practitioners. This function, which later Heidegger calls "truth setting itself to work," can be performed by what he calls a work of art. Heidegger takes the Greek temple as his illustration of an artwork working. The temple held up to the Greeks what was important, and so let there be heroes and slaves, victory and disgrace, disaster and blessing, and so on. People whose practices were manifested and focused by the temple had guidelines for leading good lives and avoiding bad ones. In the same way, the medieval cathedral made it possible to be a saint or a sinner by showing people the dimensions of salvation and damnation. In either case, one knew where one stood and what one had to do. Heidegger holds that "there must always be some being in the open [the clearing], something that is, in which the openness takes its stand and attains its constancy."[33]

We could call such special objects cultural paradigms. A cultural paradigm focuses and collects the scattered practices of a culture, unifies them into coherent possibilities for action, and holds them up to the people who can then act and relate to each other in terms of the shared exemplar.

When we see that for later Heidegger only those practices focused in a paradigm can establish what things can show up as and what it makes sense to do, we can see why he was pessimistic about salvaging aspects of the Enlightenment or reviving practices focused in the past. Heidegger would say that we should, indeed, try to preserve such practices, but they can save us only if they are radically transformed and integrated into a new understanding of reality. In addition we must learn to appreciate marginal practices—what Heidegger calls the saving power of insignificant things—practices such as friendship, backpacking into the wilderness, and drinking the local wine with friends. All these practices are marginal precisely because they are not efficient. They can, of course, be engaged in for the sake of health and greater efficiency. This expanding of technological efficiency is the greatest danger. But these saving practices could come together in a new cultural paradigm that held up to us a new way of doing things, thereby focusing a world in which formerly marginal practices were central and efficiency marginal. Such a new object or event that grounded a new understanding of reality Heidegger would call a new god. This is why he holds that "only another god can save us."[34]

Once one sees what is needed, one also sees that there is not much we can do to bring it about. A new sense of reality is not something that can be made the goal of a crash program like the moon flight—a paradigm of modern technological power. A hint of what such a new god might look like is offered by the music of the sixties. The Beatles, Bob Dylan, and other rock groups became for many the articulation of new understanding of what really mattered. This new understanding almost coalesced into a cultural paradigm in the Woodstock Music Festival, where people actually lived for a few days in an understanding of being in which mainline contemporary concern with rationality, sobriety, willful activity, and flexible, efficient control were made marginal and subservient to Greek virtues such as openness, enjoyment of nature, dancing, and Dionysian ecstasy along with a neglected Christian concern with peace, tolerance, and love of one's neighbor without desire and exclusivity. Technology was not smashed or denigrated but all the power of the electronic media was put at the service of the music which focused all the above concerns.

If enough people had found in Woodstock what they most cared about, and recognized that all the others shared this recognition, a new understanding of being might have coalesced and been stabilized. Of course, in retrospect we see that the concerns of the Woodstock generation were not broad and deep enough to resist technology and to sustain a culture. Still we are left with a hint of how a new cultural paradigm would work, and the realization that we must foster human receptivity and preserve the endangered species of pre-technological practices that remain in our culture, in the hope that one day they will be pulled together into a new paradigm, rich enough and resistant enough to give new meaningful directions to our lives.

To many, however, the idea of *a* god which will give us a unified but open community—one set of concerns which everyone shares if only as a focus of disagreement—sounds either unrealistic or dangerous. Heidegger would probably agree that its open democratic version looks increasingly unobtainable and that we have certainly seen that its closed totalitarian form can be disastrous. But Heidegger holds that given our historical essence—the kind of beings we have become during the history of our culture—such a community is necessary to us. This raises the question of whether our need for one community is, indeed, dictated by our historical essence, or whether the claim that we can't live without a centered and rooted culture is simply romantic nostalgia.

It is hard to know how one could decide such a question, but Heidegger has a message even for those who hold that we, in this pluralized modern world, should not expect and do not need one all-embracing community. Those who, from Dostoievsky, to the hippies, to Richard Rorty, think of communities as local enclaves in an otherwise impersonal society still owe us an account of what holds these local communities together. If Dostoievsky and Heidegger are right, each local community still needs its local god—its particular incarnation of what the community is up to. In that case we are again led to the view that releasement is not enough, and to the modified Heideggerian slogan that only some new *gods* can save us.

Notes

1. Martin Heidegger, "The Question Concerning Technology," *The Question Concerning Technology* (New York: Harper Colophon, 1977), pp. 25–26.

2. Ibid.

3. Heidegger, "Overcoming Metaphysics," *The End of Philosophy* (New York: Harper and Row, 1973), p. 107.

4. Heidegger, *Discourse on Thinking* (New York: Harper and Row, 1966), p. 50.

5. Ibid., p. 48.

6. Heidegger, "The Turning," *The Question Concerning Technology*, p. 48.

7. Heidegger, "The Question Concerning Technology," *The Question Concerning Technology*, p. 5.

8. Heidegger, *Discourse on Thinking*, p. 52.

9. Martin Heidegger, "What Are Poets For?" *Poetry, Language, Thought* (New York: Harper and Row, 1971), p. 116.

10. Heidegger, *Discourse on Thinking*, p. 56.

11. Heidegger, "The Origin of the Work of Art," *Poetry, Language, Thought*, p. 53.

12. Heidegger, "The Question Concerning Technology," p. 5.

13. Ibid.

14. Ibid., p. 15.

15. Ibid., p. 17.

16. Heidegger, "Science and Reflection," *The Question Concerning Technology*, p. 173.

17. Heidegger, "The Question Concerning Technology," p. 17.

18. Ibid.

19. Heidegger, "Overcoming Metaphysics," *The End of Philosophy*, p. 104.

20. Heidegger, *Discourse on Thinking*, p. 53.

21. Ibid., p. 54.

22. Heidegger, "The Turning," *The Question Concerning Technology*, p. 39.

23. Heidegger, *Poetry, Language, Thought*, pp. 152–53.

24. Ibid., p. 44.

25. Heidegger, "The Turning," *The Question Concerning Technology*, p. 39.

26. Heidegger, *Discourse on Thinking*, p. 54.

27. Ibid., p. 55.

28. Ibid. (My italics.)

29. Ibid., p. 56.

30. Ibid.

31. Heidegger, "The Turning," *The Question Concerning Technology*, p. 47.

32. "Nur noch ein Gott kann uns retten," *Der Spiegel*, May 31, 1976.

33. Heidegger, "The Origin of the Work of Art," *Poetry, Language, Thought*, p. 61.

34. This is an equally possible translation of the famous phrase from *Der Spiegel*.

Study Questions

1. What is the main point of Dreyfus's article? Why do you think it might be correct? Why do you think it might be wrong? Explain and defend your answers.

2. What assumptions does Heidegger make about technology? about ethics? Do you think these assumptions are correct? Explain and defend your answers.

3. If society accepts Heidegger's main conclusions, what consequences might follow? Would these consequences be desirable? Explain and defend your answers.

4. Dreyfus recognizes that Heidegger distinguishes between the danger of technological catastrophe and the supreme danger of losing an understanding of being. What value consequences does this distinction mark?

5. Does Heidegger's call for a new cultural paradigm require us to change our life habits? Why or why not? Explain.

6. Why does Heidegger, according to Dreyfus, believe that it is not enough to understand technology in terms of instrument applications and values?

7. Why is Heidegger's distinction between calculative and meditative thinking important?

2.3

Technologies as Forms of Life

Langdon Winner

From the early days of manned space travel comes a story that exemplifies what is most fascinating about the human encounter with modern technology. Orbiting the earth aboard *Friendship 7* in February 1962, astronaut John Glenn noticed something odd. His view of the planet was virtually unique in human experience; only Soviet pilots Yuri Gagarin and Gherman Titov had preceded him in orbital flight. Yet as he watched the continents and oceans moving beneath him, Glenn began to feel that he had seen it all before. Months of simulated space shots in sophisticated training machines and centrifuges had affected his ability to respond. In the words of chronicler Tom Wolfe, "The world demanded awe, because this was a voyage through the stars. But he couldn't feel it. The backdrop of the event, the stage, the environment, the true orbit . . . was not the vast reaches of the universe. It was the simulators. *Who could possibly understand this?*"[1] Synthetic conditions generated in the training center had begun to seem more "real" than the actual experience.

It is reasonable to suppose that a society thoroughly committed to making artificial realities would have given a great deal of thought to the nature of that commitment. One might expect, for example, that the philosophy of technology would be a topic widely discussed by scholars and technical professionals, a lively field of inquiry often chosen by students at our universities and technical institutes. One might even think that the basic issues in this field would be well defined, its central controversies well worn. However, such is not the case. At this late date in the development of our industrial/technological civiliza-

Originally published in *The Whale and the Reactor*. Copyright © 1986 by University of Chicago Press. Published by University of Chicago Press, 1986.

tion the most accurate observation to be made about the philosophy of technology is that there really isn't one.

The basic task for a philosophy of technology is to examine critically the nature and significance of artificial aids to human activity. That is its appropriate domain of inquiry, one that sets it apart from, say, the philosophy of science. Yet if one turns to the writings of twentieth-century philosophers, one finds astonishingly little attention given to questions of that kind. The eight-volume *Encyclopedia of Philosophy*, a recent compendium of major themes in various traditions of philosophical discourse, contains no entry under the category "technology."[2] Neither does that work contain enough material under possible alternative headings to enable anyone to piece together an idea of what a philosophy of technology might be.

True, there are some writers who have taken up the topic. The standard bibliography in the philosophy of technology lists well over a thousand books and articles in several languages by nineteenth- and twentieth-century authors.[3] But reading through the material listed shows, in my view, little of enduring substance. The best writing on this theme comes to us from a few powerful thinkers who have encountered the subject in the midst of much broader and ambitious investigations—for example, Karl Marx in the development of his theory of historical materialism or Martin Heidegger as an aspect of his theory of ontology. It may be, in fact, that the philosophy is best seen as a derivative of more fundamental questions. For despite the fact that nobody would deny its importance to an adequate understanding of the human condition, technology has never joined epistemology, metaphysics, esthetics, law, science, and politics as a fully respectable topic for philosophical inquiry.

Engineers have shown little interest in filling this void. Except for airy pronouncements in yearly presidential addresses at various engineering societies, typically ones that celebrate the contributions of a particular technical vocation to the betterment of humankind, engineers appear unaware of any philosophical questions their work might entail. As a way of starting a conversation with my friends in engineering, I sometimes ask, "What are the founding principles of your discipline?" The question is always greeted with puzzlement. Even when I explain what I am after, namely, a coherent account of the nature and significance of the branch of engineering in which they are involved, the question still means nothing to them. The scant few who raise important first questions about their technical professions are usually seen by their colleagues as dangerous cranks and radicals. If Socrates' suggestion that the "unexamined life is not worth living" still holds, it is news to most engineers.[4]

Technological Somnambulism

Why is it that the philosophy of technology has never really gotten under way? Why has a culture so firmly based upon countless sophisticated instruments, techniques, and systems remained so steadfast in its reluctance to examine its own foundations? Much of the answer can be found in the astonishing hold the idea of "progress" has exercised on social thought during the industrial age. In the twentieth century it is usually taken for granted that the only reliable sources for improving the human condition stem from new machines, techniques, and chemicals. Even the recurring environmental and social ills that have accompanied technological advancement have rarely dented this faith. It is still a prerequisite that the person running for public office swear his or her unflinching confidence in a positive link between technical development and human well-being and affirm that the next wave of innovations will surely be our salvation.

There is, however, another reason why the philosophy of technology has never gathered much steam. According to conventional views, the human relationship to technical things is too obvious to merit serious reflection. The deceptively reasonable notion that we have inherited from much earlier and less complicated times divides the range of possible concerns about technology into two basic categories: *making* and *use*. In the first of these our attention is drawn to the matter of "how things work" and of "making things work." We tend to think that this is a fascination of certain people in certain occupations, but not for anyone else. "How things work" is the domain of inventors, technicians, engineers, repairmen, and the like who prepare artificial aids to human activity and keep them in good working order. Those not directly involved in the various spheres of "making" are thought to have little interest in or need to know about the materials, principles, or procedures found in those spheres.

What the others do care about, however, are tools and uses. This is understood to be a straightforward matter. Once things have been made, we interact with them on occasion to achieve specific purposes. One picks up a tool, uses it, and puts it down. One picks up a telephone, talks on it, and then does not use it for a time. A person gets on an airplane, flies from point A to point B, and then gets off. The proper interpretation of the meaning of technology in the mode of use seems to be nothing more complicated than an occasional, limited, and nonproblematic interaction.

The language of the notion of "use" also includes standard terms that enable us to interpret technologies in a range of moral contexts. Tools can be "used well or poorly" and for "good or bad purposes"; I can use my knife to slice a loaf of bread or to stab the next person that walks by. Because technological objects and processes have a promiscuous utility, they are taken to be fundamentally neutral as regards their moral standing.

The conventional idea of what technology is and what it means, an idea powerfully reinforced by familiar terms used in everyday language, needs to be overcome if a critical philosophy of technology is to move ahead. The crucial weakness of the conventional idea is that it disregards the many ways in which technologies provide structure for human activity. Since, according to accepted wisdom, patterns that take shape in the sphere of "making" are of interest to practitioners alone, and since the very essence of "use" is its occasional, innocuous, nonstructuring occurrence, any further questioning seems irrelevant.[5]

If the experience of modern society shows us anything, however, it is that technologies are not merely aids to human activity, but also powerful forces acting to reshape that activity and its meaning. The introduction of a robot to an industrial workplace not only increases productivity, but often radically changes the process of production, redefining what "work" means in that setting. When a sophisticated new technique or instrument is adopted in medical practice, it transforms not only what doctors do, but also the ways people think about health, sickness, and medical care. Widespread alterations of this kind in techniques of communication, transportation, manufacturing, agriculture, and the like are largely what distinguishes our times from early periods of human history. The kinds of things we are apt to see as "mere" technological entities become much more interesting and problematic if we begin to observe how broadly they are involved in conditions of social and moral life.

It is true that recurring patterns of life's activity (whatever their origins) tend to become unconscious processes taken for granted. Thus, we do not pause to reflect upon how we speak a language as we are doing so or the motions we go through in taking a shower. There is, however, one point at which we may become aware of a pattern taking shape—the very first time we encounter it. An opportunity of that sort occurred several years ago at the conclusion of a class I was teaching. A student came to my office on the day term papers were due and told me his essay would be late. "It crashed this morning," he explained. I immediately interpreted this as a "crash" of the conceptual variety, a flimsy array of arguments and observations that eventually collapses under the weight of its own ponderous absurdity. Indeed, some of my own papers have "crashed" in exactly that manner. But this was not the kind of mishap that had befallen this particular fellow. He went on to explain that his paper had been composed on a computer terminal and that it had been stored in a time-sharing minicomputer. It some-times happens that the machine "goes down" or "crashes," making everything that happens in and around it stop until the computer can be "brought up," that is, restored to full functioning.

As I listened to the student's explanation, I realized that he was tell-ing me about the facts of a particular form of activity in modern life in which he and others similarly situated were already involved and that I had better get ready for. I remembered J. L. Austin's little essay "A Plea for Excuses" and noticed that the student and I were negotiating one of the boundaries of contemporary moral life—where and how one gives and accepts an excuse in a particular technology-mediated situation.[6] He was, in effect, asking me to recognize a new world of parts and pieces and to acknowledge appropriate practices and expec-tations that hold in that world. From then on, a knowledge of this situation would be included in my understanding of not only "how things work" in that generation of computers, but also how we do things as a consequence, including which rules to follow when the machines break down. Shortly thereafter I got used to computers crashing, disrupting hotel reservations, banking, and other everyday transactions; eventually, my own papers began crashing in this new way.

Some of the moral negotiations that accompany technological change eventually become matters of law. In recent times, for example, a number of activities that employ computers as their operating me-dium have been legally defined as "crimes." Is unauthorized access to a computerized data base a criminal offense? Given the fact that elec-tronic information is in the strictest sense intangible, under what con-ditions is it "property" subject to theft? The law has had to stretch and reorient its traditional categories to encompass such problems, creating whole new classes of offenses and offenders.

The ways in which technical devices tend to engender distinctive worlds of their own can be seen in a more familiar case. Picture two men traveling in the same direction along a street on a peaceful, sunny day, one of them afoot and the other driving an automobile. The pedes-trian has a certain flexibility of movement: he can pause to look in a shop window, speak to passersby, and reach out to pick a flower from a sidewalk garden. The driver, although he has the potential to move much faster, is constrained by the enclosed space of the automobile, the physical dimensions of the highway, and the rules of the road. His realm is spatially structured by his intended destination, by a periph-ery of more-or-less irrelevant objects (scenes for occasional side glances), and by more important objects of various kinds—moving and parked cars, bicycles, pedestrians, street signs, etc., that stand in his way. Since the first rule of good driving is to avoid hitting things, the immediate environment of the motorist becomes a field of obstacles.

Imagine a situation in which the two persons are next-door neighbors. The man in the automobile observes his friend strolling along the street and wishes to say hello. He slows down, honks his horn, rolls down the window, sticks out his head, and shouts across the street. More likely than not the pedestrian will be startled or annoyed by the sound of the horn. He looks around to see what's the matter and tries to recognize who can be yelling at him across the way. "Can you come to dinner Saturday night?" the driver calls out over the street noise. "What?" the pedestrian replies, straining to understand. At that moment another car to the rear begins honking to break up the temporary traffic jam. Unable to say anything more, the driver moves on.

What we see here is an automobile collision of sorts, although not one that causes bodily injury. It is a collision between the *world* of the driver and that of the pedestrian. The attempt to extend a greeting and invitation, ordinarily a simple gesture, is complicated by the presence of a technological device and its standard operating conditions. The communication between the two men is shaped by an incompatibility of the form of locomotion known as walking and a much newer one, automobile driving. In cities such as Los Angeles, where the physical landscape and prevailing social habits assume everyone drives a car, the simple act of walking can be cause for alarm. The U.S. Supreme Court decided one case involving a young man who enjoyed taking long walks late at night through the streets of San Diego and was repeatedly arrested by police as a suspicious character. The Court decided in favor of the pedestrian, noting that he had not been engaged in burglary or any other illegal act. Merely traveling by foot is not yet a crime.[7]

Knowing how automobiles are made, how they operate, and how they are used and knowing about traffic laws and urban transportation policies does little to help us understand how automobiles affect the texture of modern life. In such cases a strictly instrumental/functional understanding fails us badly. What is needed is an interpretation of the ways, both obvious and subtle, in which everyday life is transformed by the mediating role of technical devices. In hindsight the situation is clear to everyone. Individual habits, perceptions, concepts of self, ideas of space and time, social relationships, and moral and political boundaries have all been powerfully restructured in the course of modern technological development. What is fascinating about this process is that societies involved in it have quickly altered some of the fundamental terms of human life without appearing to do so. Vast transformations in the structure of our common world have been undertaken with little attention to what those alterations mean. Judgments about technology have been made on narrow grounds, paying attention to such matters as whether a new device serves a particular need, performs more efficiently than its predecessor, makes a profit, or provides a convenient service. Only later does the broader significance of the choice become clear, typically as a series of surprising "side effects" or "secondary consequences." But it seems characteristic of our culture's involvement with technology that we are seldom inclined to examine, discuss, or judge pending innovations with broad, keen awareness of what those changes mean. In the technical realm we repeatedly enter into a series of social contracts, the terms of which are revealed only after the signing.

It may seem that the view I am suggesting is that of technological determinism: the idea that technological innovation is the basic cause of changes in society and that human beings have little choice other than to sit back and watch this ineluctable process unfold. But the concept of determinism is much too strong, far too sweeping in its implications to provide an adequate theory. It does little justice to the genuine choices that arise, in both principle and practice, in the course of technical and social transformation. Being saddled with it is like attempting to describe all instances of sexual intercourse based only on the concept of rape. A more revealing notion, in my view, is that of technological somnambulism. For the interesting puzzle in our times is that we so willingly sleepwalk through the process of reconstituting the conditions of human existence.

Beyond Impacts and Side Effects

Social scientists have tried to awaken the sleeper by developing methods of technology assessment. The strength of these methods is that they shed light on phenomena that were previously overlooked. But an unfortunate shortcoming of technology assessment is that it tends to see technological change as a "cause" and everything that follows as an "effect" or "impact." The role of the researcher is to identify, observe, and explain these effects. This approach assumes that the causes have already occurred or are bound to do so in the normal course of events. Social research boldly enters the scene to study the "consequences" of the change. After the bulldozer has rolled over us, we can pick ourselves up and carefully measure the treadmarks. Such is the impotent mission of technological "impact" assessment.

A somewhat more farsighted version of technology assessment is sometimes used to predict which changes are likely to happen, the "social impacts of computerization" for example. With these forecasts at its disposal, society is, presumably, better able to chart its course. But, once again, the attitude in which the predictions are offered usually suggests that the "impacts" are going to happen in any case. Assertions of the sort "Computerization will bring about a revolution in the way we educate our children" carry the strong implication that those who will experience the change are obliged simply to endure it. Humans must adapt. That is their destiny. There is no tampering with the source of change, and only minor modifications are possible at the point of impact (perhaps some slight changes in the fashion contour of this year's treadmarks).

But we have already begun to notice another view of technological development, one that transcends the empirical and moral shortcomings of cause-and-effect models. It begins with the recognition that as technologies are being built and put to use, significant alterations in patterns of human activity and human institutions are already taking place. New worlds are being made. There is nothing "secondary" about this phenomenon. It is, in fact, the most important accomplishment of any new technology. The construction of a technical system that involves human beings as operating parts brings a reconstruction of social roles and relationships. Often this is a result of a new system's own operating requirements: it simply will not work unless human behavior changes to suit its form and process. Hence, the very act of using the kinds of machines, techniques, and systems available to us generates patterns of activities and expectations that soon become "second nature." We do indeed "use" telephones, automobiles, electric lights, and computers in the conventional sense of picking them up and putting them down. But our world soon becomes one in which telephony, automobility, electric lighting, and computing are forms of life in the most powerful sense: life would scarcely be thinkable without them.

My choice of the term "forms of life" in this context derives from Ludwig Wittgenstein's elaboration of that concept in *Philosophical Investigations*. In his later writing Wittgenstein sought to overcome an extremely narrow view of the structure of language then popular among philosophers, a view that held language to be primarily a matter of naming things and events. Pointing to the richness and multiplicity of the kinds of expression or "language games" that are a part of everyday speech, Wittgenstein argued that "the speaking of language is a part of an activity, or of a form of life."[8] He gave a variety of examples—the giving of orders, speculating about events, guessing riddles, making up stories, forming and testing hypotheses, and so forth—to indicate the wide range of language games involved in various "forms of life." Whether he meant to suggest that these are patterns that occur naturally to all human beings or that they are primarily cultural conventions that can change with time and setting is a question open to dispute.[9] For the purposes here, what matters is not the ultimate philosophical status of Wittgenstein's concept but its suggestiveness in helping us to overcome another widespread and extremely narrow conception: our normal understanding of the meaning of technology in human life.

As they become woven into the texture of everyday existence, the devices, techniques, and systems we adopt shed their tool-like qualities to become part of our very humanity. In an important sense we become the beings who work on assembly lines, who talk on telephones, who do our figuring on pocket calculators, who eat processed foods, who clean our homes with powerful chemicals. Of course, working, talking, figuring, eating, cleaning, and such things have been parts of human activity for a very long time. But technological innovations can radically alter these common patterns and on occasion generate entirely new ones, often with surprising results. The role television plays in our society offers some poignant examples. None of those who worked to perfect the technology of television in its early years and few of those who brought television sets into their homes ever intended the device to be employed as the universal babysitter. That, however, has become one of television's most common functions in the modern home. Similarly, if anyone in the 1930s had predicted people would eventually be watching seven hours of television each day, the forecast would have been laughed away as absurd. But recent surveys indicate that we Americans do spend that much time, roughly one-third of our lives, staring at the tube. Those who wish to reassert freedom of choice in the matter sometimes observe, "You can always turn off your TV." In a trivial sense that is true. At least for the time being the on/off button is still included as standard equipment on most sets (perhaps someday it will become optional). But given how central television has become to the content of everyday life, how it has become the accustomed topic of conversation in workplaces, schools, and other social gatherings, it is apparent that television is a phenomenon that, in the larger sense, cannot be "turned off" at all. Deeply insinuated into people's perceptions, thoughts, and behavior, it has become an indelible part of modern culture.

Most changes in the content of everyday life brought on by technology can be recognized as versions of earlier patterns. Parents have always had to entertain and instruct children and to find ways of keeping the little ones out of their hair. Having youngsters watch several hours of television cartoons is, in one way of looking at the matter, merely a new method for handling this age-old task, although the "merely" is of no small significance. It is important to ask, Where, if at all, have modern technologies added *fundamentally new* activities to the range of things human beings do? Where and how have innovations in science and technology begun to alter the very *conditions of life* itself? Is computer programming only a powerful recombination of forms of life known for ages—doing mathematics, listing, sorting, planning, organizing, etc.—or is it something unprecedented? Is industrialized agribusiness simply a renovation of older ways of farming, or does it amount to an entirely new phenomenon?

Certainly, there are some accomplishments of modern technology, manned air flight, for example, that are clearly altogether novel. Flying in airplanes is not just another version of modes of travel previously known; it is something new. Although the hope of humans flying is as old as the myth of Daedalus and Icarus or the angels of the *Old Testament*, it took a certain kind of modern machinery to realize the dream in practice. Even beyond the numerous breakthroughs that have pushed the boundaries of human action, however, lie certain kinds of changes now on the horizon that would amount to a fundamental change in the conditions of human life itself. One such prospect is that of altering human biology through genetic engineering. Another is the founding of permanent settlements in outer space. Both of these possibilities call into question what it means to be human and what constitutes "the human condition."[10] Speculation about such matters is now largely the work of science fiction, whose notorious perversity as a literary genre signals the troubles that lie in wait when we begin thinking about becoming creatures fundamentally different from any the earth has seen. A great many futuristic novels are blatantly technopornographic.

But, on the whole, most of the transformations that occur in the wake of technological innovation are actually variations of very old patterns. Wittgenstein's philosophically conservative maxim "What has to be accepted, the given, is—so one could say—*forms of life*" could well be the guiding rule of a phenomenology of technical practice.[11] For instance, asking a question and awaiting an answer, a form of interaction we all know well, is much the same activity whether it is a person we are confronting or a computer. There are, of course, significant differences between persons and computers (although it is fashionable in some circles to ignore them). Forms of life that we mastered before the coming of the computer shape our expectations as we begin to use the instrument. One strategy of software design, therefore, tries to "humanize" the computers by having them say "Hello" when the user logs in or having them respond with witty remarks when a person makes an error. We carry with us highly structured anticipations about entities that appear to participate, if only minimally, in forms of life and associated language games that are parts of human culture. Those anticipations provide much of the persuasive power of those who prematurely claim great advances in "artificial intelligence" based on narrow but impressive demonstrations of computer performance. But then children have always fantasized that their dolls were alive and talking.

The view of technologies as forms of life I am proposing has its clearest beginnings in the writings of Karl Marx. In Part I of *The German Ideology*, Marx and Engels explain the relationship of human individuality and material conditions of production as follows: "The way in which men produce their means of subsistence depends first of all on the nature of the means of subsistence they actually find in existence and have to reproduce. This mode of production must not be considered simply as being the reproduction of the physical existence of the individuals. Rather it is a definite form of activity of these individuals, a definite form of expressing their life, a definite *mode of life* on their part. As individuals express their life, so they are."[12]

Marx's concept of production here is a very broad and suggestive one. It reveals the total inadequacy of any interpretation that finds social change a mere "side effect" or "impact" of technological innovation. While he clearly points to means of production that sustain life in an immediate, physical sense, Marx's view extends to a general understanding of human development in a world of diverse natural resources, tools, machines, products, and social relations. The notion is clearly not one of occasional human interaction with devices and material conditions that leave individuals unaffected. By changing the shape of material things, Marx observes, we also change ourselves. In this process human beings do not stand at the mercy of a great deterministic punch press that cranks out precisely tailored persons at a certain rate during a given historical period. Instead, the situation Marx describes is one in which individuals are actively involved in the daily creation and recreation, production and reproduction of the world in which they live. Thus, as they employ tools and techniques, work in social labor arrangements, make and consume products, and adapt their behavior to the material conditions they encounter in their natural and artificial environment, individuals realize possibilities for human existence that are inaccessible in more primitive modes of production.

Marx expands upon this idea in "The Chapter on Capital" in the *Grundrisse*. The development of forces of production in history, he argues, holds the promise of the development of a many-sided individuality in all human beings. Capital's unlimited pursuit of wealth leads it to develop the productive powers of labor to a state "where the possession and preservation of general wealth require a lesser labour time of society as a whole, and where the labouring society relates scientifically to the process of its progressive reproduction, its reproduction in constantly greater abundance." This movement toward a general form of wealth "creates the material elements for the development of the rich individuality which is all-sided in its production as in its consumption, and whose labour also therefore appears no longer as labour, but as the full development of activity itself."[13]

If one has access to tools and materials of woodworking, a person can develop the human qualities found in the activities of carpentry. If one is able to employ the instruments and techniques of music making, one can become (in that aspect of one's life) a musician. Marx's ideal here, a variety of materialist humanism, anticipates that in a properly structured society under modern conditions of production, people would engage in a very wide range of activities that enrich their individuality along many dimensions. It is that promise which, he argues, the institutions of capitalism thwart and cripple.[14]

As applied to an understanding of technology, the philosophies of Marx and Wittgenstein direct our attention to the fabric of everyday existence. Wittgenstein points to a vast multiplicity of cultural practices that comprise our common world. Asking us to notice "what we say when," his approach can help us recognize the way language reflects the content of technical practice. It makes sense to ask, for example, how the adoption of digital computers might alter the way people think of their own faculties and activities. If Wittgenstein is correct, we would expect that changes of this kind would appear, sooner or later, in the language people use to talk about themselves. Indeed, it has now become commonplace to hear people say "I need to access your data." "I'm not programmed for that." "We must improve our interface." "The mind is the best computer we have."

Marx, on the other hand, recommends that we see the actions and interactions of everyday life within an enormous tapestry of historical developments. On occasion, as in the chapter on "Machinery and Large-Scale Industry" in *Capital*, his mode of interpretation also includes a place for a more microscopic treatment of specific technologies in human experience.[15] But on the whole his theory seeks to explain very large patterns, especially relationships between different social classes, that unfold at each stage in the history of material production. These developments set the stage for people's ability to survive and express themselves, for their ways of being human.

Return to Making

To invoke Wittgenstein and Marx in this context, however, is not to suggest that either one or both provide a sufficient basis for a critical philosophy of technology. Proposing an attitude in which forms of life must be accepted as "the given," Wittgenstein decides that philosophy "leaves everything as it is."[16] Although some Wittgensteinians are eager to point out that this position does not necessarily commit the philosopher to conservatism in an economic or political sense, it does seem that as applied to the study of forms of life in the realm of technology, Wittgenstein leaves us with little more than a passive traditionalism. If one hopes to interpret technological phenomena in a way that suggests positive judgments and actions, Wittgensteinian philosophy leaves much to be desired.

In a much different way Marx and Marxism contain the potential for an equally woeful passivity. This mode of understanding places its hopes in historical tendencies that promise human emancipation at some point. As forces of production and social relations of production develop and as the proletariat makes its way toward revolution, Marx and his orthodox followers are willing to allow capitalist technology, for example, the factory system, to develop to its farthest extent. Marx and Engels scoffed at the utopians, anarchists, and romantic critics of industrialism who thought it possible to make moral and political judgments about the course a technological society ought to take and to influence that path through the application of philosophical principles. Following this lead, most Marxists have believed that while capitalism is a target to be attacked, technological expansion is entirely good in itself, something to be encouraged without reservation. In its own way, then, Marxist theory upholds an attitude as nearly lethargic as the Wittgensteinian decision to "leave everything as it is." The famous eleventh thesis on Feuerbach—"The philosophers have only interpreted the world in various ways; the point, however, is to change it"—conceals an important qualification: that judgment, action, and change are ultimately products of history. In its view of technological development Marxism anticipates a history of rapidly evolving material productivity, an inevitable course of events in which attempts to propose moral and political limits have no place. When socialism replaces capitalism, so the promise goes, the machine will finally move into high gear, presumably releasing humankind from its age-old miseries.

Whatever their shortcomings, however, the philosophies of Marx and Wittgenstein share a fruitful insight: the observation that social activity is an ongoing process of world-making. Throughout their lives people come together to renew the fabric of relationships, transactions, and meanings that sustain their common existence. Indeed, if they did not engage in this continuing activity of material and social production, the human world would literally fall apart. All social roles and frameworks—from the most rewarding to the most oppressive—must somehow be restored and reproduced with the rise of the sun each day.

From this point of view, the important question about technology becomes, As we "make things work," what kind of *world* are we making? This suggests that we pay attention not only to the making of physical instruments and processes, although that certainly remains important, but also to the production of psychological, social, and political conditions as a part of any significant technical change. Are we going to design and build circumstances that enlarge possibilities for growth in human freedom, sociability, intelligence, creativity, and self-government? Or are we headed in an altogether different direction?

It is true that not every technological innovation embodies choices of great significance. Some developments are more-or-less innocuous; many create only trivial modifications in how we live. But in general, where there are substantial changes being made in what people are doing and at a substantial investment of social resources, then it always pays to ask in advance about the qualities of the artifacts, institutions, and human experiences currently on the drawing board.

Inquiries of this kind present an important challenge to all disciplines in the social sciences and humanities. Indeed, there are many historians, anthropologists, sociologists, psychologists, and artists whose work sheds light on long-overlooked human dimensions of technology. Even engineers and other technical professionals have much to contribute here when they find courage to go beyond the narrow-gauge categories of their training.

The study of politics offers its own characteristic route into this territory. As the political imagination confronts technologies as forms of life, it should be able to say something about the choices (implicit or explicit) made in the course of technological innovation and the grounds for making those choices wisely. That is a task I take up elsewhere. Through technological creation and many other ways as well, we make a world for each other to live in. Much more than we have acknowledged in the past, we must admit our responsibility for what we are making.

Notes

1. Tom Wolfe, *The Right Stuff* (New York: Bantam Books, 1980), 270.
2. *The Encyclopedia of Philosophy,* 8 vols., Paul Edwards (editor-in-chief) (New York: Macmillan: 1967).

3. *Bibliography of the Philosophy of Technology*, Carl Mitcham and Robert Mackey (eds.) (Chicago: University of Chicago Press, 1973).

4. There are, of course, exceptions to this general attitude. See Stephen H. Unger, *Controlling Technology: Ethics and the Responsible Engineer* (New York: Holt, Rinehart, and Winston, 1982).

5. An excellent corrective to the general thoughtfulness about "making" and "use" is to be found in Carl Mitcham, "Types of Technology," in *Research in Philosophy and Technology*, Paul Durbin (ed.) (Greenwich, Conn.: JAI Press, 1978), 229–294.

6. J. L. Austin, *Philosophical Papers* (Oxford: Oxford University Press, 1961), 123–153.

7. See William Kolender et al., "Petition v. Edward Lawson," *Supreme Court Reporter* 103: 1855–1867, 1983. Edward Lawson had been arrested approximately fifteen times on his long walks and refused to provide identification when stopped by the police. Lawson cited his rights guaranteed by the Fourth and Fifth Amendments of the U.S. Constitution. The Court found the California vagrancy statute requiring "credible and reliable" identification to be unconstitutionally vague. See also Jim Mann, "State Vagrancy Law Voided as Overly Vague," *Los Angeles Times*, May 3, 1983, 1, 19.

8. Ludwig Wittgenstein, *Philosophical Investigations*, ed. 3, translated by G. E. M. Anscombe, with English and German indexes (New York: Macmillan, 1958), 11e.

9. Hanna Pitkin, *Wittgenstein and Justice: On the Significance of Ludwig Wittgenstein for Social and Political Thought* (Berkeley: University of California Press, 1972), 293.

10. For a thorough discussion of this idea, see Hannah Arendt, *The Human Condition* (Chicago: University of Chicago Press, 1958); and Hannah Arendt, *Willing*, vol. II of *The Life of the Mind* (New York: Harcourt Brace Jovanovich, 1978).

11. *Philosophical Investigations*, 226e.

12. Karl Marx and Friedrich Engels, "The German Ideology," in *Collected Works*, vol. 5 (New York: International Publishers, 1976), 31.

13. Karl Marx, *Grundrisse*, translated with a foreword by Martin Nicolaus (Harmondsworth, England: Penguin Books, 1973), 325.

14. An interesting discussion of Marx in this respect is Kostas Axelos' *Alienation, Praxis and Technē in the Thought of Karl Marx*, translated by Ronald Bruzina (Austin: University of Texas Press, 1976).

15. Karl Marx, *Capital*, vol. 1, translated by Ben Fowkes, with an introduction by Ernest Mandel (Harmondsworth, England: Penguin Books, 1976), chap. 15.

16. *Philosophical Investigations*, 49e.

Study Questions

1. What is the main point of Winner's article? Why do you think it might be correct? Why do you think it might be wrong? Explain and defend your answers.

2. What assumptions does Winner make about technology? about ethics? Do you think these assumptions are correct? Explain and defend your answers.
3. If society accepts Winner's main conclusions, what consequences might follow? Would these consequences be desirable? Explain and defend your answers.
4. Why does Winner say that technologies are "forms of life"?
5. Why is the influence of technological representations of life problematic from a moral point of view? Discuss this response in terms of different ethical theories.
6. What are the foundational questions about technology that Winner raises? What do these have to do with questions of moral responsibility?

2.4

The Role of Technology in Society

Emmanuel G. Mesthene

Emmanuel Mesthene's essay, "The Role of Technology in Society," and the piece that follows, "Technology: The Opiate of the Intellectuals" by John Mc-Dermott, constitute a classic debate over the role of technology in society. Both articles date from the late 1960s, when the war in Vietnam was in full swing and intellectual life in the United States was torn by bitter conflicts between the "establishment" and the "New Left."

Mesthene's perspective is that of the establishment. The article originated as the overview section of the fourth annual report of the Harvard Program on Technology and Society, an interdisciplinary program of academic studies funded by a $5 million grant from IBM. Mesthene was the program's director, and this essay was his general statement of what the program had learned about the implications of technological change for society.

According to Mesthene, technology appears to induce social change in two ways: by creating new opportunities and by generating new problems for individuals and for societies. "It has both positive and negative effects, and it usually has the two at the same time and in virtue of each other." By enlarging the realm of goal choice, or by altering the relative costs associated with different values, technology can induce value change. The essay deals briefly with the value implications of economic change, the impact of technology on religion, and the impact of technology on the individual. In all these areas, technology is seen to have two faces, one positive and one negative.

Emmanuel G. Mesthene directed the Harvard Program on Technology and Society from 1964 through 1974, following 11 years with the Rand Corpora-

Originally published in *Technology and Culture* 10:4 (1969), pp. 489–536. Reprinted by permission of the author and The University of Chicago Press. Further published as chapter 6 in *Technology and the Future.* Copyright © 1997 by St. Martin's Press. Published by St. Martin's Press, 1997.

tion. He joined Rutgers University in 1974, serving as the dean of Livingston College for several years, then as distinguished professor of philosophy and professor of management. Mesthene died in 1990. Among his books were Technological Change: Its Impact on Man and Society *(1970) and* How Language Makes Us Know *(1964).*

Social Change

Three Unhelpful Views about Technology

While a good deal of research is aimed at discerning the particular effects of technological change on industry, government, or education, systematic inquiry devoted to seeing these effects together and to assessing their implications for contemporary society as a whole is relatively recent and does not enjoy the strong methodology and richness of theory and data that mark more established fields of scholarship. It therefore often has to contend with facile or one-dimensional views about what technology means for society. Three such views, which are prevalent at the present time, may be mildly caricatured somewhat as follows.

The first holds that technology is an unalloyed blessing for man and society. Technology is seen as the motor of all progress, as holding the solution to most of our social problems, as helping to liberate the individual from the clutches of a complex and highly organized society, and as the source of permanent prosperity; in short, as the promise of utopia in our time. This view has its modern origins in the social philosophies of such 19th-century thinkers as Saint-Simon, Karl Marx, and Auguste Comte. It tends to be held by many scientists and engineers, by many military leaders and aerospace industrialists, by people who believe that man is fully in command of his tools and his destiny, and by many of the devotees of modern techniques of "scientific management."

A second view holds that technology is an unmitigated curse. Technology is said to rob people of their jobs, their privacy, their participation in democratic government, and even, in the end, their dignity as human beings. It is seen as autonomous and uncontrollable, as fostering materialistic values and as destructive of religion, as bringing about a technocratic society and bureaucratic state in which the individual is increasingly submerged, and as threatening, ultimately, to poison nature and blow up the world. This view akin to historical "back-to-nature" attitudes toward the world and is propounded mainly by artists, literary commentators, popular social critics, and existentialist philosophers. It is becoming increasingly attractive to many of our youth, and it tends to be held, understandably enough, by segments of the population that have suffered dislocation as a result of technological change.

The third view is of a different sort. It argues that technology as such is not worthy of special notice, because it has been well recognized as a factor in social change at least since the Industrial Revolution, because it is unlikely that the social effects of computers will be nearly so traumatic as the introduction of the factory system in 18th-century England, because research has shown that technology has done little to accelerate the rate of economic productivity since the 1880s, because there has been no significant change in recent decades in the time periods between invention and widespread adoption of new technology, and because improved communications and higher levels of education make people much more adaptable than heretofore to new ideas and to new social reforms required by technology.

While this view is supported by a good deal of empirical evidence, however, it tends to ignore a number of social, cultural, psychological, and political effects of technological change that are less easy to identify with precision. It thus reflects the difficulty of coming to grips with a new or broadened subject matter by means of concepts and intellectual categories designed to deal with older and different subject matters. This view tends to be held by historians, for whom continuity is an indispensable methodological assumption, and by many economists, who find that their instruments measure some things quite well while those of the other social sciences do not yet measure much of anything.

Stripped of caricature, each of these views contains a measure of truth and reflects a real aspect of the relationship of technology and society. Yet they are oversimplifications that do not contribute much to understanding. One can find empirical evidence to support each of them without gaining much knowledge about the actual mechanism by which technology leads to social change or significant insight into its implications for the future. All three remain too uncritical or too partial to guide inquiry. Research and analysis lead to more differentiated conclusions and reveal more subtle relationships.

Some Countervailing Considerations

. . . Whether modern technology and its effects constitute a subject matter deserving of special attention is largely a matter of how technology is defined. The research studies of the Harvard Program on Technology and Society reflect an operating assumption that the meaning of technology includes more than machines. As most serious investigators have found, understanding is not advanced by concentrating single-mindedly on such narrowly drawn yet imprecise questions as "What are the social implications of computers, or lasers, or space technology?" Society and the influences of technology upon it are much too complex for such artificially limited approaches to be meaningful. The opposite error, made by some, is to define technology too broadly by identifying it with rationality in the broadest sense. The term is then operationally meaningless and unable to support fruitful inquiry.

We have found it more useful to define technology as tools in a general sense, including machines, but also including linguistic and intellectual tools and contemporary analytic and mathematical techniques. That is, we define technology as the organization of knowledge for practical purposes. It is in this broader meaning that we can best see the extent and variety of the effects of technology on our institutions and values. Its pervasive influence on our very culture would be unintelligible if technology were understood as no more than hardware.

It is in the pervasive influence of technology that our contemporary situation seems qualitatively different from that of past societies, for three reasons. (1) Our tools are more powerful than any before. The rifle wiped out the buffalo, but nuclear weapons can wipe out man. Dust storms lay whole regions waste, but too much radioactivity in the atmosphere could make the planet uninhabitable. The domestication of animals and the invention of the wheel literally lifted the burden from man's back, but computers could free him from all need to labor. (2) This quality of finality of modern technology has brought our society, more than any before, to explicit awareness of technology as an important determinant of our lives and institutions. (3) As a result, our society is coming to a deliberate decision to understand and control technology to good social purpose and is therefore devoting significant effort to the search for ways to measure the full range of its effects rather than only those bearing principally on the economy. It is this prominence of technology in many dimensions of modern life that seems novel in our time and deserving of explicit attention.

How Technological Change Impinges on Society

It is clearly possible to sketch a more adequate hypothesis about the interaction of technology and society than the partial views outlined above. Technological change would appear to induce or "motor" social change in two principal ways. New technology creates new opportunities for men and societies, and it also generates new problems for them. It has both positive and negative effects, and it usually has the two *at the same time and in virtue of each other.* Thus, industrial technology strengthens the economy, as our measures of growth and productivity show. . . . However, it also induces changes in the relative importance of individual supplying sectors in the economy as new techniques of production alter the amounts and kinds of materials, parts and components, energy, and service inputs used by each industry to produce its output. It thus tends to bring about dislocations of businesses and people as a result of changes in industrial patterns and in the structure of occupations.

The close relationship between technological and social change itself helps to explain why any given technological development is likely to have both positive and negative effects. The usual sequence is that (1) technological advance creates a new opportunity to achieve some desired goal; (2) this requires (except in trivial cases) alterations in social organization if advantage is to be taken of the new opportunity, (3) which means that the functions of existing social structures will be interfered with, (4) with the result that other goals which were served by the older structures are now only inadequately achieved.

As the Meyer-Kain[1] study has shown, for example, improved transportation technology and increased ownership of private automobiles have increased the mobility of businesses and individuals. This has led to altered patterns of industrial and residential location, so that older unified cities are being increasingly transformed into larger metropolitan complexes. The new opportunities for mobility are largely defined to the poor and black populations of the core cities, however, partly for economic reasons, and partly as a result of restrictions on choice of residence by blacks, thus leading to persistent black unemployment despite a generally high level of economic activity. Cities are thus increasingly unable to perform their traditional functions of providing employment opportunities for all segments of their populations and an integrated social environment that can temper ethnic and racial differences. The new urban complexes are neither fully viable economic units nor effective political organizations able to upgrade and integrate their core populations into new economic and social structures. The resulting instability is further aggravated by modern mass communications technology, which heightens the expectations of the poor and the fears of the well-to-do and add frustration and bitterness to the urban crisis. . . .

In all such cases, technology creates a new opportunity and a new problem at the same time. That is why isolating the opportunity or the problem and construing it as the whole answer is ultimately obstructive of rather than helpful to understanding.

How Society Reacts to Technological Change

 The heightened prominence of technology in our society makes the interrelated tasks of profiting from its opportunities and containing its dangers a major intellectual and political challenge of our time.

Failure of society to respond to the opportunities created by new technology means that much actual or potential technology lies fallow, that is, not used at all or is not used to its full capacity. This can mean that potentially solvable problems are left unsolved and potentially achievable goals unachieved, because we waste our technological resources or use them inefficiently. A society has at least as much stake in the efficient utilization of technology as in that of its natural or human resources.

There are often good reasons, of course, for not developing or utilizing a particular technology. The mere fact that it can be developed is not sufficient reason for doing so. The costs of development may be too high in the light of the expected benefits, as in the case of the project to develop a nuclear-powered aircraft. Or, a new technological device may be so dangerous in itself or so inimical to other purposes that it is never developed, as in the cases of Herman Kahn's "Doomsday Machine" and the proposal to "nightlight" Vietnam by reflected sunlight.

But there are also cases where technology lies fallow because existing social structures are inadequate to exploit the opportunities it offers. . . . Community institutions wither for want of interest and participation by residents. City agencies are unable to marshal the skills and take the systematic approach needed to deal with new and intensified problems of education, crime control, and public welfare. Business corporations, finally, which are organized around the expectation of private profit, are insufficiently motivated to bring new technology and management know-how to bear on urban projects where the benefits will be largely social. All these factors combine to dilute what may otherwise be a genuine desire to apply our best knowledge and adequate resources to the resolution of urban tensions and the eradication of poverty in the nation. . . .

Containing the Negative Effects of Technology

The kinds and magnitude of the negative effects of technology are no more independent of the institutional structures and cultural attitudes of society than is realization of the new opportunities that technology offers. In our society, there are individuals or individual firms always on the lookout for new technological opportunities, and large corporations hire scientists and engineers to invent such opportunities. In deciding whether to develop a new technology, individual entrepreneurs engage in calculations of expected benefits and expected costs to themselves, and proceed if the former are likely to exceed the latter. Their calculations do not take adequate account of the probable benefits and costs of the new developments to others than themselves or to society generally. These latter are what economists call external benefits and costs.

The external benefits potential in new technology will thus not be realized by the individual developer and will rather accrue to society as a result of deliberate social action, as has been argued above. Similarly with the external costs. In minimizing only expected costs to himself, the individual decision maker helps to contain only some of the potentially negative effects of the new technology. The external costs and therefore the negative effects on society at large are not of principal concern to him and, in our society, are not expected to be.

Most of the consequences of technology that are causing concern at the present time—pollution of the environment, potential damage to the ecology of the planet, occupational and social dislocations, threats to the privacy and political significance of the individual, social and psychological malaise—are negative externalities of this kind. They are with us in large measure because it has not been anybody's explicit business to foresee and anticipate them. They have fallen between the stools of innumerable individual decisions to develop individual technologies for individual purposes without explicit attention to what all these decisions add up to for society as a whole and for people as human beings. This freedom of individual decision making is a value that we have cherished and that is built into the institutional fabric of our society. The negative effects of technology that we deplore are a measure of what this traditional freedom is beginning to cost us. They are traceable, less to some mystical autonomy presumed to lie in technology, and much more to the autonomy that our economic and political institutions grant to individual decision making. . . .

Measures to control and mitigate the negative effects of technology, however, often appear to threaten freedoms that our traditions still take for granted as inalienable rights of men and good societies, however much they may have been tempered in practice by the social pressures of modern times: the freedom of the market, the freedom of private enterprise, the freedom of the scientist to follow truth wherever it may lead, and the freedom of the individual to pursue his fortune and decide his fate. There is thus set up a tension between the need to control technology and our wish to preserve our values, which leads some people to conclude that technology is inherently inimical to human values. The political effect of this tension takes the form of inability to adjust our decision-making structures to the realities of technology so as to take maximum advantage of the opportunities it offers and so that we can act to contain its potential ill effects before they become so pervasive and urgent as to seem uncontrollable.

To understand why such tensions are so prominent a social consequence of technological change, it becomes necessary to look explicitly at the effects of technology on social and individual values.

Values

Technology as a Cause of Value Change

Technology has a direct impact on values by virtue of its capacity for creating new opportunities. By making possible what was not possible before, it offers individuals and society new options to choose from. For example, space technology makes it possible for the first time to go to the moon or to communicate by satellite and thereby adds those two new options to the spectrum of choices available to society. By adding new options in this way, technology can lead to changes in values in the same way that the appearance of new dishes on the heretofore standard menu of one's favorite restaurant can lead to changes in one's tastes and choices of food. Specifically, technology can lead to value change either (1) by bringing some previously unattainable goal within the realm of choice or (2) by making some values easier to implement than heretofore, that is, by changing the costs associated with realizing them. . . .

One example related to the effect of technological change on values is implicit in our concept of democracy. The ideal we associate with the old New England town meeting is that each citizen should have a direct voice in political decisions. Since this has not been possible, we have elected representatives to serve our interests and vote our opinions. Sophisticated computer technology, however, now makes possible rapid and efficient collection and analysis of voter opinion and could eventually provide for "instant voting" by the whole electorate on any issue presented to it via television a few hours before. It thus raises the possibility of instituting a system of direct democracy and gives rise to tensions between those who would be violently opposed to such a prospect and those who are already advocating some system of participatory democracy.

This new technological possibility challenges us to clarify what we mean by democracy. Do we constitute it as the will of an undifferentiated majority, as the resultant of transient coalitions of different interest groups representing different value commitments, as the considered judgment of the people's elected representatives, or as by and large the kind of government we actually have in the United States, minus the flaws in it that we would like to correct? By bringing us face to face with such questions, technology has the effect of calling society's bluff and thereby preparing the ground for changes in its values.

In the case where technological change alters the relative costs of implementing different values, it impinges on inherent contradictions in our value system. To pursue the same example, modern technology can enhance the values we associate with democracy. But it can also enhance another American value—that of "secular rationality," as sociologists call it—by facilitating the use of scientific and technical expertise in the process of political decision making. This can in turn further reduce citizen participation in the democratic process. Technology thus has the effect of facing us with contradictions in our value system and of calling for deliberate attention to their resolution.

Religion and Values

Much of the unease that our society's emphasis on technology seems to generate among various sectors of society can perhaps be explained in terms of the impact that technology has on religion. The formulations and institutions of religion are not immune to the influences of technological change, for they too tend toward an accommodation to changes in the social milieu in which they function. But one way in which religion functions is as an ultimate belief system that provides legitimation, that is, a "meaning" orientation, to moral and social values. This ultimate meaning orientation, according to Professor Harvey Cox, is even more basic to human existence than the value orientation. When the magnitude or rapidity of social change threatens the credibility of that belief system, therefore, and when the changes are moreover seen as largely the results of technological change, the meanings of human existence that we hold most sacred seem to totter and technology emerges as the villain. . . .

Technology, as noted, creates new possibilities for human choice and action but leaves their disposition uncertain. What its effects will be and what ends it will serve are not inherent in the technology, but depend on what man will do with technology. Technology thus makes possible a future of open-ended options that seems to accord well with the presuppositions of the prophetic tradition. It is in that tradition above others, then, that we may seek the beginnings of a religious synthesis that is both adequate to our time and continuous with what is most relevant in our religious history. But this requires an effort at deliberate religious innovation for which Cox finds insufficient theological ground at the present time. Although it is recognized that religions have changed and developed in the past, conscious innovation in religion has been condemned and is not provided for by the relevant theologies. The main task that technological change poses for theology in the next decades, therefore, is that of deliberate religious innovation and symbol reformulation to take specific account of religious needs in a technological age. . . .

Individual Man in a Technological Age

What do technological change and the social and value changes that it brings with it mean for the life of the individual today? It is not clear that their effects are all one-way. For example, we are often told that today's individual is alienated by the vast proliferation of technical expertise and complex bureaucracies, by a feeling of impotence in the face of "the machine," and by a decline in personal privacy. It is probably true that the social pressures placed on individuals today are more complicated and demanding than they were in earlier times. Increased geographical and occupational mobility and the need to function in large organizations place difficult demands on the individual to conform or "adjust." It is also evident that the privacy of many individuals tends to be encroached upon by sophisticated eavesdropping and surveillance devices, by the accumulation of more and more information about individuals by governmental and many private agencies, and by improvements in information-handling technologies such as the proposed institution of centralized statistical data banks. There is little doubt, finally, that the power, authority, influence, and scope of government are greater today than at any time in the history of the United States.

But, as Professor Edward Shils points out in his study on technology and the individual, there is another, equally compelling side of the coin. First, government seems to be more shy and more lacking in confidence today than ever before. Second, while privacy may be declining in the ways indicated above, it also tends to decline in a sense that most individuals are likely to approve. The average man in Victorian times, for example, probably "enjoyed" much more privacy than today. No one much cared what happened to him, and he was free to remain ignorant, starve, fall ill, and die in complete privacy; that was the "golden age of privacy," as Shils puts it. Compulsory universal education, social security legislation, and public health measures—indeed, the very idea of a welfare state—are all antithetical to privacy in this sense, and it is the rare individual today who is loath to see that kind of privacy go. . . .

Recognition that the impact of modern technology on the individual has two faces, both negative and positive, is consistent with the double effect of technological change that was discussed above. It also suggests that appreciation of that impact in detail may not be achieved in terms of old formulas, such as more or less privacy, more or less government, more or less individuality.

Economic and Political Organization

The Enlarged Scope of Public Decision Making

When technology brings about social changes (as described in the first section of this essay) which impinge on our existing system of values (in ways reviewed in the second section), it poses for society a number of problems that are ultimately political in nature. The term "political" is used here in the broadest sense: it encompasses all of the decision-making structures and procedures that have to do with the allocation and distribution of wealth and power in society. The political organization of society thus includes not only the formal apparatus of the state but also industrial organizations and other private institutions that play a role in the decision-making process. It is particularly important to attend to the organization of the entire body politic when technological change leads to a blurring of once clear distinctions between the public and private sectors of society and to changes in the roles of its principal institutions.

It was suggested above that the political requirements of our modern technological society call for a relatively greater public commitment on the part of individuals than in previous times. The reason for this, stated most generally, is that technological change has the effect of enhancing the importance of public decision making in society, because technology is continually creating new possibilities for social action as well as new problems that have to be dealt with.

A society that undertakes to foster technology on a large scale, in fact, commits itself to social complexity and to facing and dealing with new problems as a normal feature of political life. Not much is yet known with any precision about the political imperatives inherent in technological change, but one may nevertheless speculate about the reasons why an increasingly technological society seems to be characterized by enlargement of the scope of public decision making.

For one thing, the development and application of technology seems to require large-scale, and hence increasingly complex, social concentrations, whether these be large cities, large corporations, big universities, or big government. In instances where technological advance appears to facilitate reduction of such first-order concentrations, it tends to instead enlarge the relevant *system* of social organization, that is, to lead to increased centralization. Thus, the physical dispersion made possible by transportation and communications technologies, as Meyer and Kain have shown, enlarges the urban complex that must be governed as a unit.

A second characteristic of advanced technology is that its effects cover large distances, in both the geographical and social senses of the term. Both its positive and negative features are more extensive. Horsepowered transportation technology was limited in its speed and capacity, but its nuisance value was also limited, in most cases to the owner and to the occupant of the next farm. The supersonic transport can carry hundreds across long distances in minutes, but its noise and vibration damage must also be suffered willy-nilly by everyone within the limits of a swath 3,000 miles long and several miles wide.

The concatenation of increased density (or enlarged system) and extended technological "distance" means that technological applications have increasingly wider ramifications and that increasingly large concentrations of people and organizations become dependent on technological systems. . . . The result is not only that more and more decisions must be social decisions taken in public ways, as already noted, but that, once made, decisions are likely to have a shorter useful life than heretofore. That is partly because technology is continually altering the spectrum of choices and problems that society faces, and partly because any decision taken is likely to generate a need to take ten more.

These speculations about the effects of technology on public decision making raise the problem of restructuring our decision-making mechanisms—including the system of market incentives—so that the increasing number and importance of social issues that confront us can be resolved equitably and effectively.

The Promise and Problems of Scientific Decision Making

There are two further consequences of the expanding role of public decision making. The first is that the latest information-handling devices and techniques tend to be utilized in the decision-making process. This is so (1) because public policy can be effective only to the degree that it is based on reliable knowledge about the actual state of the society, and thus requires a strong capability to collect, aggregate, and analyze detailed data about economic activities, social patterns, popular attitudes, and political trends, and (2) because it is recognized increasingly that decisions taken in one area impinge on and have consequences for other policy areas often thought of as unrelated, so that it becomes necessary to base decisions on a model of society that sees it as a system and that is capable of signaling as many as possible of the probable consequences of a contemplated action.

As Professor Alan F. Westin points out, reactions to the prospect of more decision making based on computerized data banks and scientific management techniques run the gamut of optimism to pessimism mentioned in the opening of this essay. Negative reactions take the form of rising political demands for greater popular participation in decision making, for more equality among different segments of the population, and for greater regard for the dignity of individuals. The increasing dependence of decision making on scientific and technological devices and techniques is seen as posing a threat to these goals, and pressures are generated in opposition to further "rationalization" of decision-making processes. These pressures have the paradoxical effect, however, not of deflecting the supporters of technological decision making from their course, but of spurring them on to renewed effort to save the society before it explodes under planlessness and inadequate administration.

The paradox goes further, and helps to explain much of the social discontent that we are witnessing at the present time. The greater complexity and the more extensive ramifications that technology brings about in society tend to make social processes increasingly circuitous and indirect. The effects of actions are widespread and difficult to keep track of, so that experts and sophisticated techniques are increasingly needed to detect and analyze social events and to formulate policies adequate to the complexity of social issues. The "logic" of modern decision making thus appears to require greater and greater dependence on the collection and analysis of data and on the use of technological devices and scientific techniques. Indeed, many observers would agree that there is an "increasing relegation of questions which used to be matters of political debate to professional cadres of technicians and experts which function almost independently of the democratic political process."[2] In recent times, that process has been most noticeable, perhaps, in the areas of economic policy and national security affairs.

This "logic" of modern decision making, however, runs counter to that element of traditional democratic theory that places high value on direct participation in the political processes and generates the kind of discontent referred to above. If it turns out on more careful examination that direct participation is becoming less relevant to a society in which the connections between causes and effects are long and often hidden—which is an increasingly "indirect" society, in other words—elaboration of a new democratic ethos and of new democratic processes more adequate to the realities of modern society will emerge as perhaps the major intellectual and political challenge of our time.

The Need for Institutional Innovation

The challenge is, indeed, already upon us, for the second consequence of the enlarged scope of public decision making is the need to develop new institutional forms and new mechanisms to replace established ones that can no longer deal effectively with the new kinds of problems with which we are increasingly faced. Much of the political ferment of the present time—over the problems of technology assessment, the introduction of statistical data banks, the extension to domestic problems of techniques of analysis developed for the military services, and the modification of the institutions of local government—is evidence of the need for new institutions. . . .

Conclusion

As we review what we are learning about the relationship of technological and social change, a number of conclusions begin to emerge. We find, on the one hand, that the creation of new physical possibilities and social options by technology tends toward and appears to require the emergence of new values, new forms of economic activity, and new political organizations. On the other hand, technological change also poses problems of social and psychological displacement.

The two phenomena are not unconnected, nor is the tension between them new: man's technical prowess always seems to run ahead of his ability to deal with and profit from it. In America, especially, we are becoming adept at extracting the new techniques, the physical power, and the economic productivity that are inherent in our knowledge and its associated technologies. Yet we have not fully accepted the fact that our progress in the technical realm does not leave our institutions, values, and political processes unaffected. Individuals will be fully integrated into society only when we can extract from our knowledge not only its technological potential but also its implications for a system of values and a social, economic, and political organization appropriate to a society in which technology is so prevalent. . . .

Notes

1. Unless otherwise noted, studies referred to in this article are described in the Fourth Annual Report (1967–68) of the Harvard University Program on Technology and Society.
2. Harvey Brooks, "Scientific Concepts and Cutural Change," in G. Holton, ed., *Science and Culture* (Boston: Houghton Mifflin, 1965), p. 71.

Study Questions

1. What is the main point of Mesthene's article? Why do you think it might be correct? Why do you think it might be wrong? Explain and defend your answers.
2. What assumptions does Mesthene make about technology? about ethics? Do you think these assumptions are correct? Explain and defend your answers.
3. If society accepts Mesthene's main conclusions, what consequences might follow? Would these consequences be desirable? Explain and defend your answers.
4. What does Mesthene mean by "externalities"? How does he suggest that society cope with these externalities?
5. Why does Mesthene use the example of rapid transit technologies away from and into cities? Do you think his example makes an important point? Explain.
6. Do you agree with Mesthene's assessment of advances in medical technologies? Why or why not? Explain.

Technology: The Opiate
of the Intellectuals

John McDermott

Several months after the report containing Emmanuel Mesthene's article was published by Harvard, a sharply critical review-essay by John McDermott appeared in the New York Review of Books. *McDermott's piece, which follows here, is not a point-by-point analysis or rebuttal of the Mesthene work. Rather, it is McDermott's attempt to critique the entire point of view that he sees as epitomized by Mesthene—"a not new but . . . newly aggressive right-wing ideology in this country." McDermott focuses on a notion he calls* laissez-innover, *which holds that technology is a self-correcting system. Mesthene, he claims, finds this principle acceptable because he defines technology abstractly. McDermott himself, however, rejects* laissez-innover *because he claims to see specific characteristics in contemporary technology that contradict the abstraction.*

Concentrating on the application of technology to the war in Vietnam, McDermott examines it nature and concludes that "technology in its concrete, empirical meaning, refers fundamentally to systems of rationalized control over large groups of men, events, and machines by small groups of technically skilled men operating through organized hierarchy." Using this definition, he proceeds to discuss the social effect of modern technology in America, concluding that the ideology of laissez-innover *is attractive to those in power since they are in a position to reap technology's benefits while avoiding its costs.*

John McDermott has served on the faculty of the State University of New York at Old Westbury, in the Department of Labor Studies.

I

. . . If religion was formerly the opiate of the masses, then surely technology is the opiate of the educated public today, or at least of its favorite authors. No other single subject is so universally invested with high hopes for the improvement of mankind generally and of Americans in particular. . . .

These hopes for mankind's, or technology's, future, however, are not unalloyed. Technology's defenders, being otherwise reasonable men, are also aware that the world population explosion and the nuclear missile race are also the fruit of the enormous advances made in technology during the past half century or so. But here too a cursory reading of their literature would reveal widespread though qualified optimism that these scourges too will fall before technology's might. Thus population (and genetic) control and permanent peace are sometimes added to the already imposing roster of technology's promises. What are we to make of such extravagant optimism?

[In early 1968] Harvard University's Program on Technology and Society, ". . . an inquiry in depth into the effect of technological change on the economy, on public policies, and on the character of society, as well as into the reciprocal effects of social progress on the nature, dimension, and directions of scientific and technological development," issued its Fourth Annual Report to the accompaniment of full front-page coverage in *The New York Times* (January 18). Within the brief (fewer than 100) pages of that report and most clearly in the concluding essay by the Program's Director, Emmanuel G. Mesthene, one can discern some of the important threads of belief which bind together much current writing on the social implications of technology. Mesthene's essay is worth extended analysis because these beliefs are of interest in themselves and, of greater importance, because they form the basis not of a new but of a newly aggressive right-wing ideology in this country, an ideology whose growing importance was accurately measured by the magnitude of the *Times*'s news report.

At the very beginning of Mesthene's essay, which attempts to characterize the relationships between technological and social change, the author is careful to dissociate himself from what he believes are several extreme views of those relationships. For example, technology is neither the relatively "unalloyed blessing" which, he claims, Marx, Comte, and the Air Force hold it to be, nor an unmitigated curse, a view he attributes to "many of our youth." (This is but the first of several reproofs Mesthene casts in the direction of youth.) Having denounced straw men to the right and left of him he is free to pursue that middle or moderate course favored by virtually all political writers of the day. This middle course consists of an extremely abstract and—politically speaking—sanitary view of technology and technological progress.

For Mesthene, it is characteristic of technology that it

> . . . creates new possibilities for human choice and action but leaves their disposition uncertain. What its effects will be and what ends it will serve are not inherent in the technology, but depend on what man will do with technology. Technology thus makes possible a future of open-ended options. . . .

This essentially optimistic view of the matter rests on the notion that technology is merely ". . . the organization of knowledge for practical purposes . . ." and therefore cannot be purely boon or wholly burden. The matter is somewhat more complex:

> New technology creates new opportunities for men and societies and it also generates new problems for them. It has both positive and negative effects, and it usually has the two *at the same time and in virtue of each other.*

. . . Mesthene believes there are two distinct problems in technology's relation to society, a positive one of taking full advantage of the opportunities it offers and the negative one of avoiding unfortunate consequences which flow from the exploitation of those opportunities. Positive opportunities may be missed because the costs of technological development outweigh likely benefits (e.g., Herman Kahn's "Doomsday Machine"). Mesthene seems convinced, however, that a more important case is that in which

> . . . technology lies fallow because existing social structures are inadequate to exploit the opportunities it offers. This is revealed clearly in the examination of institutional failure in the ghetto carried on by [the Program]. . . .

His diagnosis of these problems is generous in the extreme:

> All these factors combine to dilute what may be otherwise a genuine desire to apply our best knowledge and adequate resources to the resolution of urban tensions and the eradication of poverty in the nation.

Moreover, because government and the media ". . . are not yet equipped for the massive task of public education that is needed . . ." if we are to exploit technology more fully, many technological opportunities are lost because of the lack of public support. This too is a problem primarily of "institutional innovation."

Mesthene believes that institutional innovation is no less important in combatting the negative effects of technology. Individuals or individual firms which decide to develop new technologies normally do not take "adequate account" of their likely social benefits or costs. His critique is anticapitalist in spirit, but lacks bite, for he goes on to add that

> . . . [most of the negative] consequences of technology that are causing concern at the present time—pollution of the environment, potential damage to the ecology of the planet, occupational and social dislocations, threats to the privacy and political significance of the individual, social and psychological malaise—are *negative externalities of this kind*. They are with us in large measure because it has not been anybody's explicit business to foresee and anticipate them. [Italics added.]

Mesthene's abstract analysis and its equally abstract diagnosis in favor of "institutional innovation" places him in a curious and, for us, instructive position. If existing social structures are inadequate to exploit technology's full potential, or if, on the other hand, so-called negative externalities assail us because it is nobody's business to foresee and anticipate them, doesn't this say that we should apply technology to this problem too? That is, we ought to apply and organize the appropriate *organizational* knowledge for the practical purpose of solving the problems of institutional inadequacy and "negative externalities." Hence, in principle, Mesthene is in the position of arguing that the cure for technology's problems, whether positive or negative, is still more technology. This is the first theme of the technological school of writers and its ultimate First Principle.

Technology, in their view, is a self-correcting system. Temporary oversight or "negative externalities" will and should be corrected by technological means. Attempts to restrict the free play of technological innovation are, in the nature of the case, self-defeating. Technological innovation exhibits a distinct tendency to work for the general welfare in the long run. *Laissez-innover!*

I have so far deliberately refrained from going into any greater detail than does Mesthene on the empirical character of contemporary technology for it is important to bring out the force of the principle of *laissez-innover* in its full generality. Many writers on technology appear to deny in their definition of the subject—organized knowledge for practical purposes—that contemporary technology exhibits distinct trends which can be identified or projected. Others, like Mesthene, appear to accept these trends, but then blunt the conclusion by attributing to technology so much flexibility and "scientific" purity that it becomes an abstraction infinitely malleable in behalf of good, pacific, just, and egalitarian purposes. Thus the analogy to the laissez-faire principle of another time is quite justified. Just as the market or the free play of competition provided in theory the optimum long-run solution for virtually every aspect of virtually every social and economic problem, so too does the free play of technology, according to its writers. Only if technology or innovation (or some other synonym) is allowed the freest possible reign, they believe, will the maximum social good be realized.

What reasons do they give to believe that the principle of *laissez-innover* will normally function for the benefit of mankind rather than, say, merely for the belief of the immediate practitioners of technology, their managerial cronies, and for the profits accruing to their corporations? As Mesthene and other writers of his school are aware, this is a very real problem, for they all believe that the normal tendency of technology is, and ought to be, the increasing concentration of decision-making power in the hands of larger and larger scientific-technical bureaucracies. *In principle*, their solution is relatively simple, though not often explicitly stated.[1]

Their argument goes as follows: the men and women who are elevated by technology into commanding positions within various decision-making bureaucracies exhibit no generalized drive for power such as characterized, say, the landed gentry of preindustrial Europe or the capitalist entrepreneur of the last century. For their social and institutional position and its supporting culture as well are defined solely by the fact that these men are problem solvers. (Organized knowledge for practical purposes again.) That is, they gain advantage and reward only to the extent that they can bring specific technical knowledge to bear on the solution of specific technical problems. Any more general drive for power would undercut the bases of their usefulness and legitimacy.

Moreover their specific training and professional commitment to solving technical problems creates a bias against ideologies in general which inhibits any attempts to formulate a justifying ideology for the group. Consequently, they do not constitute a class and have no general interests antagonistic to those of their problem-beset clients. We may refer to all of this as the disinterested character of the scientific-technical decision-maker, or, more briefly and cynically, as the principle of the Altruistic Bureaucrat.

As if not satisfied by the force of this (unstated) principle, Mesthene like many of his school fellows spends many pages commenting around the belief that the concentration of power at the top of technology's organizations is a problem, but that like other problems technology should be able to solve it successfully through institutional innovation. You may trust in it; the principle of *laissez-innover* knows no logical or other hurdle.

This combination of guileless optimism with scientific tough-mindedness might seem to be no more than an eccentric delusion were the American technology it supports not moving in directions that are strongly antidemocratic. To show why this is so we must examine more closely Mesthene's seemingly innocuous distinction between technology's positive opportunities and its "negative externalities." In order to do this I will make use of an example drawn from the very frontier of American technology, the war in Vietnam.

II

At least two fundamentally different bombing programs [have been] carried out in South Vietnam. There are fairly conventional attacks against targets which consist of identified enemy troops, fortifications, medical center, vessels, and so forth. The other program is quite different and, at least since March 1968, infinitely more important. With some oversimplification it can be described as follows:

Intelligence data is gathered from all kinds of sources, of all degrees of reliability, on all manner of subjects, and fed into a computer complex located, I believe, at Bien Hoa. From this data and using mathematical models developed for the purpose, the computer then assigns probabilities to a range of potential targets, probabilities which represent the likelihood that the latter contain enemy forces or supplies. These potential targets might include: a canal-river crossing known to be used occasionally by the NLF [National Liberation Front]; a section of trail which would have to be used to attack such and such an American base, now overdue for attack; a square mile of plain rumored to contain enemy troops; a mountainside from which camp fire smoke was seen rising. Again using models developed for the purpose, the computer divides pre-programmed levels of bombardment among these potential targets which have the highest probability of containing actual targets. Following the raids, data provided by further reconnaissance is fed into the computer and conclusions are drawn (usually optimistic ones) on the effectiveness of the raids. This estimate of effectiveness then becomes part of the data governing current and future operations, and so on.

Two features must be noted regarding this program's features, which are superficially hinted at but fundamentally obscured by Mesthene's distinction between the abstractions of positive opportunity and "negative externality." First, when considered from the standpoint of its planners, the bombing program is extraordinarily rational, for it creates previously unavailable "opportunities" to pursue their goals in Vietnam. It would make no sense to bomb South Vietnam simply at random, and no serious person or air force general would care to mount the effort to do so. So the system employed in Vietnam significantly reduces, though it does not eliminate, that randomness. That canal-river crossing which is bombed at least once every eleven days or so is a very poor target compared to an NLF battalion observed in a village. But it is an infinitely more promising target than would be selected by throwing a dart at a grid map of South Vietnam. In addition to bombing the battalion, why not bomb the canal crossing to the frequency and extent that it *might* be used by enemy troops?

Even when we take into account the crudity of the mathematical models and the consequent slapstick way in which poor information is evaluated, it is a "good" program. No single raid will definitely kill an enemy soldier but a whole series of them increases the "opportunity" to kill a calculable number of them (as well, of course, as a calculable but not calculated number of nonsoldiers). This is the most rational bombing system to follow if American lives are very expensive and American weapons and Vietnamese lives very cheap. Which, of course, is the case.

Secondly, however, considered from the standpoint of goals and values not programmed in by its designers, the bombing program is incredibly irrational. In Mesthene's terms, these "negative externalities" would include, in the present case, the lives and well-being of various Vietnamese as well as the feelings and opinions of some less important Americans. Significantly, this exclusion of the interests of people not among the managerial class is based quite as much on the so-called technical means being employed as on the political goals of the system. In the particular case of the Vietnamese bombing system, the political goals of the bombing system clearly exclude the interests of certain Vietnamese. After all, the victims of the bombardment are communists or their supporters, they are our enemies, they resist US intervention. In short, their interests are fully antagonistic to the goals of the program and simply must be excluded from consideration. The technical reasons for this exclusion require explanation, being less familiar and more important, especially in the light of Mesthene's belief in the malleability of technological systems.

Advanced technological systems such as those employed in the bombardment of South Vietnam make use not only of extremely complex and expensive equipment but, quite as important, of large numbers of relatively scarce and expensive-to-train technicians. They have immense capital costs; a thousand aircraft of a very advanced type, literally hundreds of thousands of spare parts, enormous stocks of rockets, bombs, shells and bullets, in addition to tens of thousands of technical specialists: pilots, bombardiers, navigators, radar operators, computer programmers, accountants, engineers, electronic and mechanical technicians, to name only a few. In short, they are "capital intensive."

Moreover, the coordination of this immense mass of esoteric equipment and its operators in the most effective possible way depends upon an extremely highly developed technique both in the employment of each piece of equipment by a specific team of operators and in the management of the program itself. Of course, all large organizations standardize their operating procedures, but it is peculiar to advanced technological systems that their operating procedures embody a very high degree of information drawn from the physical sciences, while their managerial procedures are equally dependent on information drawn from the social sciences. We may describe this situation by saying that advanced technological systems are both "technique intensive" and "management intensive."

It should be clear, moreover, even to the most casual observer that such intensive use of capital, technique, and management spills over into almost every area touched by the technological system in question. An attack program delivering 330,000 tons of munitions more or less selectively to several thousand different targets monthly would be an anomaly if forced to rely on sporadic intelligence data, erratic maintenance systems, or a fluctuating and unpredictable supply of heavy bombs, rockets, jet fuel, and napalm tanks. Thus it is precisely because the bombing program requires an intensive use of capital, technique, and management that the same properties are normally transferred to the intelligence, maintenance, supply, coordination and training systems which support it. Accordingly, each of these supporting systems is subject to sharp pressures to improve and rationalize the performance of its machines and men, the reliability of its techniques, and the efficiency and sensitivity of the management controls under which it operates. Within integrated technical systems, higher levels of technology drive out lower, and the normal tendency is to integrate systems.

From this perverse Gresham's Law of Technology follow some of the main social and organizational characteristics of contemporary technological systems: the radical increase in the scale and complexity of operations that they demand and encourage; the rapid and widespread diffusion of technology to new areas; the great diversity of activities which can be directed by central management; an increase in the ambition of management's goals; and, as a corollary, especially to the last, growing resistance to the influence of so-called negative externalities.

Complex technological systems are extraordinarily resistant to intervention by persons or problems operating outside or below their managing groups, and this is so regardless of the "politics" of a given situation. Technology creates its own politics. The point of such advanced systems is to minimize the incidence of personal or social behavior which is erratic or otherwise not easily classified, of tools and equipment with poor performance, of improvisory techniques, and of unresponsiveness to central management. . . .

To define technology so abstractly that it obscures these observable characteristics of contemporary technology—as Mesthene and his school have done—makes no sense. It makes even less sense to claim some magical malleability for something as undefined as "institutional innovation." Technology, in its concrete, empirical meaning, refers fundamentally to systems of rationalized control over large groups of men, events, and machines by small groups of technically skilled men operating through organizational hierarchy. The latent "opportunities" provided by that control and its ability to filter out discordant "negative externalities" are, of course, best illustrated by extreme cases. Hence the most instructive and accurate example should be of a technology able to suppress the humanity of its rank-and-file and to commit genocide as a by-product of its rationality. The Vietnam bombing program fits technology to a "T".

III

It would certainly be difficult to attempt to translate in any simple and direct way the social and organizational properties of highly developed technological systems from the battlefields of Vietnam to the different cultural and institutional setting of the US. Yet before we conclude that any such attempt would be futile or even absurd, we might consider the following story.

In early 1967 I stayed for several days with one of the infantry companies of the US Fourth Division whose parent battalion was then based at Dau Tieng. From the camp at Dau Tieng the well-known Black Lady Mountain, sacred to the Cao Dai religious sect, was easily visible and in fact dominated the surrounding plain and the camp itself. One afternoon when I began to explain the religious significance of the mountain to some GI friends, they interrupted my somewhat academic discourse to tell me a tale beside which even the strange beliefs of the Cao Dai sect appeared prosaic.

According to GI reports which the soldiers had heard and believed, the Viet Cong had long ago hollowed out most of the mountain in order to install a very big cannon there. The size of the cannon was left somewhat vague—"huge, fucking . . ."—but clearly the GIs imagined that it was in the battleship class. In any event, this huge cannon had formerly taken a heavy toll of American aircraft and had been made impervious to American counterattacks by the presence of two— "huge, fucking"—sliding steel doors, behind which it retreated whenever the Americans attacked. Had they seen this battleship cannon, and did it ever fire on the camp, which was easily within its range? No, they answered, for a brave flyer, recognizing the effectiveness of the cannon against his fellow pilots, had deliberately crashed his jet into those doors one day, jamming them, and permitting the Americans to move into the area unhindered.

I have never been in the army, and at the time of my trip to Vietnam had not yet learned how fantastic GI stories can be. Thus I found it hard to understand how they could be convinced of so improbable a tale. Only later, after talking to many soldiers and hearing many other wild stories from them as well, did I realize what the explanation for this was. Unlike officers and civilian correspondents who are almost daily given detailed briefings on a unit's situation capabilities and objectives, GIs are told virtually nothing of this sort by the army. They are simply told what to do, where, and how, and it is a rare officer, in my experience anyway, who thinks they should be told any more than this. Officers don't think soldiers are stupid; they simply assume it, and act accordingly. For the individual soldier's personal life this doesn't make too much difference; he still has to deal with the facts of personal feelings, his own well-being, and that of his family.

But for the soldier's group life this makes a great deal of difference. In their group life, soldiers are cut off from sources of information about the situation of the group and are placed in a position where their social behavior is governed largely by the principle of blind obedience. Under such circumstances, reality becomes elusive. Because the soldiers are not permitted to deal with facts in their own ways, facts cease to discipline their opinions. Fantasy and wild tales are the natural outcome. In fact, it is probably a mark of the GI's intelligence to fantasize, for it means that he has not permitted his intellectual capacity to atrophy. The intelligence of the individual is thus expressed in the irrationality of the group.

It is this process which we may observe when we look to the social effect of modern technological systems in America itself. Here the process is not so simple and clear as in Vietnam, for it involves not simply the relations of today's soldiers to their officers and to the army but the historical development of analogous relations between the lower and upper orders of our society. Moreover, these relations are broadly cultural rather than narrowly social in nature. It is to a brief review of this complex subject that I now wish to turn.

IV

Among the conventional explanations for the rise and spread of the democratic ethos in Europe and North America in the seventeenth, eighteenth, and nineteenth centuries, the destruction of the gap in political culture between the mass of the population and that of the ruling classes is extremely important. There are several sides to this explanation. For example, it is often argued that the invention of the printing press and the spread of Protestant Christianity encouraged a significant growth in popular literacy. . . . The dating of these developments is, in the nature of the case, somewhat imprecise. But certainly by the middle of the eighteenth century, at least in Britain and North America, the literacy of the population was sufficient to support a variety of newspapers and periodicals not only in the large cities but in the smaller provincial towns as well. . . . Common townsmen had closed at least one of the cultural gaps between themselves and the aristocracy of the larger cities.

Similarly, it is often argued that with the expansion and improvement of road and postal systems, the spread of new tools and techniques, the growth in the number and variety of merchants, the consequent invigoration of town life, and other numerous and familiar related developments, the social experiences of larger numbers of people became richer, more varied, and similar in fact to those of the ruling class. This last, the growth in similarity of the social experiences of the upper and lower classes, is especially important. Social skills and experiences which underlay the monopoly of the upper classes over the processes of law and government were spreading to important segments of the lower orders of society. For carrying on trade, managing a commercial—not a subsistence—farm, participating in a vestry or workingmen's guild, or working in an up-to-date manufactory or business, unlike the relatively narrow existence of the medieval serf or artisan, were experiences which contributed to what I would call the social rationality of the lower orders.

Activities which demand frequent intercourse with strangers, accurate calculation of near means and distant ends, and a willingness to devise collective ways of resolving novel and unexpected problems demand and reward a more discriminating attention to the realities and deficiencies of social life, and provide thereby a rich variety of social experiences analogous to those of the governing classes. As a result not only were the processes of law and government, formerly treated with semireligious veneration, becoming demystified but, equally important, a population was being fitted out with sufficient skills and interests to contest their control. Still another gap between the political cultures of the upper and lower ends of the social spectrum was being closed.

The same period also witnessed a growth in the organized means of popular expression. . . .

These same developments were also reflected in the spread of egalitarian and republican doctrines such as those of Richard Price and Thomas Paine, which pointed up the arbitrary character of what had heretofore been considered the rights of the higher orders of society, and thus provided the popular ideological base which helped to define the legitimate lower-class demands.

This description by no means does justice to the richness and variety of the historical process underlying the rise and spread of what has come to be called the democratic ethos. But it does, I hope, isolate some of the important structural elements and, moreover, it enables us to illuminate some important ways in which the new technology, celebrated by Mesthene and his associates for its potential contributions to democracy, contributes instead to the erosion of that same democratic ethos. For if, in an earlier time, the gap between the political cultures of the higher and lower orders of society was being widely attacked and closed, this no longer appears to be the case. On the contrary, I am persuaded that the direction has been reversed and that we now observe evidence of a growing separation between ruling and lower-class culture in America, a separation which is particularly enhanced by the rapid growth in technology and the spreading influence of its *laissez-innover* ideologues.

Certainly, there has been a decline in popular literacy, that is to say, in those aspects of literacy which bear on an understanding of the political and social character of the new technology. Not one person in a hundred is even aware of, much less understands, the nature of technologically highly advanced systems such as are used in the Vietnam bombing program. . . .

Secondly, the social organization of this new technology, by systematically denying to the general population experiences which are analogous to those of its higher management, contributes very heavily to the growth of social irrationality in our society. For example, modern technological organization defines the roles and values of its members, not vice versa. An engineer or a sociologist is one who does all those things but only those things called for by the "table of organization" and the "job description" used by his employer. Professionals who seek self-realization through creative and autonomous behavior without regard to the defined goals, needs, and channels of their respective departments have no more place in a large corporation or government agency than squeamish soldiers in the army. . . .

However, those at the top of technology's most advanced organizations hardly suffer the same experience. For reasons which are clearly related to the principle of the Altruistic Bureaucracy the psychology of an individual's fulfillment through work has been incorporated into management ideology. As the pages of *Fortune, Time,* or *Business Week* . . . serve to show, the higher levels of business and government are staffed by men and women who spend killing hours looking after the economic welfare and national security of the rest of us. The rewards of this life are said to be very few: the love of money would be demeaning and, anyway, taxes are said to take most of it; its sacrifices are many, for failure brings economic depression to the masses or gains for communism as well as disgrace to the erring managers. Even the essential high-mindedness or altruism of our managers earns no reward, for the public is distracted, fickle, and, on occasion, vengeful. . . . Hence for these "real revolutionaries of our time," as Walt Rostow has called them, self-fulfillment through work and discipline is the only reward. The managerial process is seen as an expression of the vital personalities of our leaders and the right to it an inalienable right of the national elite.

In addition to all this, their lonely and unrewarding eminence in the face of crushing responsibility, etc., tends to create an air of mystification around technology's managers. . . .

It seems fundamental to the social organization of modern technology that the quality of the social experience of the lower orders of society declines as the level of technology grows no less than does their literacy. And, of course, this process feeds on itself, for with the consequent decline in the real effectiveness and usefulness of local and other forms of organization open to easy and direct popular influence their vitality declines still further, and the cycle is repeated.

The normal life of men and women in the lower and, I think, middle levels of American society now seems cut off from those experiences in which near social means and distant social ends are balanced and rebalanced, adjusted, and readjusted. But it is from such widespread experience with effective balancing and adjusting that social rationality derives. To the degree that it is lacking, social irrationality becomes the norm, and social paranoia a recurring phenomenon. . . .

Mesthene himself recognizes that such "negative externalities" are on the increase. His list includes ". . . pollution of the environment, potential damage to the ecology of the planet, occupational and social dislocations, threats to the privacy and political significance of the individual, social and psychological malaise. . . ." Minor matters all, however, when compared to the marvelous opportunities *laissez-innover* holds out to us: more GNP, continued free world leadership, supersonic transports, urban renewal on a regional basis, institutional innovation, and the millennial promises of his school.

This brings us finally to the ideologies and doctrines of technology and their relation to what I have argued is a growing gap in political culture between the lower and upper classes in American society. Even more fundamentally than the principles of *laissez-innover* and the altruistic bureaucrat, technology in its very definition as the organization of knowledge for practical purposes assumes that the primary and really creative role in the social processes consequent on technological change is reserved for a scientific and technical elite, the elite which presumably discovers and organizes that knowledge. But if the scientific and technical elite and their indispensable managerial cronies are the really creative (and hardworking and altruistic) elements in American society, what is this but to say that the common mass of men are essentially drags on the social weal? This is precisely the implication which is drawn by the *laissez-innover* school. Consider the following quotations from the article which appeared in *The New Republic* in December 1967, written by Zbigniew Brzezinski, one of the Intellectual leaders of the school.

Brzezinski is describing a nightmare which he calls the "technetronic society" (the word like the concept is a pastiche of technology and electronics). This society will be characterized, he argues, by the application of ". . . the principle of equal opportunity for all but . . . special opportunity for the singularly talented few." It will thus combine ". . . continued *respect* for the popular will with an increasing *role* in the key decision-making institutions of individuals with special intellectual and scientific attainments." (Italics added.) Naturally, "The educational and social systems [will make] it increasingly attractive and easy for those meritocratic few to develop to the fullest of their special potential."

However, while it will be ". . . necessary to require everyone at a sufficiently responsible post to take, say, two years of [scientific and technical] retraining every ten years . . . ," the rest of us can develop a new ". . . interest in the cultural and humanistic aspects of life, *in addition to purely hedonistic preoccupations.*" (Italics added.) The latter, he is careful to point out, "would serve as a social valve, reducing tensions and political frustration."

It is not fair to ask how much *respect* we carefree pleasure lovers and culture consumers will get from the hard-working bureaucrats, going to night school two years in every ten, while working like beavers in the "key decision-making institutions"? The altruism of our bureaucrats has a heavy load to bear.

Stripped of their euphemisms these are simply arguments which enhance the social legitimacy of the interests of new technical and scientific elites and detract from the interests of the rest of us. . . .

As has already been made clear the *laissez-innover* school accepts as inevitable and desirable the centralizing tendencies of technology's social organization, and they accept as well the mystification which comes to surround the management process. Thus equality of opportunity, as they understand it, has precious little to do with creating a more egalitarian society. On the contrary, it functions as an indispensable feature of the highly stratified society they envision for the future. For in their society of meritocratic hierarchy, equality of opportunity assures that talented young meritocrats (the word is no uglier than the social system it refers to) will be able to climb into the "key decision-making" slots reserved for trained talent, and thus generate the success of the new society, and its cohesion against popular "tensions and political frustration."

The structures which formerly guaranteed the rule of wealth, age, and family will not be destroyed (or at least not totally so). They will be firmed up and rationalized by the perpetual addition of trained (and, of course, acculturated) talent. In technologically advanced societies, equality of opportunity functions as a hierarchical principle, in opposition to the egalitarian social goals it pretends to serve. To the extent that it has already become the kind of "equality" we seek to institute in our society, it is one of the main factors contributing to the widening gap between the cultures of upper- and lower-class America.

V

. . . *Laissez-innover* is now the premier ideology of the technological impulse in American society, which is to say, of the institutions which monopolize and profit from advanced technology and of the social classes which find in the free exploitation of *their* technology the most likely guarantee of their power, status, and wealth.

This said, it is important to stress both the significance and limitations of what has in fact been said. Here Mesthene's distinction between the positive opportunities and negative "externalities" inherent in technological change is pivotal; for everything else which I've argued follows inferentially from the actual social meaning of that distinction. As my analysis of the Vietnam bombing program suggested, those technological effects which are sought after as positive opportunities and those which are dismissed as negative externalities are decisively influence by the fact that this distinction between positive and negative within advanced technological organizations tends to be made among the planners and managers themselves. Within these groups there are, as was pointed out, extremely powerful organizational, hierarchical, doctrinal, and other *"technical"* factors, which tend by design to filter out "irrational" demands from below, substituting for them the "rational" demands of technology itself. As a result, technological rationality is as socially neutral today as market rationality was a century ago.

Turning from the inner social logic of advanced technological organizations and systems to their larger social effect, we can observe a significant convergence. For both the social tendency of technology and the ideology (or rhetoric) of the *laissez-innover* school converge to encourage a political and cultural gap between the upper and lower ends of American society. As I have pointed out, these can now be characterized as those who manage and those who are managed by advanced technological systems.

This analysis lends some weight (though perhaps no more than that) to a number of wide-ranging and unorthodox conclusions about American society today and the directions in which it is tending. It may be useful to sketch out the most important of those conclusions in the form of a set of linked hypotheses, not only to clarify what appear to be the latent tendencies of America's advanced technological society but also to provide more useful guides to the investigation of the technological impulse than those offered by the obscurantism and abstractions of the school of *laissez-innover*.

First, and most important, technology should be considered as an institutional system, not more and certainly not less. Mesthene's definition of the subject is inadequate, for it obscures the systematic and decisive social changes, especially their political and cultural tendencies, that follow the widespread application of advanced technological systems. At the same time, technology is less than a social system per se, though it has many elements of a social system, viz., an elite, a group of linked institutions, an ethos, and so forth. Perhaps the best summary statement of the case resides in an analogy—with all the vagueness and imprecision attendant on such things: today's technology stands in relation to today's capitalism as, a century ago, the latter stood to the free market capitalism of the time. . . .

A second major hypothesis would argue that the most important dimension of advanced technological institutions is the social one, that is, the institutions are agencies of highly centralized and intensive social control. Technology conquers nature, as the saying goes. But to do so it must first conquer man. More precisely, it demands a very high degree of control over the training, mobility, and skills of the work force. The absence (or decline) of direct controls or of coercion should not serve to obscure from our view the reality and intensity of the social controls which are employed (such as the internalized belief in equality of opportunity, indebtedness through credit, advertising, selective service channeling, and so on).

Advanced technology has created a vast increase in occupational specialties, many of them requiring many, many years of highly specialized training. It must motivate this training. It has made ever more complex and "rational" the ways in which these occupational specialties are combined in our economic and social life. It must win passivity and obedience to this complex activity. Formerly, technical rationality had been employed only to organize the production of rather simple physical objects, for example, aerial bombs. Now technical rationality is increasingly employed to organize all of the processes necessary to the utilization of physical objects, such as bombing systems. For this reason it seems a mistake to argue that we are in a "postindustrial" age, a concept favored by the *laissez-innover* school. On the contrary, the rapid spread of technical rationality in organizational and economic life and, hence, into social life is more aptly described as a second and much more intensive phase of the industrial revolution. One might reasonably suspect that it will create analogous social problems.

Accordingly, a third major hypothesis would argue that there are very profound social antagonisms or contradictions not less sharp or fundamental than those ascribed by Marx to the development of nineteenth-century industrial society. The general form of the contradictions might be described as follows: a society characterized by the employment of advanced technology requires an ever more socially disciplined population, yet retains an ever declining capacity to enforce the required discipline. . . .

These are brief and, I believe, barely adequate reviews of extremely complex hypotheses. But, in outline, each of these contradictions appears to bear on roughly the same group of the American population, a technological underclass. If we assume this to be the case, a fourth hypothesis would follow, namely that technology is creating the basis for new and sharp class conflict in our society. That is, technology is creating its own working and managing classes just as earlier industrialization created its working and owning classes. Perhaps this suggests a return to the kind of class-based politics which characterized the US in the last quarter of the nineteenth century, rather than the somewhat more ambiguous politics which was a feature of the second quarter of this century. I am inclined to think that this is the case, though I confess the evidence for it is as yet inadequate.

This leads to a final hypothesis, namely that *laissez-innover* should be frankly recognized as a conservative or right-wing ideology. This is an extremely complex subject, for the hypothesis must confront the very difficult fact that the intellectual genesis of *laissez-innover* is traceable much more to leftist and socialist theorizing on the wonders of technical rationality and social planning than it is to the blood politics of a De Maistre or the traditionalism of a Burke. So be it. Much more important is the fact that *laissez-innover* is now the most powerful and influential statement of the demands and program of the technological impulse of our society, an impulse rooted in its most powerful institutions. More than any other statement, it succeeds in identifying and rationalizing the interests of the most authoritarian elites within this country, and the expansionism of their policies overseas. . . .

The point of this final hypothesis is not primarily to reimpress the language of European politics on the American scene. Rather it is to summarize the fact that many of the forces in American life hostile to the democratic ethos have enrolled under the banner of *laissez-innover*. Merely to grasp this is already to take the first step toward a politics of radical reconstruction and against the malaise, irrationality, powerlessness, and official violence that characterize American life today.

Note

1. For a more complete statement of the argument which follows, see Suzanne Keller, *Beyond the Ruling Class* (New York: Random House, 1963).

Study Questions

1. What is the main point of McDermott's article? Why do you think it might be correct? Why do you think it might be wrong? Explain and defend your answers.

2. What assumptions does McDermott make about technology? about ethics? Do you think these assumptions are correct? Explain and defend your answers.
3. If society accepts McDermott's main conclusions, what consequences might follow? Would these consequences be desirable? Explain and defend your answers.
4. What is the significance of the Vietnam War in McDermott's article?
5. What is the major disagreement between Mesthene and McDermott? Explain and argue who is more correct.
6. Which ethical and moral theory would best support McDermott's argument? Explain.

2.6

The Social Construction of Safety

Rachelle D. Hollander

This article concentrates on safety. Focusing on safety is important for many reasons. One is because safety is a goal for even risky activities. Risky activities can be made more safe. This article uses illustrations to clarify how safety is a social construction. It includes activities of science, engineering, business, government, law, individuals, communities—all these behaviors shape safety. Focusing on safety also issues a challenge to scientists and scholars who have limited their attention to risk. Proceeding in this way helps to overcome views that inappropriately dichotomize and reify notions of objective and subjective risk, thereby overlooking ways to improve human safety.

My purpose is to focus attention on the need for research on issues and processes by which safety can be improved. Philosophers and other researchers doing research on risk assessment and risk management will benefit from turning their attention to questions of safety from questions of risk. They can recognize "safety research" as a topic in applied ethics, requiring attention to complex issues of collective moral responsibility. My approach recommends using notions of legitimate expectations, feasible control, and due care in order to structure the inquiry into ways to develop collective moral responsibility. This approach needs further consideration and development, in conjunction with as well as testing against other approaches.

The first three examples that I use to clarify safety as a social construction concern occupational safety. These examples and the later discussion of their implications and those of other examples come, in

Originally published in *The International Journal of Applied Ethics* 8:2 (1994), pp. 15–18.

part, from a paper that I co-authored with Jacob Abel, a professor of mechanical engineering at the University of Pennsylvania.[1]

[Example 1.] In this case, a German manufacturer of a large industrial saw for building materials knew that it was dangerous for workers to stand near the point where the large stacks of material were automatically pushed onto the table which fed the material into the saw. It painted the "danger zone" with a stripe of alternating yellow and black bands, expecting workers to understand that the striped zone was dangerous and to stay away from it. The machine was installed in a plant in the US. Seeing that some minor mishap concerning the building material was about to occur, an American worker reached into the danger zone and suffered a serious injury to his arm. The manufacturer's expectations concerning worker behavior, ability to translate the meaning of the stripe, and obedience to its implied instruction were mistaken. The worker's expectations concerning the manufacturer's care and the employer's ability and willingness to improve a dangerous machine were also unmet. Learning of the accident, the manufacturer expressed outrage over the conduct of the employee, contrasting it with worker obedience in Germany.

[Example 2.] The manufacturer of a mechanical power press devised and installed a system of guards and interlocks which required the machine operator to use both hands to actuate the press and which interrupted power to the press if any door to the dangerous area was opened. But the purchaser wished to increase profits by speeding production. To accomplish this goal the purchaser removed the safeguards that the manufacturer had installed. This allowed two workers to operate the press, which had been designed to be operated by only one. During operation, one worker inserted a hand into the dangerous area through an opening which was no longer protected. The other caused the press to operate and the first was injured. The manufacturer expected his safe machine to be used in its original condition. The workers expected their employer to value their well-being above the incremental profits that could be earned by compromising their safety. The employer expected the workers to accept the more hazardous working conditions, which they did, and to assume and discharge more responsibility for their own safety in the face of the heightened danger. These expectations were only partially fulfilled.

[Example 3.] In a small metal stamping plant in Pennsylvania containing two dozen dangerous presses, an injunction concerning punching-in, posted over the time clock, was written in Spanish. All of the signs on the presses were in English. The expectation that the workforce could understand and comply with the safety instructions was mistaken. Usually the materials accompanying a machine or attached to it are written in the language of the country of origin, which may or may not be helpful when it reaches its destination. Instructions on machines for US use in production or consumption are often written in an English that is so unidiomatic that its value as communication is seriously diminished. This deficiency may be a minor irritant to the purchaser of a piece of consumer electronics, but it can be a cause of action to injured workers. To receive workers' compensation in some situations, they may lose the legal right to sue their employers, but not the manufacturer of the machines. In other settings, both employer and manufacturer can be sued.[2]

Real life teaches us that failures of congruence with respect to these matters account for many serious accidents involving injury and loss of life. Often the expectations which are of paramount importance are displayed only implicitly. Actors are unaware of what they are and how they influence their behavior until someone else points out that unjustified assumptions influenced the design of a machine or workplace and these erroneous assumptions caused harm.

These examples demonstrate that the workplace is an extraordinarily complicated cultural phenomenon. In it, socio-technical behaviors shape entities (both physical and abstract) and relationships (both physical and abstract). Thus, social behaviors shape more or less safety in the workplace.

The workplace inevitably includes machines of varying degrees of power and complexity. The level and experience of safety in the workplace depend on the interrelated set of expectations concerning this machinery and the conduct, the assumption of duties, and the propensity to act with due care, that each actor has of the other actors in the setting. Thus, employees' expectations of their fellow workers and their employers' concern for worker well-being modulate the employees' behavior. At the same time, employers' assumptions about employee conduct, willingness to follow orders, and concern for their own health influence the employers' conduct. The manufacturers of even the simplest machine or tool, sometimes without realizing it, act on expectations they have about its purchasers and users. In particular, cultural differences among employees, employers, and tool manufacturers play a role in the causation of injury.

Research about improving workplace safety must take into account the expectations of all the stakeholders and the care that they are likely to exercise. Feasible control is a function of human habits of behavior and training, not simply of engineering design or statistical analysis. If promoting workers' safety is a legitimate goal, expectations that practices to promote that goal will continue, or will be augmented, are also legitimate. To reach this goal, engineers and other humans in these organizational contexts need to consider and improve the complex social behaviors, including technologies, that can promote or frustrate it.

All of these actors' assumptions and behaviors also depend on the broader setting of law and regulation and informal cultural expectations which influence the assumption of responsibility and the outcomes for the stakeholders. Research at the macro-level needs to examine such phenomena as the implications of education, worker training, laws, and regulations within and between countries and states on behavior that affects safety. The role of innovation in improving or degrading safety is an important topic.

Injury is hardly limited to workplaces, be they for-profit and not-for-profit, public or private sector organizations. The roles of scientists and engineers, among others, cannot be limited to workplaces either. My next example makes this quite clear.

[Example 4.] *The Baltimore Sun* reports that the parents of an 11-year-old killed while crossing the street to his school bus are suing the Baltimore County government.[3] These parents and others had objected to having their children cross the 50-mph thoroughfare at dawn and had asked that the bus pull up first and put on its flashing red lights. County policy is that children be at their stops before the buses arrive. Officials said they developed their policy because too many motorists violate the law and do not stop on seeing the lights. Officials say that it is safer for children to cross ahead of time, when they see there is no traffic.

Like worker safety, pedestrian safety is a social construct with ethical dimensions. It is a positive value, something human beings seek. Social behaviors, including technologies, can foster safety, or interfere with safety. Human beings come into conflict about safety matters, as social behaviors expose people to harms or the potential of harms. Improving pedestrian safety requires cooperative efforts from many individuals and organizations in US communities. All of their behaviors will shape safety.

Whenever human agency exposes people to harms, questions of moral responsibility arise. Moral responsibility requires foresight; foresight depends on knowledge or coming to know about the future.[4] There are legitimate expectations about foresight, expectations limited by feasible control, or due care. If I know that my children have to cross a very busy street to get to school, I know that they could be hurt and I need to take precautions to help to assure that they do not. If I do not, I can be asked why I did not live up to my responsibility in a way that I cannot if there is a plane crash into my house at 2 a.m.

Moral responsibility is also a non-exclusive matter. Just because I have responsibility to take precautions for my children does not mean that no one else does. For instance, there will be "slow down, school crossing" signs which drivers will be expected to obey. There will be crossing guards.

In the Baltimore County example, the parents and the county are at odds about what legitimate expectations about safety are. In one sense in which the term legitimate is used, the sense of being accurate or not mistaken, the parents' expectations are not legitimate. But in another sense, the sense in which legitimate means that due care will be taken, the parents are raising a normative issue about how that legitimate expectation might be met. To make the expectations accurate, all parties involved have to consider how changes in the technological and social features of the transportation system can contribute to more safety.

[Example 5.] *The Sunpapers* of November 29, 1993, contains an editorial complimenting the towns of Havre de Grace and Westminster in Hartford and Carroll counties in Maryland for going for more than a decade without a pedestrian traffic fatality.[5] The editorial notes that the towns "have made a concerted effort to raise public awareness and to engineer their towns so as to enhance pedestrian safety." The American Automobile Association has been sponsoring the Safety Commendation/Achievement Awards, which the towns just won, for 54 years. In the first year, 1939, the number of pedestrian deaths in the US exceeded 12,400; in 1993, the number was 5,546. The AAA indicates that noticeable reductions in pedestrian deaths and injuries requires good education, engineering, and law enforcement. Baltimore County has formed a transportation safety committee, which may help to prevent future tragedies like that noted in my fourth example.

All cultures are concerned about safety. Different cultures may develop different practices by which children's safety is preserved. Expectations about how things will be handled to promote safety arise. When promoting children's safety is taken as a legitimate goal, expectations that practices to promote that goal will continue, or will be augmented, are also legitimate. The *Sun* editorial shows an example of a social system that promotes safety. The system builds legitimate expectations that attention will be paid to promote and augment feasible control and due care.

The next example I use shows how current scientific and engineering constructions of safety can work at cross purposes with the promotion of that goal. These constructions can be misleading and maybe even dangerous.

[Example 6.] In the *New Scientist* of November 13, 1993,[6] in a comment titled "Real deaths, real choices," the editorialists write about a just released report from the National Radiological Protection Board.

The report estimates that if a nuclear reprocessing plant, Thorp, is allowed to operate, about 200 people will die from radioactive emissions over its 25-year lifetime. The editorial notes that this report will make it harder for the British government to go ahead on Thorp. The writers then ask us to imagine this kind of analysis of a new road, calculating the number of children likely to die, and asking whether townspeople would then support construction. They conclude that such "objective" (their quotes) comparative exercises would have the result of making nuclear fuel reprocessing look safe.

But would it? The important point to recognize here is the contrast between a situation in which people to be affected, at least apparently, have no control, and one in which they can take steps to mitigate risks to themselves and others. The two examples that precede this discussion make the differences between these two types of situation quite clear. If people recognize the difference, they can ask how citizens and facilities developers, managers, and operators might develop oversight, control, and procedures for a nuclear processing plant adequate to assure that the facility is safe. It might not be possible to do so, but it is very important to ask the question and attempt, seriously, to answer it.

The problem with the scientific and engineering construction of risk as it is currently practiced is that it often seems to assume that risk is impervious to human influence. This results in science and engineering that takes on both too much and too little. The too much is the attempt to get rid of risk, to engineer it down to disappearance or to some magical "de minimus" level which society will or should accept. The too little is the assumption that once this has been done, once the scientists and engineers have done their best, people will or should accept the risks. Nothing further can or should be expected, and the people who do not accept this are being unreasonable. This construction of risk may not promote safety.

Risks are real, but they are real social constructions involving scientists and engineers and other human actors and non-human entities. Risk assessment, not just management, requires understanding how human expectations, control, care, among other things, shape the nature and impacts of risk. Collective moral responsibility requires actions to eliminate, mitigate, prevent, or compensate harms, in circumstances where the actions are expressions of legitimate expectations, feasible control, and due care.

Human beings expect that social systems, including those with scientific and technological components, will be designed and maintained so as to promote safety. This cannot happen unless human beings take responsibility for examining the systems they create with that goal in mind. Safe enough requires vigilance, and people are made uneasy by an attitude of complacency and self-satisfaction that in the light of historical events is often seen to be unfounded. This attitude will not promote safety.

Developing a better understanding about how human expectations shape safety is a topic in need of attention in the fields of applied ethics and in science and technology studies. Both conceptual and empirical work are needed, in order to see how social systems (including those of science and engineering) have improved and can improve legitimate expectations, feasible control, and due care with respect to safety and to other quality of life concerns.

Notes

1. This article grows out of a presentation at the American Philosophical Association Eastern Division Meeting, Atlanta, December 28, 1994, published as "Is Engineering Safety Just Business Safety? in the *International Journal of Applied Philosophy* 8:2, 15–18, Winter/Spring 1994. The first three examples come from a presentation of Jacob Abel and Rachelle D. Hollander, "Ethics, Safety and Workplace Cultures, or Expectations and Safety," developed for the Joint Meeting of the Society for Social Studies of Science and the European Association for Studies of Science and Technology, Gothenberg, Sweden, August 1992.

2. Lawrence K. Bell, PE, "The Care and Feeding of PLC-Controlled Machinery," *Safety Brief* 8:3, June 1993, 1–17.

3. Sheridan Lyons, "Dead boy's parents sue officials for $20 million," *The Baltimore Sunpapers,* May 29, 1993, p. 3B.

4. This and the following paragraphs use notions of foresight and non-exclusivity as components of moral responsibility, taken from John Ladd, "Collective and Individual Moral Responsibility in Engineering: Some Questions," *IEEE Technology and Society,* June 1982, pp. 3-10.

5. Editorial, "Safer Walking, Safer Driving," *The Baltimore Sunpapers,* November 29, 1993, p. A6.

6. Comment, "Real deaths, real choices?" *New Scientist* 1899, November 13, 1993, p. 3.

Study Questions

1. What is the main point of Hollander's article? Why do you think it might be correct? Why do you think it might be wrong? Explain and defend your answers.

2. What assumptions does Hollander make about technology? about

ethics? Do you think these assumptions are correct? Explain and defend your answers.

3. If society accepts Hollander's main conclusions, what consequences might follow? Would these consequences be desirable? Explain and defend your answers.

4. What is the point of Hollander's first example, about German manufacture of large industrial saws?

5. From the standpoint of safety, what considerations should engineers and scientists keep in mind as they are developing technologies?

6. Can technical measures assure safety? Explain.

7. Is moral responsibility for technological processes different from moral responsibility for technological products? Explain.

2.7

The Political Construction of Technology: A Call for Constructive Technology Assessment

Jesse S. Tatum

It is one of the fundamental contentions of philosophy since Thomas Kuhn (1962) and of the more recent literature on the "social construction" of science and technology (Bloor, 1976; Latour, 1987; Bijker, Hughes, & Pinch, 1987) that modern patterns of science and technology are not simply inherent in nature. These patterns do not, it is broadly suggested, emerge of necessity, but bear the indelible imprint of human agency, emerging at least in part as a product of the inevitable selectivity of human attention.[1] Modern patterns, furthermore, are not obviously the culmination or even the most advanced product of any fixed "method" or notion of "progress" (e.g., Kuhn, 1962; Feyerabend, 1975, 1978; Marx, 1987). Science and technology, it is argued, are "underdetermined" (Latour, 1987) by external realities and thus are always also contingent on human actions and/or choices.

This paper begins with the contention that more vigorous attention to the developmental path of science and technology, specifically in the form of "constructive technology assessment" (CTA) efforts, is necessitated in a democratic system by recognition of the contingent nature of science and technology. The focus of the argument is on the exercise of power implicit in the *political* construction of technology and on the often subtle, even unrecognized, nature of that exercise.

Originally published as *Research in Philosophy and Technology*, Volume 15, "Social and Philosophical Constructions of Technology," pages 103–115. Copyright © 1995 by JAI Press Inc.

After developing this argument, specific examples will be used to suggest means by which CTA efforts might be furthered and to indicate where they might most critically be needed.

Political Construction of Technology

As the contingent or underdetermined nature of science and technology has become increasingly accepted in recent years, much has been made of practical and theoretical evidence for their "social construction." In many cases, this "social" construction may seem an unexceptional process. The precise configuration of the modern bicycle, for example, may be of relatively little concern even if "there is nothing 'natural' or logically necessary about [the process of] closure" that brought us the air tire and eliminated the old "high-wheelers," as Pinch and Bijker (1987) argue. Speed apparently became the ultimate design criterion in this case, and "It could be argued that speed is not the most important characteristic of the bicycle or that existing cycle races were not appropriate tests of a cycle's 'real' speed" (Pinch & Bijker, 1987, p. 46). yet the design details of modern bicycles and how they came to predominate may seem both a matter of little consequence.

A slightly different term can, however, be employed to label much the same process of selecting particular technical configurations among multiple alternatives: "political construction" of technology. This is not to disparage the notion of "social construction."[2] The bicycle illustration of "social construction" offered by Pinch and Bijker does have the unique and, in its context, laudable advantage of focusing on the "underdetermination" of the design by natural facts and the associated contingency of the technological outcome on human participation, without the possible distraction of sharply differing (political) commitments as to how the design "should" have come out.

The term "political construction" of technology, however, has the advantage of a different focus. While it works reasonably well in calling attention to the contingent nature of technology, it tends also to call our attention to the exercise of political *power* that may be implicit in the selection of one configuration for technology over others. In the science, technology, and society (STS) studies literature, there are many excellent illustrations of what might more aptly be termed "political" rather than "social" construction of technology. David Noble (1984) argues, for example, that the determining factor in the development of "numerical control" rather than "record playback" automation may have been a reshaping of the balance of power between management and workers, rather than any realistically defensible belief in the superior technical efficiency of numerical control. Langdon Winner's (1986, pp. 19–39) discussion of Robert Moses' low bridges as barriers to buses, and hence to low income people's access to suburbs and beaches in the New York area, also has the ring of political construction of technology. Winner's reference in the same essay to the development of automatic tomato picking machines usable only on large farms, and to the implementation of pneumatic molding machines in the 1880s as a way to break a union, offer further illustrations of what appear to be clear and consequential exercises of power in the selective construction of technology.

Subtle Exercise of Power

In focusing attention on the exercise of power often implicit in the selective development of science and technology, allegations of conspiracy and even of malevolent or self-serving intent may be advanced. However, no such allegation is necessary to justify a focus on power.

> As far as I know, no one argued that the development of the tomato harvester was the result of a plot. Two students of the controversy, William Friedland and Amy Barton, specifically exonerate the original developers of the machine and the hard tomato from any desire to facilitate economic concentration in that industry. What we see here instead is an ongoing social process in which scientific knowledge, technological invention, and corporate profit reinforce each other in deeply entrenched patterns, patterns that bear the unmistakable stamp of political and economic power. (Winner, 1986, p. 27)

Even where an exercise of power may be only implicit—that is, entirely unrecognized and, as such, often not defined as an exercise of power at all—it may well remain a matter worthy of careful attention in the development of technology, just as it is in other areas of life.

Where there is a concern for the *responsible* exercise of power and for *legitimacy* within democratic traditions, *awareness* of the exercise of power and *accountability* for that exercise are logically matters of rudimentary significance. There may, of course, be a wide range of opinion regarding the degree to which human actors *actually choose* or have the capacity to choose in matters of science and technology. In the world described by Ellul (1964), Mumford (1967, 1970), or Marcuse (1964), we may be virtually powerless to choose, or we may be so deeply conditioned to the rule of technology as to require vigorous shaking, at the very least, if we are to become active choice makers.

In practical terms, however, if we accept as fact the underdetermination of science and technology, there is room for attention to the process by which choices are made. What alternatives are there (or have there been) to the chosen path? How have those alternatives been selectively eliminated? The importance of such questions is asserted, implicitly or explicitly, by virtually every participant in the STS community who raises the issues of the political construction of technology, whether or not they employ this term. Concern with the choice process, and the often suspect or regrettable consequences of widespread default in that process, may well be one of the fundamental origins of the STS movement (Cutcliffe, 1990).

Concern with the political construction of technology is very much heightened by an appreciation for the subtler workings of power. How do we know, after all, that any particular "advance" in science or technology is responsive to human needs or desires at either an individual or collective level? How do we know what people "really" want? Is it enough that people do in fact purchase the VCR or the microwave oven, for example, when it comes on the market? Is this enough to guarantee us that resource investments in these new technologies have been appropriate and that the public good has indeed been best served by the choice of this among all possible routes for technology development? Is this market behavior a sufficient guarantee that power, if you will, has been legitimately and responsibly exercised in choosing this evolutionary direction over others?

For many, of course, these are idle questions. Where, after all, is the exercise of power in the development of a VCR? Surely there has been no significant *conflict* over whether or not a VCR should be developed. And is it not the proper right of those taking the risk and making the investment in new technology development to dispose of their own property in this way if they so choose?

Subtler notions of power, however, suggest that power may have been exercised in the decision to develop and market the VCR even without overt conflict or opposition to this technological thrust. Without proposing the abandonment of property rights, we may yet imagine conflict between a democratic shaping of science and technology, and a developmental path shaped by the interests of private property. The fact that "consumers" buy VCRs, furthermore, is no a priori guarantee that anyone's "real interests" are served by this new device; we have no data, after all, on what the outcome would be if people have been offered an "anti-VCR" as an alternative situated in a "nonconsumer" world.

This may still sound like idle speculation. To make the matter more concrete, consider Langdon Winner's discussion of technology as an impediment to access and participation on the part of handicapped people (Winner, 1986, p. 25). While there surely was no deliberate intent to prevent handicapped people from participating as other citizens do in public life, the use of curbs, stairs, and other technologies in traditional construction in this country long had precisely this effect. The (effective) exercise of power that prevented their participation was not intentional and long went essentially unrecognized, yet we collectively went about shaping technology in particular ways (to the exclusion of obvious alternatives that have been pursued since) that had a significant political effect.

In his handling of the subtler instances of the exercise of power, Steven Lukes offers a definitional refinement:

> Can *A* properly be said to exercise power over *B* where knowledge of the effects of *A* upon *B* is just not available to *A*? If *A*'s ignorance of those effects is due to his (remediable) failure to find out, the answer appears to be yes. (Lukes, 1974, p. 51)

He further explains that

> The point . . . of locating power is to fix responsibility for consequences held to flow from the action, or inaction, of certain specifiable agents. (p. 56)

In the context of this definition, those who have designed and built curbs and stairways are at least potentially responsible for an exercise of power that long excluded handicapped people from public life. If the designers or builders *could have found out* about the impacts of their technology choices on handicapped people, then it may be that they should be held responsible for this effect.

If we adopt Lukes' notions of power, the social construction of science and technology may well be political in a rather pervasive sense. *Every* choice among avenues of development has the potential, at least in theory, of selectively foreclosing or disadvantaging particular patterns of life that could be of interest to certain members of society. As such, *every* such choice could be regarded as an exercise of power of real significance. The fact that no one appears to raise serious objections to a specific selection or developmental path, the fact that no vociferous advocate of an alternative path comes forward, still leaves the prospect that we may simply have failed to "find out" how those who remain silent might approach the choices afforded within the realm of scientific and technological possibility.

The Legitimate Exercise of Power:
Constructive Technology Assessment

Does science and technology, as presently constructed, best serve human needs or aspirations, and has the selection process that brought us the particular science and technology now in place been democratic? There would appear to be room for concern that the political construction of science and technology, especially to the degree that it may involve subtle and less than overt exercises of power, indeed may *not* occur in a way that is entirely in keeping with the principles of our democratic tradition. If particular avenues of development are being selectively advanced, others effectively undermined, without an even-handed exploration of alternatives or in the absence of active efforts to get at the preferences and perspectives of all citizens, the exercise of power involved may not prove fully legitimate within a democratic tradition.

Acceptance of the underdetermined nature of science and technology, coupled with an appreciation for the potential subtleties inherent in the exercise of power, appear to call for a more conscientious approach to the development of science and technology, an approach that might be referred to as "constructive technology assessment" (CTA).

Constructive technology assessment (as the term will be used here) strives to envision alternative technological futures that might serve alternative configurations of human values and commitments.[3] It is concerned not simply with specific alternatives to particular technological developments or with alternative configurations for particular developments. The question it addresses is not simply one of weighing the costs and benefits, even broadly conceived, of particular developments, but of probing alternative notions of "cost" and "benefit" and of framing assessments in such a way as to address real choices among alternative "forms of life" (Winner, 1986, pp. 3–18). If we have become, as Winner suggests, "the beings who work on assembly lines, who talk on telephones, who do our figuring on pocket calculators, who eat processed foods, who clean our homes with powerful chemicals" (Winner, 1986, p. 12), then we need CTA to begin addressing the question, *who else could we be*?

Legitimate (i.e., democratic) exercise of power in the development of science and technology calls for active efforts to *construct* a range of alternatives *for* choice, along with *active* efforts to comprehend and discuss the full range of reactions to those alternatives that might arise. CTA efforts along these lines will call for a good deal more than the usual "market data" or questionnaires targeting people's "values." A far more probing concern with human "cares, commitments, responsibilities, preferences, tastes, religious convictions, personal aspirations, and so forth" (Winner, 1986, p. 156) will be required.

Ethnography and Constructive Technology Assessment

Schot (1992) has recently proposed three approaches to the actual conduct of CTA: "stimulating alternative [technological] variations, changing the [technology] selection environment, and creating or utilizing technological nexus." (The last of these focuses on institutional links and real world interactions between the variation process and the selection process.) While these approaches are developed in the context of examples taken from efforts to develop environmentally clean technologies, they appear to provide, as Schot suggests, good foundations for CTA in general. They are, in fact, indicative of interesting possibilities for extracting what amounts to CTA from ethnographic examinations of ongoing technology-related activities, and for developing CTA from almost purely theoretical exercises as well.

A variety of alternative variations and selection environments may already be available as elements of CTA if they can only be captured and somewhat formalized.[4] Ethnographic examination of ongoing departures from traditional norms may be especially useful in this regard, as may be seen from recent studies of the "home power" movement in the United States (Tatum, 1992, 1994).

Although not widely recognized by government or the popular media, the home power movement has generated a host of alternative variations in the area of residential energy systems. In a technical sense, the core of the movement includes more than 40,000 home electric supply systems (mostly photovoltaic or "solar cell" systems), augmented by small home hydroelectric and wind systems, as well as substantially improved and more reliable 110 volt inverters and battery systems to store energy and convert to household current. New technological variations also include uniquely efficient refrigerators, lighting systems, well pumps, and other appliances, specifically suited to (more expensive) renewable electricity supplies.

Participants in the home power movement appear to be motivated by a desire for a reformulation of work roles (more varied, less specialized content, and shared rather than hierarchically structured responsibility). They also appear to seek a strengthened sense of community and different and less damaging relations with the natural environment.

Together, the alternative variations and differently constituted selection environment of the home power movement arise from what might be termed a newly created technological nexus. Emerging from this new nexus is a very different "form of life" from those given consideration in the context of traditional economic or government policy analyses. There is, in the movement, a strong sense of "the coemergence of values and practices" (Bijker, 1993, p. 130) and the usefulness of ethnographic methods as a means of escaping conventional mindsets is much in evidence.

In efforts to initiate CTA from ethnography of ongoing departures from traditional norms, it will be important to remember that the intent is *not* to argue that people should not speak directly for themselves. The argument is rather that it may be useful at least to try to know people better than they know themselves, or more accurately, to attempt to *explain* behavior, attitudes, and values in words that communicate better with outsiders than the words that a particular group may be willing or able to come up with on its own.

It should be emphasized as well that there will be no guarantee that the descriptions arrived at through ethnographic methods will more accurately reflect "real" interests than simple market data on the "consumption patterns" of the people under study. Certain knowledge of "real" interests inevitably remains beyond our grasp. The objective, instead, would be to contribute to a broadened conversation about the goals, objectives, and effects of scientific and technological "advance." In undertaking ethnographic analysis, we simply take the first step in admitting that *we do not know* (no individual, professional expert, or group knows) what people "want." In adopting an ethnographic approach, we admit that what people want may change over time and with changing circumstance—and even that we do not fully know what we ourselves want. In undertaking such analysis, we assert only that what we and others want may benefit from examination and discussion in a context of choice among practical alternatives for action.[5]

Constructive Technology Assessment by Theoretical Means

While ethnographic studies of departures from the mainstream of technology development would, in essence, tap existing *practical* efforts to do what amounts to CTA, assessment efforts can also proceed from theoretical speculations. Consider, for example, the possibility of "success" in the development of an electric automobile with performance characteristics comparable to present gasoline-fueled cars. Substantial efforts are now underway to accomplish this advance in technology with relatively few very different alternatives apparently under consideration. Supposing, then, that we succeed. As a matter of theoretical speculation, what developments might we expect to follow this success? If the cost and mechanical energy requirements of the electric auto are comparable to those of present autos, can we expect substantial easing of the environmental problems associated with present autos, or more nearly a shift from auto to power plant emissions? Can we expect an easing of pressures on increasingly scarce fossil energy supplies? Where will the new power plants likely to be associated with the approximate doubling of present electric power production required by a fully electric fleet be sited? What fuels will they utilize?

Strictly theoretical speculations such as these may lead us to a consideration of other alternatives for future technology development. We may probe possibilities for much lighter, lower energy, shorter range electric vehicles to be used only in local commuting. We may consider possibilities for demographic shifts, living closer to work and to essential shopping, that would simply eliminate the need for many trips.

Purely theoretical considerations, in other words, may effectively lead us to a critical examination of traditional selection criteria and to a collection of alternative variations. Where we are at a loss to come up with new theoretical directions, we may benefit by interactions with others who may be in a position to see things differently. In the transportation example, we might address the question of alternative transportation systems to a backyard tinkerer or an engineering student, or to someone who doesn't have the money to buy a car or to feed it gasoline. Conversations with people who may bring new perspectives to technology development can provide access, in effect, to possibilities for changes in the selection environment that can be pursued in purely speculative theoretical terms.

Much may be possible in the way of theoretical as well as practical CTA with relatively little government support. Both theoretical and practical approaches, on the other hand, by the very nature of their departure from, or questioning of, established patterns of science and technology development, are likely to require some degree of public funding. At a minimal level, a great deal might be learned simply by examining and reporting on efforts like the home power movement. At the opposite extreme, public funding could be (and to a degree already is being) used in the actual experimental exploration of alternatives such as those that might emerge from theoretical speculations. A certain amount of publicly funded CTA is already being done, either within governmental agencies, or through activist organizations[6]— with other countries apparently in the lead (Vig, 1992).

An overall reduction in government intervention in the shaping of science and technology may, however, also be worthy of careful consideration in advancing a program of constructive technology assessment. It can be argued that democratic decision making is best assured when ordinary people directly confront actual alternatives for the development of science and technology. To the degree that people are naturally included to practice what amounts to CTA on their own, it may in some cases be well simply to get out of the way.

In a sense, "doing" CTA is nothing new. Surely most technology development efforts begin with more than one possible alternative and with some consideration of how the selection environment is or should be defined. The argument here again is simply that a full recognition of the underdetermined nature of technological advance and of the subtleties in the exercise of power implicit in that advance necessitate a more conscious and a more conscientious dedication to the process. In order to be sure that we have "found out" what directions might be preferred by particular individuals or groups, CTA needs to be pursued more actively. It needs to be pursued from a wider range of perspectives and to carry those perspectives farther in terms of their possible expression in alternative technologies. And it needs more closely to involve citizens to ensure both that a wider range of perspectives is represented, and that alternative developmental paths actually achieve broad consideration by those who ultimately are supposed to be represented in the choices that are made.

Where Do We Need Constructive Technology Assessment?

In an ideal world, CTA would be pursued continuously and on a broad front, not simply in response to specific technological proposals. Critical areas of need, however, might be located by certain signals. In areas where there appear to be unusual opportunities for commercial *profit*, for example, special attention to CTA efforts may be justified by the concern that broadly democratic decision making may be overwhelmed by particularized economic interests. Areas in which appeals of an apparently unchallenged *ideological* nature are made may also be critical targets for CTA efforts.

The case of bioengineering may be a good illustration in the first category and technology development targeting "global competitiveness" may be exemplary in the second.

The potential for profit from commercial developments in the field of bioengineering requires little explanation. The problem is that that potential for profit may overwhelm less organized and less focused alternatives for technology development that could arise in a more broadly democratic decision-making setting. Once particular bioengineering alternatives are proposed, whether they involve organisms engineered to help keep strawberries from freezing in the field or biological systems engineered to produce bovine growth hormone to enhance milk production, the potential for profit gives those proposals a life of their own. With strong arguments (i.e., profits) from well-organized groups (i.e., corporations) in favor of these developments, the burden of proof shifts to opponents who in practice are generally required to come up with fairly specific arguments against particular products. In the case of "recombinant" bovine growth hormone (rBGH), for example, we hear objections that include the expectation of minimal direct benefit to consumers (prices are already low, given a surplus of milk on the market), possible concerns about human health effects, and possible increases in udder infections and other tissue reactions in cows. Four major corporations have, however, developed one or another version of rBGH. For one of them, Monsanto, rBGH is only the first of more than a dozen biotech products in a pipeline with $1 billion a year in projected sales and royalties, and research and development costs already amounting to $800 million (Roush, 1991).

Specific objections to rBGH or other bioengineered products may well not carry the day, and perhaps they should not. Yet there can be little doubt but that genetic engineering as a whole has the potential for "reconstituting the conditions of human existence" (Winner, 1986, p. 10) in terms of continuing ecosystem transformations, human/nature relationships, and even the structure and content of human associations. It is difficult to regard that process of reconstitution as "democratic" in the absence of a countervailing balance to the highly focused profit incentive, and in the absence of equally aggressive efforts to explore and define alternative courses of development. Such a situation is expressive of a critical need for CTA efforts.

In the area of global competitiveness, there is the appearance of an appeal to unexamined ideological commitments indicative of a critical need for CTA. The phrase "global competitiveness" seems at times to have acquired a mesmerizing quality, silencing once insurmountable resistance to layoffs and other labor concessions, for example, where it is argued that global competition necessitates leaner and more efficient operations. Similarly, we seem to proceed without a blink in the apparent abandonment of traditional separations between publicly and privately supported research and development (separations once carefully guarded in the national lab system, for example), if that abandonment is proffered in the name of global competitiveness (cf. Hill, 1989). As developments of this sort take on an "automated" character, advancing without the apparent possibility of resistance or debate, there may be a critical need for aggressive CTA efforts. While the arguments for the obvious course of action may in fact be strong, their very strength may serve to eclipse alternatives with significant appeal from marginal or marginalized perspectives.

In continuing development of science and technology, we too often confront what amounts to one and only one alternative. We accept it or, implicitly, we oppose it and embrace the social stigma and practical impotence of siding with a "non-alternative." It is as if we were presented with the choice of Mr. X for president, or no president at all, as if we were to choose between global competitiveness and nothing at all. In such a setting, and especially in the areas of critical need identified earlier, constructive technology assessment is essential as a tool to general enfranchisement and democratic choice.

Conclusion

In ongoing genetic engineering, "global competitiveness," and innumerable other technology development efforts, we are unquestionably engaged in "reconstituting the conditions of human existence" (Winner, 1986, p. 10). Without vigorous constructive technology assessment, we are at best doing this as "somnambulists" (Winner, 1986). At worst we are reshaping the lives of others without their knowledge, participation, or permission. Recognition of the underdetermined nature of science and technology, an understanding of the subtle nature of power and its exercise in the selection of developmental paths, and a concern with democratic decision making, together constitute a compelling argument for new and more active practical and theoretical efforts to explore, define, and discuss a broadened range of technological alternatives for the future.

Notes

1. The term "selective attention" is derived from David Rose (1981).
2. Bijker himself appears to work more in the area of political construction in recent work (Bijker, 1992, 1993).
3. For comparison, see Schot (1992) and his references. Bijker (1993, p. 129) offers a characterization of "orthodox 'technology assessment,' as exemplified since 1972 by the work of the U.S. Office of Technology Assessment," which he contrasts nicely with a constructivist approach, though without use of the term CTA. See also Martin Rein (1993) and the notion of "value-critical" policy analysis, which in many ways parallels that of CTA.
4. Support for such a claim is widely available even in such classic studies as *Small Town in Mass Society* (Vidisch & Bensman, 1968).

5. What is envisioned here is beginning again to cultivate what Albert Borgmann (1984) has termed "diectic discourse."

6. For example, the National Resources Defense Council, Washington, DC, has constructed arguments for transportation alternatives along lines similar to those outlined previously.

References

Bijker, W.E., T.P. Hughes, & T.J. Pinch (Eds.) (1987). *The social construction of technological systems.* Cambridge, MA: MIT Press.

Bijker, W.E. (1992). The social construction of fluorescent lighting, or how an artifact was invented in its diffusion state. In W.E. Bijker & J. Law (Eds.), *Shaping technology/building society.* Cambridge, MA: MIT Press, 1992.

Bijker, W.E. (1993). Do not despair: There is life after constructivism. *Science, Technology, and Human Values, 18* (1), 113–138.

Bloor, D. (1976). *Knowledge and social imagery.* Boston: Routledge and Kegan Paul.

Borgmann, A. (1984). *Technology and the character of contemporary life: A philosophy inquiry.* Chicago: University of Chicago Press.

Cutcliffe, S. (1990). The STS curriculum: What have we learned in twenty years? *Science, Technology, and Human Values, 15* (3), 360–372.

Ellul, J. (1964). *The technological society.* J. Wilkinson (trans.). New York: Knopf.

Feyerabend, P. (1975). *Against method.* London: Verso.

Feyerabend, P. (1978). *Science in a free society.* London: Verso.

Hill, C.T. (1989). Technology and international competitiveness: Metaphor for progress. In S.L. Goldman (Ed.), *Science, technology, and social progress* (pp. 33–47). Bethlehem, PA: Lehigh University Press).

Kuhn, T.S. (1962). *The structure of scientific revolutions.* Chicago: University of Chicago Press.

Latour, B. (1978). *Science in action.* Cambridge, MA: Harvard University Press.

Lukes, S. (1974). *Power: A radical view.* London: Macmillan.

Marcuse, H. (1964). *One dimensional man: Studies in the ideology of advanced industrial society.* Boston: Beacon Press.

Marx, L. (1987). Does improved technology mean progress?" *Technology Review, 90* (1), 32–41, 71.

Mumford, L. (1967). *The myth of the machine* (volume 1: *Technics and human development*). New York: Harcourt Brace Jovanovich.

Mumford, L. (1970). *The myth of the machine* (volume 2: *The Pentagon of Power*). New York: Harcourt Brace Jovanovich.

Noble, D. (1984). *Forces of production: A social history of industrial automation.* New York: Knopf.

Pinch, T.J., & W.E. Bijker. (1989). The social construction of facts and artifacts: Or how the sociology of science and the sociology of technology might benefit each other. In Bijker et al. (pp. 17–50).

Rein, M. (1983). Value-critical policy analysis. In D. Callahan & B. Jennings (Eds.), *Ethics, the social sciences, and policy analysis.* New York: Plenum Press.

Rose, D.J. (1981). Continuity and change: Thinking in new ways about large and persistent problems. *Technology Review, 84* (2), 53–67.

Roush, W. (1991). Who decides about biotech? The clash over bovine growth hormone. *Technology Review, 94* (5), 28–36.

Schot, J.W. (1992). Constructive technology assessment and technology dynamics: The case of clean technologies. *Science, Technology, and Human Values, 17* (1), 36–56.

Tatum, J.S. (1992). The home power movement and the assumption of energy policy analysis. *Energy: The International Journal, 17* (2), 99–107.

Tatum, J.S. (1994). Technology and values: Getting beyond the 'device paradigm' impasse. *Science, Technology, and Human Values, 19* (1), 99–108.

Vidich, A.J., & J. Besman. (1968). *Small town in mass society.* (rev. ed.). Princeton, NJ: Princeton University Press.

Vig, N.J. (1992). Parliamentary technology assessment in Europe: Comparative evolution. Paper delivered at the Annual Meeting of the American Political Science Association, Sept. 3–6.

Winner, L. (1977). *Autonomous technology: Technics-out-of-control as a theme in political theory.* Cambridge, MA: MIT Press.

Winner, L. (1986). *The whale and the reactor: The search for limits in an age of high technology.* Chicago: University of Chicago Press.

Study Questions

1. What is the main point of Tatum's article? Why do you think it might be correct? Why do you think it might be wrong? Explain and defend your answers.
2. What assumptions does Tatum make about technology? about ethics? Do you think these assumptions are correct? Explain and defend your answers.
3. If society accepts Tatum's main conclusions, what consequences might follow? Would these consequences be desirable? Explain and defend your answers.
4. Why does Tatum speak of the "political" construction of technology? Explain.
5. Why does Tatum use the two examples from biotechnology, the "hard tomato" and the BGH? What is the point of each example?
6. What is a "constructive technology assessment (CTA)? Why does Tatum refer to it as a legitimate exercise of power? What ethical or moral reasons are there for doing a CTA?
7. How would CTAs redress the problems caused by present approaches that place the "burden of proof" on victims of technology?

2.8

Further Reading

Balestra, Dominic J. 1990. "Technology in a Free Society: The New Frankenstein," *Thought* 65, 257 (June): 155–68.

Bluhm, Louis H. 1987. "Trust, Terrorism, and Technology," *Journal of Business Ethics* 6 (July): 333–41.

Borgmann, Albert, 1987. "The Question of Heidegger and Technology: A Critical Review of the Literature," *Philosophy Today* 31, 2/4 (Summer): 98–194.

Brown, Lester. 1981. *Building a Sustainable Society.* New York: Norton.

Byrne, Edmund F., and Pitt, Joseph C. (eds.). 1989. *Technological Transformation: Contextual and Conceptual Implications.* Dordrecht: Kluwer.

Durbin, Paul T. 1983. *Philosophy and Technology.* Dordrecht: Reidel.

Feenberg, Andrew. 1995a. *Alternative Modernity: The Technical Turn in Philosophy and Social Theory.* Berkeley: University of California Press.

———. 1995b. *Technology and the Politics of Knowledge.* Bloomington: Indiana University Press.

———. 1991. *Critical Theory of Technology.* New York: Oxford University Press.

Glimm, James G. (ed.). 1991. *Mathematical Sciences, Technology, and Economic Competitiveness.* Washington, D.C.: National Academy Press.

Gomory, Ralph E. 1980. *Technology Development.* Washington, D.C.: National Academy Press.

Hickman, Larry A. 1994. "John Dewey: Philosopher of Technology," *Free Inquiry* 14, 4 (Fall): 41–43.

Ihde, Don. 1990. *Technology and the Lifeworld: From Garden to Earth.* Bloomington: Indiana University Press.

———. 1983. *Existential Technics.* Albany: SUNY Press.

King, Johnathan B. 1994. "Tools–'R'–Us," *Journal of Business Ethics* 13, 4 (April): 243–57.

McGinn, Robert E. 1994. "Technology, Demography, and the Anachronism of Traditional Right," *Journal of Applied Philosophy* 11, 1: 57–70.

McLaughlin, Andrew C. 1993. *Regarding Nature: Industrialism and Deep Ecology*. Albany: SUNY Press.

Milbrath, L. 1989. *Envisioning a Sustainable Society: Learning Our Way Out*. Albany: SUNY Press.

Mitcham, Carl, and Mackey, Robert. 1983. *Philosophy and Technology: Readings in the Philosophical Problems of Technology*. New York: Free Press.

Noss, Reed F., and Copperrider, A. Y. 1994. *Saving Nature's Legacy*. Washington, D.C.: Island Press.

Rubenstein, A. H., and Hanna, A. M. (eds.). 1991. *Research on the Management of Technology: Unleashing the Hidden Competitive Advantage*. Washington, D.C.: National Academy Press.

Shrader-Frechett, Kristin. 1995. "Evaluating the Expertise of Experts," *Risk: Environment, Health, and Safety* 6, 2 (Spring): 115–26.

———. 1991. *Risk and Rationality*. Berkeley: University of California Press.

Strong, David. 1992. "The Technological Subversion of Environmental Ethics," *Research in Philosophy and Technology* 12: 33–66.

Part Three

How to Evaluate Technology

3.1

Introduction and Overview

Kristin Shrader-Frechette and Laura Westra

Forecasting and evaluating the effects of technological development are extraordinarily difficult. In the middle of the nineteenth century, technologists investigating horse-and-buggy technology worried that the major problem of cities in the next century would be disposing of horse manure in the streets. Automobile technology, however, outstripped this impact and created quite different benefits and costs than those associated with the horse and buggy.

This part of the book examines ethical and value problems associated with evaluating technology. One of the most basic ethical problems arising in connection with the evaluation of technology concerns the evaluation itself. Many scientists and engineers often believe that they can assess a technology, even its second- and higher-order impacts, while other experts are more skeptical about long-term evaluation. A recent report by the National Research Council, for example, affirmed the adequacy of million-year estimates of the safety and technology for storing high-level nuclear waste.[1] A peer-review committee of the Department of Energy, however, said that such a long-term evaluation was impossible because of the many subjective judgments involved and because of uncertainty about future geological activity, including seismicity and volcanism.[2] Much of the debate between the National Research Council group and the peer-review group concerns the desirability of using subjective estimates of long-term technological behavior. Some evaluators would rather have a number or a model, even a bad one, than no number or no model at all. Other evaluators are more able to admit the uncertainty that besets many technological evaluations. To the degree that technologists admit the uncertainty in their assessments, citizens can claim that they have a right to make the ethical decision about how to behave in a situation of uncertainty, for example, whether to use a risky technology whose effects are partially unknown. To the degree that technologists claim there is little uncertainty in their assessments and that particular technologies are safe, citizens have fewer rights to decision making because they have less to lose and fewer threats to their welfare.

Besides disputes over the reliability of expert judgments about the risks or the consequences of a particular technology, many assessors disagree over how technologies should be evaluated. Some authors believe that economic evaluations of technology are the most defensible, and that society should adopt those technologies whose overall benefits outweigh the costs. Other authors believe that economic evaluations miss some of the values that are most crucial to people, because not everything has a price, and because not all values can be reduced to prices. Still other authors evaluate technologies from social, political, religious, or environmental points of views. Crucial value questions concern not only how reliable the evaluations are but also what kinds of evaluations are desirable.

In discussing evaluations and the kinds of values that can arise in technological risk assessment, it is important to distinguish at least three types of values that often occur in science and technology. On Helen Longino's classification, these include bias values, contextual values, and methodological values.[3] Bias values occur when one deliberately includes or excludes data for the purpose of serving one's own ends or goals. Sexist and racist beliefs are examples of bias values. Contextual values are personal, social, or philosophical emphases that can influence science or assessment practice. Financial constraints are examples of contextual values. Although it is possible to avoid bias values in evaluating science and technology, it is not possible to avoid contextual values, because all activity is accomplished in some context. Likewise it is not possible to avoid methodological values in science and technology assessment. Methodological values are judgments about how to deal with uncertainty, how to interpret data or conclusions, such as how to select sample size or how to select which statistical test to use.

Chapter 3.2 by Langdon Winner argues that bias and contextual values influence the technologies that society develops and uses. He says that many technologies are like Frankensteins: although society has created them, they are not under societal control. For Winner, the first stage of technological evaluation is to realize that humans do not control technology, but that technology often controls humans. To illustrate his point, Winner takes the reader back to the original novel about Frankenstein, and he uses insights from the book to develop many lessons about contemporary technology. Winner claims that there are two main camps evaluating technology, those from the utilitarian-pluralist school and those from the counterculture-activist-utopian school. The first group demands changes in legislation to control technology, while the second group demands a change in life and lifestyle to control technology. Different ideals of social and political life generate different technologies, says Winner. In order to create the technologies that are appropriate to our democratic social and political life, Winner says that people must employ Luddism as epistemology. They must learn how and what technologies to build by letting current technologies die or fall into disrepair.

Robert McGinn, in chapter 3.3 "Technology, Demography, and the Anachronism of Traditional Rights," argues that ethical analysis of technology reveals that creators of technology design it to be maximalist in product, size, scale, performance, speed, volume, and so on. This maximality of technology is a problem, however, says McGinn, because society gives humans the right to create technology and to maximize it in a variety of ways. Many people using their property rights to do things bigger, better, and faster, says McGinn, means technology is destroying the societal and environmental quality of life. The only solution, he says, is to redefine human rights in contextualist ways that do not allow people to engage in widespread technological maximalization.

In chapter 3.4 Kristin Shrader-Frechette discusses economic evaluations of technology by uncovering the assumptions underlying benefit-cost analyses and the consequences that follow from using benefit-cost techniques to evaluate various technologies. She shows, for example, that the assumptions of benefit-cost analysis require evaluators to equate price with value, to identify welfare with preferences, and to confuse utility with morality. Her essay also reveals that using benefit-cost analysis to evaluate technology results in discrimination against the poor. Applying her analyses of the assumptions and consequences of economic evaluations to different contemporary technology assessments, Shrader-Frechette shows that a variety of poor policy choices arise because of the way assessors uncritically use economic analyses of technology.

Many scientists and technologists believe that their methods of assessment are value free, while many philosophers and social scientists, as chapter 3.4 illustrates, believe that neither science nor technological evaluation can be value free.[4] Deborah Mayo, in chapter 3.5, investigates the view that social, ethical, and policy values can be separated from the methods of assessing technological risks. She points out that those who believe values are always present (the nonseparatists) fall into two quite different groups. The sociological group believes that values are always present because of the societal and nonscientific values (like the profit motive) that influence technology assessment. The metascientific group believes that values are always present because of the methodological values that influence technology assessment. Methodological values include, for example, assumptions about how to collect data, how to test hypotheses, and how to interpret theories and findings. Mayo believes that it is important to distinguish the sociological from the metascientific separatists because they emphasize different values present in science and technological risk assessment. The sociological separatists emphasize social values to such an extent that they often argue that neither science nor technology assessment can be rational or objective. The metascientific separatists emphasize methodological values but typically believe that despite these unavoidable values, science and technology assessments can be rational (in the sense of being defensible) or objective (in the sense of not being deliberately biased). One asset of Mayo's essay is that she shows how science and technology assessment can be value laden without also being hopelessly biased and subjective. Her point is crucial because many scientists and technologists are reluctant to recognize the values implicit in their work because they fear that such values would compromise its objectivity. Obviously, if objectivity is close to certainty and infallibility, then no scientific or technological work is objective in this unrealistic sense. However, if objectivity means being free of bias, then any technology assessment can be objective in this sense.

Mayo accepts the metascientific view, but she believes that ethical and policy judgments can enter science and technological risk assessment because, in the face of scientific uncertainty, often there is more than one scientifically acceptable methodological value judgment or inference option. Thus scientists and technologists could advance their own hidden biases—their own ethical and policy values—simply by choosing one rather than another methodological value judgment (for example, about how to interpret data). Using a case study of technological risk assessment of formaldehyde, Mayo shows how bias values can influence assessment. In the formaldehyde case, scientists and assessors disagreed about how to interpret negative epidemiological results (that formaldehyde presented no statistically significant health risk). The industry group said these results showed there was no health risk; the government regulators said these results were inconclusive, because failure to find a result does not mean there is no result, but merely that one may not yet have found it. Mayo's article shows how to avoid bias values and how to tell the truth about negative statistical or epidemiological results in technological risk assessment.

Notes

1. National Research Council, *Technical Bases for Yucca Mountain Standards* (Washington, D.C.: National Academy Press, 1995).

2. J. L. Albrecht et al., *Report of the Peer Review Panel on the Early Site Suitability Evaluation of the Potential Repository Site at Yucca Mountain, Nevada,* SAIC 91/8001 (Washington, D.C.: U.S. Department of Energy, 1991).

3. Helen Longino, *Science as Social Knowledge* (Princeton: Princeton University Press, 1990).

4. See K. S. Shrader-Frechette, *Risk and Rationality: Philosophical Foundations for Populist Reforms* (Berkeley: University of California Press, 1991), chs. 1–4, for a discussion of these two positions.

3.2

Frankenstein's Problem: Autonomous Technology

Langdon Winner

Our inquiry begins with the simple recognition that ideas and images of technology-out-of-control have been a persistent obsession in modern thought. Rather than dismiss this notion out of hand, I ask the reader to think through some ways in which the idea could be given reasonable form. The hope is that such an enterprise could help us reexamine and revise our conceptions about the place of technology in the world. In offering this perspective, I try to indicate that many of our present conceptions about technics are highly questionable, misleading, and sometimes positively destructive. I also try to lay some of the early groundwork for a new philosophy of technology, one that begins in criticism of existing forms but aspires to the eventual articulation of genuine, practical alternatives.

A possible objection to the notions I develop here is that they are altogether Frankensteinian. Some may suppose that in choosing this approach, the inquiry enters into an old and discredited myth about the age of science and technology. It is easy, indeed, for the imagination to get carried away with images of man-made monstrosities. And it is not difficult to get snagged on the linguistic hook which allows us to talk about inanimate objects with transitive verbs, as if they were alive. Some may even conclude that such traps fulfill a certain need and that autonomous technology is nothing more than an irrational construct, a psychological projection, in the minds of persons who, for whatever reason, cannot cope with the realities of the world in which they live.

To doubts of this kind I would reply: Does the point refer to the book or the motion picture? What, after all, is Frankenstein's problem? What exactly is at stake in the notorious archetypical inventor's relationship to his creation? Unfortunately, the answers to this question now commonly derive not from Mary Wollstonecraft Shelley's remarkable novel but from an endless stream of third-rate monster films whose makers give no indication of having read or understood the original work. Here is a case in which the book is truly superior to the movie. The Hollywood retelling fails to notice the novel's subtitle, "A Modern Prometheus," much less probe its meaning. The filmmakers totally ignore the essence of the story written by a nineteen-year-old woman, daughter of the radical political theorist William Godwin and one of the earliest of militant feminists, Mary Wollstonecraft. As a consequence, no justice is done to a work that it seems to me is still the closest thing we have to a definitive modern parable about mankind's ambiguous relationship to technological creation and power.

In the familiar Hollywood version, the story goes something as follows. A brilliant but deranged young scientist constructs a hideous creature from human parts stolen from graveyards. On a stormy evening in the dead of winter the doctor brings his creature to life and celebrates his triumph. But there is a flaw in the works. The doctor's demented assistant, Igor, has mistakenly stolen a criminal brain for the artificial man. When the monster awakes, he tears up the laboratory, smashes Doctor Frankenstein, and escapes into the countryside killing people right and left. The doctor is horrified at this development and tries to recapture the deformed beast. But before he can do so, the local townspeople chase down the monster and exterminate him. This ending is, of course, variable and never certain, lending itself to the need for a plot—over a forty-year period—for Boris Karloff, Bela Lugosi, Lon Chaney, Jr., Peter Cushing, and Mel Brooks.

The fact of the matter is that the film scenarios have virtually nothing to do with *Frankenstein* the novel. In the original there is no crazed assistant, no criminal brain mistakenly transplanted, no violent rampage of random terror, no final extermination of the creature to bring safety and reassurance (although there is mention of a graveyard theft). In the place of such trash, the book contains a story offering an interesting treatment of the themes of creation, responsibility, neglect, and the ensuing consequences. Let us see what Mary Shelley's gothic tale actually has to say.

From the time of his youth the young Genevan, Victor Frankenstein, was fascinated by the causes of natural phenomena. "The world," he tells us, "was to me a secret which I desired to divine. Curiosity, earnest research to learn the hidden laws of nature, gladness akin to rapture, as they were unfolded to me, are amongst the earliest sensations I can remember."[1] As he reaches maturity, his first response to this lingering obsession is to probe alchemy and the occult, the texts of Albertus Magnus and Paracelsus, in quest of the philosopher's stone. But realizing the futility of this research, he soon turns to the new science of Bacon and Newton. He hears a professor tell how the modern masters are superior to the ancients since they "penetrate into the recesses of nature and show how she works in her hiding places,"[2] a distinctly Baconian notion of what is involved. Following the principles of mathematics and natural philosophy, he eventually comes upon "the cause of the generation of life; nay more, I became myself capable of bestowing animation upon lifeless matter."[3]

To this point the story sounds very much like the one we all think we know. But from here on the novel takes some surprising turns. One evening Victor Frankenstein does bring his artificial man to life. He sees it open its eyes and begin to breathe. But instead of celebrating his victory over the powers of nature, he is seized by a rash of misgivings. "Now that I have finished, the beauty of the dream vanished, and breathless horror and disgust filled my heart. Unable to endure the aspect of the being I had created, I rushed out of the room and continued a long time traversing my bedchamber, unable to compose my mind to sleep."[4] And what about the newborn "human" back in the laboratory? He is left to his own devices trying to figure out what in the world has happened to him. Quietly he walks to Victor's bedroom, draws back the bed curtain, smiles, and tries to speak. But Victor, in the throes of a crisis of nerve, is still not ready to accept the life that he brought into existence and simply panics. "He might have spoken, but I did not hear; one hand was stretched out, seemingly to detain me, but I escaped and rushed downstairs. I took refuge in the courtyard belonging to the house which I inhabited, where I remained during the rest of the night, walking up and down in the greatest agitation, listening attentively, catching and fearing each sound as if it were to announce the approach of the daemonical corpse to which I had so miserably given life."[5]

Thus, it is Frankenstein himself who flees the laboratory, *not* his benighted creation. The next morning Victor leaves the house altogether and goes to a nearby town to tell his troubles to an old friend. This is very clearly a flight from responsibility, for the creature still alive, still benign, left with nowhere to go, and, more important, stranded with no introduction to the world in which he must live. Victor's protestations of misery, remorse, and horror at the results of his work sound particularly feeble. It is clear, for example, that the monstrosity of his creation is in the first instance less a matter of its physical appearance than of Frankenstein's terror at his own success. He is haunted henceforth not by the creature itself but by the vision of it in his imagination. He does not return to his laboratory and makes no arrangements of any kind to look after his work of artifice. The next encounter between the father and his technological son comes more than two years later.

An important feature of *Frankenstein*, the feature of the book that makes it useful for our purposes, is that the artificial being is able to explain his own position. Fully a third of the text is either "written" by his hand or spoken by him in dialogue with his maker. After his abandonment in the laboratory, the creature leaves the place and enters the world to make his way. Eventually, he takes up residence in a forest near a cottage inhabited by a Swiss family. He eavesdrops on them, notices how they use words, and after a while masters language himself. Stumbling upon a collection of books, he teaches himself to read and soon finishes off *Paradise Lost*, Pultarch's *Lives*, and the *Sorrows of Young Werther*. Later he examines the coat he had carried with him from the laboratory and finds Frankenstein's diary describing the circumstances of the experiment and giving the true identity of his maker. When the creature finally meets Victor on an icy slope in the Alps, he is ready to state an eloquent case. Autonomous technology personified finds it voice and speaks. The argument presented emphasizes the perils of an unfinished, imperfect creation, cites the continuing obligations of the creator, and describes the consequences of further insensitivity and neglect.

> "I am thy creature, and I will be even mild and docile to my natural lord and king if thou wilt also perform thy part, that which thou owest me."[6]

> "You propose to kill me. How dare you sport thus with life? Do your duty towards me, and I will do mine towards you and the rest of mankind. If you comply with my conditions I will leave them and you at peace; but if you refuse, I will glut the maw of death, until it be satiated with the blood of your remaining friends."[7]

The monster explains that his first preference is to be made part of the human community. Frankenstein was wrong to release him into the world with no provision for his role or influence in the presence of normal men. Already his attempts to find a home have had disastrous results. He introduced himself to the Swiss family, only to find them terrified at his grotesque appearance. On another occasion he unintentionally caused the death of a young boy. He now asks Frankenstein to recognize that the invention of something powerful and novel is not enough. Thought and care must be given to its place in the sphere of human relationships. But Frankenstein is still too thick and self-interested to comprehend the message. "Abhorred monster! Fiend that thou art! . . . Begone! I will not hear you. There can be no community between you and me; we are enemies. Begone, or let us try our strength in a fight, in which one must fall."[8]

Despite this stream of invective, the creature continues to reason with Victor. It soon becomes apparent that he is, if anything, the more "human" of the two and the man with the better case. At the same time, he leaves no doubt that he means business. If no accommodation is made to his needs, he will take revenge. After a while Victor begins to yield to the logic of the monster's argument. "For the first time," he admits, "I felt what the duties of a creator towards his creature were, and that I ought to render him happy before I complained of his wickedness."[9] The two are able to agree that it is probably too late for the nameless "wretch" to enter human society, and they arrive at a compromise solution: Frankenstein will return to the laboratory and build a companion, a female, for his original masterpiece. "It is true, we shall be monsters, cut off from all the world; but on that account we shall be more attached to one another."[10] The problems caused by technology are to find a technological cure.

Of course, the scheme does not work. After a long period of procrastination, Victor sets to work on the second model of his invention, but in the middle of his labors he remembers a pertinent fact. The first creature "had sworn to quit the neighborhood of man and hide himself in deserts, but she had not; and she, who in all probability was to become a thinking and reasoning animal, might refuse to comply with a compact made before her creation."[11] The artificial female would have a life of her own. What was to guarantee that she would not make demands and extract the consequences if the demands were not properly met? Then an even more disquieting thought strikes Victor. What if the two mate and have children? "A race of evils would be propagated upon the earth who might make the very existence of the species of man a condition precarious and full of terror."[12] "I shuddered to think that future ages might curse me as their pest, whose selfishness had not hesitated to buy its own peace at the price, perhaps, of the existence of the whole human race."[13] Recognizing what he believes to be a heroic responsibility, Victor commits an act of violence. With the first creature looking on, he tears the unfinished female artifact to pieces.

From this point the story moves toward a melodramatic conclusion befitting a gothic novel. The creature reminds Victor, "You are my creator, but I am your master," and then vows, "*I will be with you on your wedding night.*"[14] He makes good his promise and eventually kills Victor's young bride Elizabeth. Frankenstein then sets out to find and destroy his creature, but after a long period without success succumbs to illness on a ship at sea. In the final scene the creature delivers a soliloquy over Victor's coffin and then floats on an ice raft, announcing that he will commit suicide by cremating himself on a funeral pyre.

In recent years it has become fashionable to take *Frankenstein* seriously. The book frequently appears as the subject of elaborate psychosexual analyses, which seize upon some colorful episodes in the relationships of Mary Shelley, her famous mother and father, as well as Percy Bysshe Shelley and friend of the family, Lord Byron.[15] There is no doubt some truth to these interpretations. The book abounds with pointed references to problems of sexual identity, child-parent conflicts, and love-death obsessions. But there is also adequate evidence that in writing her story Shelley was also interested in the possibilities of science and the problems of scientific invention. In her time, as in our own, it was not considered fantasy that the secrets of nature upon which life depends might be laid open to scrutiny and that this knowledge could be used to synthesize, in whole or in part, an artificial human being.[16] It is not unlikely in this regard that the book was meant as criticism of the Promethean ideals of her husband. Percy Shelley saw in the figure of Prometheus, rebel and life-giver, a perfect symbol to embody his faith in the perfectibility of man, the creative power of reason, and the possibility of a society made new through enlightened, radical reconstruction. His play *Prometheus Unbound* sees its hero released from the fetters imposed by the gods and freed for endless good works. In its preface Shelley explains that "Prometheus is, as it were, the type of the highest perfection of moral and intellectual nature, impelled by the purest and truest motives to the best and noblest ends."[17] To the charge that the poet himself has gotten carried away with "a passion for reforming the world," Shelley replies: "For my part I had rather be damned with Plato and Lord Bacon than go to Heaven with Paley and Malthus."[18] Plato and Lord Bacon?

Mary Shelley's novel, published at about the same time as her husband's play, may well have been an attempt to discover the tragic flaw in a vision from which Shelley hoped to eliminate any trace of tragedy. In the Baconian-Promethean side of her spouse's quest, the side that marveled at the powers that could come from the discovery and taming of nature's secrets, she found a hidden agenda for trouble. The best single statement of her view comes on the title page of the book, a quotation from Milton's *Paradise Lost*:

Did I request thee, Maker, from my clay
To mould me man? Did I solicit thee
From darkness to promote me?—

Suggested in these words is, it seems to me, the issue truly at stake in the whole of *Frankenstein*: the plight of things that have been created but not in a context of sufficient care. This problem captures the essence of the themes my inquiry has addressed.

Victor Frankenstein is a person who discovers, but refuses to ponder, the implications of his discovery. He is a man who creates something new in the world and then pours all of his energy into an effort to forget. His invention is incredibly powerful and represents a quantum jump in the performance capability of a certain kind of technology. Yet he sends it out into the world with no real concern for how best to include it in the human community. Victor embodies an artifact with a kind of life previously manifest only in human beings. He then looks on in surprise as it returns to him as an autonomous force, with a structure of its own, with demands upon which it insists absolutely. Provided with no plans for its existence, the technological creation enforces a plan upon its creator. Victor is baffled, fearful, and totally unable to discover a way to repair the disruptions caused by his half-completed, imperfect work. He never moves beyond the dream of progress, the thirst for power, or the unquestioned belief that the products of science and technology are an unqualified blessing for humankind. Although he is aware of the fact that there is something extraordinary at large in the world, it takes a disaster to convince him that the responsibility is his. Unfortunately, by the time he overcomes his passivity, the consequences of his deeds have become irreversible, and he finds himself totally helpless before an unchosen fate.

If the arguments we have examined have any validity at all, it is likely that Victor's problems have now become those of a whole culture. At the outset, the development of all technologies reflects the highest attributes of human intelligence, inventiveness, and concern. But beyond a certain point, the point at which the efficacy of the technology becomes evident, these qualities begin to have less and less influence upon the final outcome; intelligence, inventiveness, and concern effectively cease to have any real impact on the ways in which technology shapes the world.

It is at this point that a pervasive ignorance and refusal to know, irresponsibility, and blind faith characterize society's orientation toward the technical. Here it happens that men release powerful changes into the world with cavalier disregard for consequences; that they begin to "use" apparatus, technique, and organization with no attention to the ways in which these "tools" unexpectedly rearrange their lives; that they willingly submit the governance of their affairs to the expertise of others. It is here also that they begin to participate without second thought in megatechnical systems far beyond their comprehension or control; that they endlessly proliferate technological forms of life that isolate people from each other and cripple rather than enrich the human potential; that they stand idly by while vast technical systems reverse the reasonable relationship between means and ends. It is here above all that modern men come to accept an overwhelmingly passive response to everything technological. The maxim "What man has made he can also change" becomes increasingly scandalous.

Until very recently this adoption of an active image to mask the passive response seemed an entirely appropriate stance. The elementary tool-use conception of scientific technology, essentially unchanged since Francis Bacon, was universally accepted as an accurate model of all technical conduct. All one had to do was to see that the tools were in good hands. Reinforcing this view was a devout acceptance of the idea of progress, originally an ideal of improvement through enlightenment, the education of all mankind, and continuing scientific and technical advance. But eventually the technological side of the notion eclipsed the others. Progress came to be coterminous with the enlarging sphere of technological achievement. This was (and still is) widely understood to be a kind of ineluctable, self-generating process of increasing beneficence—autonomous change toward a desirable *telos*.

Beyond these dominant beliefs and attitudes, however, lies something even more fundamental, for there is a sense in which all technical activity contains an inherent tendency toward forgetfulness. Is not the point of all invention, technique, apparatus, and organization to have something and *have it over with*? One does not want to bother anymore with building, developing, or learning it again. One does not want to bother with its structure or the principles of its internal workings. One simply wants the technical thing to be present in its utility. The goods are to be obtained without having to understand the factory or the distribution network. Energy is to be utilized without understanding the myriad of connections that made its generation and delivery possible. Technology, then, allows us to ignore our own works. It is *license to forget*. In its sphere the truths of all important processes are encased, shut away, and removed from our concern. This more than anything else, I am convinced, is the true source of the colossal passivity in man's dealings with technical means.

I do not mean to overlook the fact that, on the whole, mankind has been well served in this relationship. The benefits of terms in health, mobility, material comfort, and the overcoming of the physical problems of production and communication are well known. That I have not recounted them frequently is not a sign that I have forgotten them. I live here too.

For the vast majority of persons, the simple, time-honored notions about technology are sufficient. There are still eloquent public spokesmen ready to explain the basic tenets at each suitable occasion. "When you talk about progress, about the new and the different, the possibilities are infinite. That's what is so fascinating and compelling about progress. The infinite possibilities. The potential for creating sights and sounds and feelings that have not yet been dreamed of, for achieving all that has yet to be achieved, for changing the world." "For it must be obvious to anyone with any sense of history and any awareness of human nature that there *will* be SST's. And Super SST's. And Super-Super SST's. Mankind is simply not going to sit back with the Boeing 747 and say 'This is as far as we go.' "[19]

For the many who embrace this faith, any criticism of technology is taken as vile heresy. Like Elijah defending Yahweh from the gods of Jezebel and Ahab, the choice for them is strictly either/or, monotheism or not. Those who find problems in the technological content of this culture or who seriously suggest that different kinds of sociotechnical arrangements might be preferable are portrayed as absolute nay-sayers, pessimists, or, worse, crafty seducers luring innocent victims toward the brink of nameless dread.

There are, nevertheless, indications that the conversation is beginning to widen its boundaries. True, the Elijahs are still on Mount Carmel commanding piles of wood to catch fire (usually under some new aircraft or weapons system). Enthusiastic boosters and cheerleaders are still busy trying to obscure the fact that "progress" once meant something more than novel hardware and technique.[20] But other voices are beginning to speak. It is possible that a more vital, intelligent questioning is beginning to replace docile prejudice. Many now understand why it is necessary to think and act differently in the face of technological realities and to begin the search for new paths.

This essay is intended as a contribution to the effort to reevaluate the circumstances of our involvement with technology. My aim has been to sketch in some detail problems I thought were underestimated or not sufficiently clear in other writings. The position of these perspectives is not, as the boosters may conclude, that technology is a monstrosity or an evil in and of itself. Instead, the view has been much like that of Mary Shelley's novel, that we are dealing with an unfinished creation, largely forgotten and uncared for, which is forced to make its own way in the world. This creation, like Victor's masterpiece, contains the precious stuff of human life. But in its present state it all too often returns to us as a bad dream—a grotesquely animated, autonomous force reflecting our own life, crippled, incomplete, and not fully in our control.

Is this a helpful conclusion to have drawn? Other than accounting for one recurrent problem in modern thought, what does it offer?

Very little. Very little, that is, unless those who build and maintain the technological order are willing to reconsider their work. Victor Frankenstein was blinded by two diametrically opposed beliefs: first, that he would produce an artifact of undeniable perfection and, later on, that his invention was a disaster about which nothing could be done. For those willing to go beyond both of these conclusions, the rest of the essay offers a few more steps.

Technology as Legislation

Obviously there are a great many specific issues and approaches within the general range of questions we have encountered. The ecology movement, consumerism, future studies, the technology assessors, students of innovation and social change, and what remains of the "counterculture" all have something to say about the ways in which technology presents difficulties for the modern world. Since the reader is no doubt familiar with the debates now raging over these issues, I will not review the details. But in their orientations toward politics and their conceptions of how a better state of affairs might be achieved, the issue areas sort themselves into roughly two categories.

In the first domain, far and away the most prominent, the focus comes to rest on matters of risk and safeguard, cost and benefit, distribution, and the familiar interest-centered style of politics. Technology is seen as a cause of certain problematic effects. All of the questions raised in the present essay, for example, would be interpreted as "risks taken" and "prices paid" in the course of technological advance. Once this is appreciated, the important tasks become those of (1) accurate prediction and anticipation to alleviate risk, (2) adequate evaluation of the costs that are or might be incurred, (3) equitable distribution of the costs and risks so that one portion of the populace neither gains nor suffers excessively as compared to others, and (4) shrewd evaluation of the political realities bearing upon social decisions about technology.

Under this model the business of prediction is usually meted out to the natural and social sciences. Occasionally, some hope is raised that a new art or science—futurism or something of the sort—will be developed to improve the social capacity of foresight. The essential task is to devise more intelligent ways of viewing technological changes and their possible consequences in nature and society. Ideal here would be the ability to forecast the full range of significant consequences in advance. One would then have a precise way of assigning the risk of proceeding in one way rather than another.[21]

The matter of determining costs is left to orthodox economic analysis. In areas in which "negative externalities" are experienced as the result of technological practice, the loss can be given a dollar value. The price paid for the undesirable "side effects" can then be compared to the benefit gained. An exception to this mode of evaluation can be seen in some environmental and sociological arguments in which non-dollar value costs are given some weight.[22] On the whole, however, considerations of cost follow the form Leibniz suggested for the solution of all rational disputes: "Let us calculate." Taking this approach one tends to ask questions of the sort: How much are you prepared to pay for pollution-free automobiles? What is the public prepared to tax itself for clean rivers? What are the trade-offs between having wilderness and open space as opposed to adequate roads and housing? Are the costs of jet airport noise enough to offset the advantage of having airports in the middle of town? Such questions are answered at the cash register, although the computer shows a great deal more style.

Once the risks have been assigned, the safeguards evaluated, and the costs calculated, one is then prepared to worry about distribution. Who will enjoy how much of the benefit? Who will bear the burden of the uncertainty or the price tag of the costs? Here is where normal politics—pressure groups, social and economic power, private and public interests, bargaining, and so forth—enters. We expect that those most aware, best supplied, and most active will manage to steer a larger proportion of the advantages of technological productivity their way while avoiding most of the disadvantages. But for those who have raised technology as a political problem under this conception, reforms are needed in this distributive process. Even persons who have no quarrel with the inequities of wealth and privilege in liberal society now step forth with the most trenchant criticisms of the ways in which technological "impacts" are distributed through the social system. A certain radicalism is smuggled in through the back door. The humble ideal of those who see things in this light is that risks and costs be allotted more equitably than in the past. Those who stand to gain from a particular innovation should be able to account for its consequences beforehand. They should also shoulder the major brunt of the costs of undesirable side effects. This in turn should eliminate some of the problems of gross irresponsibility in technological innovation and application of previous times. Since equalization and responsibility are to be induced through a new set of laws, regulations, penalties, and encouragements, the attention of this approach also aims at a better understanding of the facts of practical political decision making.

Most of the work with any true influence in the field of technology studies at present has its basis in this viewpoint. The ecology movement, Naderism, technology assessment, and public-interest science each have somewhat different substantive concerns, but their notions of politics and rational conduct all fit within this frame. There is little new in it. What one finds here is the utilitarian-pluralist model refined and aimed at new targets. In this form it is sufficiently young to offer spark to tired arguments, sufficiently critical of the status quo to seem almost risqué. But since it accepts the major premises and disposition of traditional liberal politics, it is entirely safe. The approach has already influenced major pieces of legislation in environmental policy and consumer protection. It promises to have a bright future in both the academic and the political realms, opening new vistas for "research," "policy analysis," and, of course, "consulting."[23]

On the whole, the questions I have emphasized here are not those now on the agendas of persons working in the first domain. But for those following this approach I have one more point to add. It is now commonly thought that what must be studied are not the technologies but their implementing and regulating systems. One must pay attention to various institutions and means of control—corporations, government agencies, public policies, laws, and so forth—to see how they influence the course our technologies follow. Fine. I would not deny that there are any number of factors that go into the original and continued employment of these technical ensembles. Obviously the "implementing" systems have a great deal to do with the eventual outcome. My question is, however, In what technological context do such systems themselves operate and what imperatives do they feel obliged to obey? In several ways I have tried to show that the hope for some "alternative implementation" is largely misguided. *That* one employs something at all far outweighs (and often obliterates) the matter of *how* one employs it. This is not sufficiently appreciated by those working within the utilitarian-pluralist framework. We may firmly believe that we are developing ways of regulating technology. But is it perhaps more likely that the effort will merely succeed in putting a more elegant administrative facade on old layers of reverse adapted rules, regulations, and practices?[24]

The second domain of issues is less easily defined, for it contains a collection of widely scattered views and spokesmen. At its center is the belief that technology is problematic not so much because it is the origin of certain undesirable side effects but rather because it enters into and becomes part of the fabric of human life and activity. The maladies technology brings—and this is not to say that it brings only maladies— derive from its tendency to structure and incorporate that which it touches. The problems of interest, therefore, do not arrive by chain reaction from some distant force. They are present and immediate, built into the everyday lives of individuals and institutions. Analyses that focus only upon risk/safeguard, cost/benefit, and distribution simply do not reveal problems of this sort. They require a much more extraordinary, deep-seeking response than the utilitarian-pluralist program can ever provide.

What, then, are the issues of this second domain? Some of the most basic of them are mirrored in our discussion of the theory of technological politics. This model represents the critical phase of a movement of thought, the attempt to do social and political analysis with technics as its primary focus. But these thoughts so far have given little care to matters of amelioration. In the present formulation of the theory, I have deliberately tried to avoid dealing in popular remedies. It is my experience that inquiries pointing to broad, easy solutions soon become cheap merchandise in the commercial or academic marketplace. They become props for the very thing criticized.[25]

For better or worse, however, most of the thinking in the second domain at present is highly specific, solution oriented, and programmatic. The school of humanist psychology, writers and activists of the counterculture, utopian and communal living experiments, the free schools, proponents of encounter groups and sensual reawakening, the hip catalogers, the peace movement, pioneers of radical software and new media, the founders and designers of alternative institutions, alternative architecture, and "appropriate" or "intermediate" technology—all of these have tackled the practical side of one or more of the issues raised in this essay.

Much of the work has begun with a sobering recognition of the psychological disorders associated with life in the technological society. The world of advanced technics is still one that makes excessive demands on human performance while offering shallow, incomplete rewards. The level of stress, repression, and psychological punishment that rational-productive systems extract from their human members is not matched by the opportunity for personal fulfillment. Men and women find their lives cut into parcels, spread out, and dissociated. While the neuroses generated are often found to be normal and productive in the sociotechnical network, there has been a strong revolt against the continuation of such sick virtues. Both professionals and amateurs in psychology have come together in a host of widely differing attempts to find the origins of these maladies and to eliminate them.

Other enterprises of this kind have their roots in a pervasive sense of personal, social, and political powerlessness. Confronted with the major forces and institutions that determine the quality of life, many persons have begun to notice that they have little real voice in most important arrangements affecting their activities. Their intelligent, creative participation is neither necessary not expected. Even those who consider themselves "well served" have cause to wonder at decisions, policies, and programs affecting them directly, over which they exercise no effective influence. In the normal state of affairs, one must simply join the "consensus." One consents to a myriad of choices made, things built, procedures followed, services rendered, in much the same way that one consents to let the eucalyptus trees continue growing in Australia. There are some, however, who have begun to question this submissive, compliant way of life. In a select few areas, some people have attempted to reclaim influence over activities they had previously let slip from their grasp. The free schools, food conspiracies and organic food stores, new arts and crafts movement, urban and rural communes, and experiments in alternative technology have all—in the beginning at least—pointed in this direction. With mixed success they have sought to overcome the powerlessness that comes from meting out the responsibility for one's daily existence to remote large-scale systems.

A closely related set of projects stems from an awareness of the ways organized institutions in society tend to frustrate rather than serve human needs. The scandal of productivity has reached astounding proportions. More and more is expended on the useless, demeaning commodities idealized in the consumer ethos (for example, vaginal deodorants), while basic social and personal needs for health, shelter, nutrition, and education fall into neglect. The working structures of social institutions that provide goods and services seem themselves badly designed. Rather than elicit the best qualities of the persons they employ or serve, they systematically evoke the smallest, the least creative, least trusting, least loving, and least lovable traits in everyone. Why and how this is so has become a topic of widespread interest. A number of attempts to build human-centered and responsive institutions, more reasonable environments for social intercourse, word, and enjoyment, are now in the hands of those who found it simply impossible to continue the old patterns.

Finally, there is a set of concerns, evident in the aftermath of Vietnam, Watergate, and revelations about the CIA, which aims at restoring the element of responsibility to situations that have tended to exclude responsible conduct. There is a point, after all, where compliance becomes complicity. The twentieth century has made it possible for a person to commit the most ghastly of domestic and foreign crimes by simply living in suburbia and doing a job. The pleas of Lieutenant Calley and Adolf Eichmann—"I just work here"—become the excuse of everyman. Yet for those who perceive the responsibility, when distant deeds are done and the casualties counted, the burdens are gigantic. As Stanley Cavell and Nedezhda Mandelstam have observed, there is a sense in which one comes to feel responsible for literally everything.[26] Evils perpetrated and the good left undone all weigh heavily on one's shoulders. Like Kafka's K. at the door of the castle, the concerned begin a search for someone or something that can be held accountable.

I admit that I have no special name for this collection of projects. *Humanist technology* has been suggested to me, but that seems wide of the mark. At a time in which the industrialization of literature demands catchy paperback titles for things soon forgotten, perhaps it is just as well to leave something truly important unnamed.

The fundamental difference between the two domains, however, can be stated: a difference in insight and commitment. The first, the utilitarian-pluralist approach, sees that technology is problematic in the sense that it now *requires legislation*. An ever-increasing array of rules, regulations, and administrative personnel is needed to maximize the benefits of technological practice while limiting its unwanted maladies. Politics is seen as the process in representative government and interest group interplay whereby such legislation takes shape.

The second approach, disjointed and feeble though it still may be, begins with the crucial awareness that technology in a true sense *is legislation*. It recognizes that technical forms do, to a large extent, shape the basic pattern and content of human activity in our time. Thus, politics becomes (among other things) an active encounter with the specific forms and processes contained in technology.

Along several lines of analysis this essay has tried to advance the idea central to all thinking in the second domain—that *technology is itself a political phenomenon*. A crucial turning point comes when one is able to acknowledge that modern technics, much more than politics as conventionally understood, now legislates the conditions of human existence. New technologies are institutional structures within an evolving constitution that gives shape to a new polity, the technopolis in which we do increasingly live. For the most part, this constitution still evolves with little public scrutiny or debate. Shielded by the conviction that technology is neutral and tool-like, a whole new order is built—piecemeal, step by step, with the parts and pieces linked together in novel ways—without the slightest public awareness or opportunity to dispute the character of the changes underway. It is somnambulism (rather than determinism) that characterizes technological politics—on the left, right, and center equally. Silence is its distinctive mode of speech. If the founding fathers had slept through the convention in Philadelphia in 1787 and never uttered a word, their response to constitutional questions before them would have been similar to our own.

Indeed, there is no denying that technological politics as I have described it is, in the main, a set of pathologies. To explain them is to give a diagnosis of how things have gone wrong. But there is no reason why the recognition of technology's intrinsic political aspect should wed us permanently to the ills of the present order. On the contrary, projects now chosen in the second domain bear a common bond with attempts made to redefine an authentic politics and reinvent conditions under which it might be practiced. As a concern for political theory this work has been admirably carried forward by such writers as Hannah Arendt, Sheldon Wolin, and Carole Pateman.[27] In the realm of historical studies it appears as a renewed interest in a variety of attempts— the Paris Communes of 1793 and 1871, nineteenth-century utopian experiments, twentieth-century Spanish anarchism, the founding of worker and community councils in a number of modern revolutions—to create decentralist democratic politics.[28] In contemporary practice it can be seen in the increasingly common efforts to establish worker self-management in factories and bureaucracies, to build self-sufficient communities in both urban and rural settings, and to experiment with modes of direct democracy in places where hierarchy and managerialism had previously ruled.[29]

Taken in this light, it is possible to see technology as legislation and then follow that insight in hopeful directions. An important step comes when one recognizes the validity of a simple yet long overlooked principle: *Different ideas of social and political life entail different technologies for their realization.* One can create systems of production, energy, transportation, information handling, and so forth that are compatible with the growth of autonomous, self-determining individuals in a democratic polity. Or one can build, perhaps unwittingly, technical forms that are incompatible with this end and then wonder how things went strangely wrong. The possibilities for matching political ideas with technological configurations appropriate to them are, it would seem, almost endless. If, for example, some perverse spirit set out deliberately to design a collection of systems to increase the general feeling of powerlessness, enhance the prospects for the dominance of technical elites, create the belief that politics is nothing more than a remote spectacle to be experienced vicariously, and thereby diminish the chance that anyone would take democratic citizenship seriously, what better plan to suggest than that we simply keep the systems we already have?

There is, of course, hope that we may decide to do better than that. The challenge of trying to do so now looms as a project open to political science and engineering equally. But the notion that technical forms are merely neutral and that "one size fits all" is a myth that no longer merits the least respect.

Luddism as Epistemology

But what next? Following the normal pattern of twentieth-century writing, I should now rush forward with suggestions and recommendations for how things might be different. What good are analyses, criticisms, and perspectives, some might say, unless they point to positive courses of action?

In view of what we have seen, however, it is not easy simply to take a deep breath and begin spewing forth plans for a better world. The issues are difficult ones. It has not been my aim to make them seem any less difficult than they are. In my experience, virtually all of the remedies proposed are little more than tentative steps in uncertain directions. Goodman's plea for the application of moral categories to technological action, Bookchin's outlines for a liberatory technology, Marcuse's rediscovery of utopian thinking, and Ellul's call to the defiant, self-assertive, free individual—all of these offer us something.[30] But when compared to the magnitude of what is to be overcome, these solutions seem trivial. I could, I suppose, fudge the matter here and seem to be zeroing in on some useful proposals. Having gone this far, the reader can probably predict how it would look.

First, I could say that there is a need to begin the search for new technological forms. Recognizing the often wrong-headed and oppressive character of existing configurations of technology, we should find new kinds of technics that avoid the human problems of the present set. This would mean, presumably, the birth of a new sort of inventiveness and innovation in the physical arrangements of this civilization.

Second, I could suggest that the development of these forms proceed through the direct participation of those concerned with their everyday employment and efforts. One major shortcoming in the technologies of the modern period is that those touched by their presence have little or no control over their design or operation. To as great an extent as possible, then, the processes of technological planning, construction, and control ought to be opened to those destined to experience the final products and full range of social consequences.

Third, I might point to the arguments presented here and offer some specific principles to guide further technological construction. One such rule would certainly be the following: *that as a general maxim, technologies be given a scale and structure of the sort that would be immediately intelligible to nonexperts.* This is to say, technological systems ought to be intellectually as well as physically accessible to those they are likely to affect. Another worthy principle would be: *that technologies be built with a high degree of flexibility and mutability.* In other words, we should seek to avoid circumstances in which technological systems impose a permanent, rigid, and irreversible imprint on the lives of the populace. Yet another conceivable rule is this: *that technologies be judged according to the degree of dependency they tend to foster, those creating a greater dependency being held inferior.* This merely recognizes a situation we have seen again and again in this essay. Those who must rely for their very existence upon artificial systems they do not understand or control are not at liberty to change those systems in any way whatsoever. For this reason, any attempt to create new technological circumstances must make certain that it does not discover freedom only to lose it again on the first step.

Finally, I could suggest a supremely important step—that we return to the original understanding of technology as a means that, like all other means available to us, must only be employed with a fully informed sense of *what is appropriate.* Here, the ancients knew, was the meeting point at which ethics, politics, and technics came together. If one lacks a clear and knowledgeable sense of which means are appropriate to the circumstances at hand, one's choice of means can easily lead to excesses and danger. This ability to grasp the appropriateness of means has, I believe, now been pretty thoroughly lost. It has been replaced by an understanding which holds that if a given means can be shown to have a narrow utility, then it ought to be adopted straight off, regardless of its broader implications.[31] For a time, perhaps from the early seventeenth century to the early twentieth, this was a fruitful way of proceeding. But we have now reached a juncture at which such a cavalier disposition will only lead us astray. A sign of the maturity of modern civilization would be its recollection of that lost sense of appropriateness in the judgment of means. We would profit from regaining our powers of selectivity and our ability to say "no" as well as "yes" to a technological prospect. There are now many cases in which we would want to say: "After all a temptation is not very tempting."[32]

I am convinced that measures of this kind point to a new beginning on the problems we have seen.[33] At the same time, these proposals have overtones of utopianism and unreality, which make them less than compelling. It may be that the only innovation I have suggested is to use my hat as a megaphone. There are excellent reasons why *any* call for the taking of a new path or new beginning now falls flat.

Not the least of these is simply the fact that while positive, utopian principles and proposals can be advanced, the real field is already taken. There are, one must admit, technologies already in existence—apparatus occupying space, techniques shaping human consciousness and behavior, organizations giving pattern to the activities of the whole society. To ignore this fact is to take flight from the reality that must be considered. One finds, for example, that in the contemporary discussions those most sanguine about the prospects for tackling the technological dilemma are those who place their confidence in *new* systems to be implemented in the future. Their hope is not that the existing state of affairs will be changed through any direct action, only that certain superior features will be added. In this manner the mass of problems now at hand is skirted.[43]

Another barrier is this: even if one seriously wanted to construct a different kind of technology appropriate to a different kind of life, one would be at a loss to know how to proceed. There is no living body of knowledge, no method of inquiry applicable to our present situation that tells us how to move any differently from the way we already do. Mumford's suggestion that society return to an older tradition of small-scale technics and craftsmanship is not convincing. The world that supported that tradition and gave it meaning has vanished. Where and how techniques of that sort could be a genuine alternative is highly problematic. Certainly a technological revivalism could *add* things to the existing technological stock. But the kind of knowledge that would make a difference is not to be found in decorating the periphery.

In no place is the force of these considerations better exemplified than in the sorry fate of the counterculture of the late 1960s. The belief of those who followed the utopian dream was that by dropping out of the dominant culture and "raising one's consciousness," a better way of living would be produced. In several areas of social fashion—clothing, music, language, drug use—there were some remarkable innovations. But behind the facade of style, a familiar reality still held sway. The basic structures of life, many of them technological structures, remained unchallenged and unchanged. Members of the movement convinced themselves that with a few gestures they had transcended all of that. But all of the networks of practical connections remained intact. The best that was done was to give the existing patterns a hip veneer. Members of the management team began to wear bell-bottoms and medallions.

The lesson, I think, is evident. Even though one commits oneself to ends radically different from those in common currency, there is no real beginning until the question of means is looked straight in the eye. One must take seriously the fact that there are already technologies occupying the available physical and social space and employing the available resources. One must also take seriously the fact that one simply does not yet know how to go ahead to find genuinely new means appropriate to the new "consciousness." No doubt some faced with this realization will simply wish to stop. They will see the virtual necessity of co-optation and the impending disappointment for anyone who tries to resist one's technological fate. Some will find it impossible to do anything else than retreat into despair and blame their plight on "those in power." But if I am not mistaken, the logic of the problem admits at least one more alternative.

In many contemporary writings the response to the idea of autonomous technology reads something like this: "Technology is not a juggernaut; being a human construction it can be torn down, augmented, and modified at will."[35] The author of this statement, Dr. Glenn T. Seaborg, would probably be the last person to suggest that any existing technology actually be "torn down." But in his mind, as in many others, the conviction that man still controls technology is rooted in the notion that at any time the whole thing could be taken apart and something better built in its place. This idea, for reasons we have seen all along, is almost pure fantasy. Real technologies do not permit such wholesale tampering. Changes here occur through "invention," "development," "progress," and "growth"—processes in which more and more additions are made to the technological store while some parts are eventually junked as obsolete. The technologies generated are understood to be more or less permanent fixtures. That they might be torn down or seriously tinkered with is unthinkable.

But perhaps Seaborg's idea has some merit. As we have already noted, is not the fundamental business of technics that of taking things apart and putting them together? One conceivable approach to tackling whatever flaws one sees in the various systems of technology might be to begin dismantling those systems. This I would propose not as a solution in itself but as a method of inquiry. The forgotten essence of technical activity, regardless of the specific purpose at hand, might well be revealed by this very basic yet, at the same time, most difficult of steps. Technologies identified as problematic would be taken apart with the expressed aim of studying their interconnections and their relationships to human need. Prominent structures of apparatus, technique, and organization would be, temporarily at least, disconnected and made unworkable in order to provide the opportunity to learn what they are doing for or to mankind. If such knowledge were available, one could then employ it in the invention of radically different configurations of technics, better suited to nonmanipulated, consciously, and prudently articulated ends.

None of this would be necessary if such information were obvious. But at present it is exactly this kind of awareness and understanding that is lacking. Our involvement in advanced technical systems resembles nothing so much as the somnambulist in Caligari's cabinet. Somewhat drastic steps must be taken to raise the important questions at all. The method of carefully and deliberately dismantling technologies, epistemological Luddism if you will, is one way of recovering the buried substance upon which our civilization rests. Once unearthed, that substance could again be scrutinized, criticized, and judged.

I can hear the outcry already. Isn't this man's Luddism simply an invitation to machine smashing? Isn't it mere nihilism with a sharp edge? How can anyone calmly suggest such an awful course of action?

Again, I must explain that I am only proposing a method. The method has nothing to do with Luddism in the traditional sense (the smashing and destroying of apparatus). The much-maligned original Luddites were, of course, merely unemployed workers with a flare for the dramatic. As they scrutinized the mechanization of the textile trade in the industrial revolution, they applied two interesting criteria. Does the new device enhance the quality of the product being manufactured? Does the machine improve the quality of work? If the answer to either question or both is "no," the innovation should not be permitted. Banned from lawful union activity, the Luddites did what they could and unwittingly brought upon themselves a lasting opprobrium.[36]

As best as I can tell, there have never been any epistemological Luddites, unless perhaps Paul Goodman was one on occasion. I am not proposing that a sledge hammer be taken to anything. Neither do I advocate any act that would endanger anyone's life or safety. The idea is that in certain instances it may be useful to dismantle or unplug a technological system in order to create the space and opportunity for learning.

The most interesting parts of the technological order in this regard are not those found in the structure of physical apparatus anyway. I have tried to suggest that the technologies of concern are actually *forms of life*—patterns of human consciousness and behavior adapted to a rational, productive design. Luddism seen in this context would seldom refer to dismantling any piece of machinery. It would seek to examine the connections of the human parts of modern social technology. To be more specific, it would try to consider at least the following: (1) the kinds of human dependency and regularized behavior centering upon specific varieties of apparatus, (2) the patterns of social activity that rationalized techniques imprint upon human relationships, and (3) the shapes given everyday life by the large-scale organized networks of technology. Far from any wild smashing, this would be a meticulous process aimed at restoring significance to the question, What are we about?

One step that might be taken, for example, is that groups and individuals would for a time, self-consciously and through advance agreement, extricate themselves from selected techniques and apparatus. This, we can expect, would create experiences of "withdrawal" much like those that occur when an addict kicks a powerful drug. These experiences must be observed carefully as prime data. The emerging "needs," habits, or discomforts should be noticed and thoroughly analyzed. Upon this basis it should be possible to examine the structure of the human relationships to the device in question. One may then ask whether those relationships should be restored and what, if any, new form those relationships should take. The participants would have a genuine (and altogether rare) opportunity to ponder and make choices about the place of that particular technology in their lives. Very fruitful experiments of this sort could now be conducted with many implements of our semiconscious technological existence, such as the automobile, television, and telephone.

Other possibilities for Luddism as methodology can be found at virtually any point in which social and political institutions depend upon advanced technologies for their effective operation. Persons who, for any reason, wish to alter or reform those institutions—the factory, school, business, public agency—have an alternative open to them that they have previously overlooked. As preparation for changes one may later wish to make, one might try disconnecting crucial links in the organized system for a time and studying the results. There is no getting around the fact that the most likely consequences will be some variety of chaos and confusion. But it is perhaps better to have this out in the open rather than endure the subliminal chaos and confusion upon which many of our most important institutions now rest. Again, these symptoms must be taken as prime data. The effects of systematic disconnection must be taken as an opportunity to inquire, to learn, and seek something better. What is the institution doing in the first place? How does its technological structure relate to the ends one would wish for it? Can one see anything more than to plug the whole back together the way it was before? The Luddite step is necessary if such questions are to be asked in any critical way. It is, perhaps, not too farfetched to suppose that some positive innovations might result from this straightforward challenge to established patterns of institutional life.

By far the most significant of Luddite alternatives, however, requires no direction action at all: the best experiments can be done simply by refusing to repair technological systems as they break down. Many of society's biggest investments at present are those that merely prop up failing technologies. This propping up is usually counted as "growth" and placed in the plus column. We build more and more freeways, larger and larger suburban developments, greater and greater systems of centralized water supply, power, sewers, and police, all in a frantic effort to sustain order and minimal comfort in the sprawling urban complex. Perhaps a better alternative would be to let dying artifice die. One might then begin the serious search, not for something superficially "better" but for totally new forms of sociotechnical existence.

Beyond these few words I have little more to say, for now. This essay has taken us on a long path to a conclusion that is actually a beginning. What one does with a beginning is to begin. I realize full well that many of these notions will be counted impractical. But that is precisely the point. I have tried to show that the practical-technical aspect of human activity has been almost totally removed from any concerned and conscious care. Autonomous technology is the part of our being that has been transferred, transformed, and separated from living needs and creative intelligence. Any effort to reclaim this part of human life must at first seem impractical and even absurd.

In this light, the suggestions at the end are not so much a call to action as an attempt to speak to logical problems that arose during the investigation. Given the power of these developments, what might possibly make a difference? My best answer at present is this: if the phenomenon of technological politics is to be overcome, a truly *political technology* must be put in its place. I have tried to give a few outlines of an experimental method that might encourage its birth.

In Mary Shelley's novel, Victor Frankenstein is portrayed as a "modern Prometheus." The young man's inevitable tragedy mirrors an ancient story in which the combined elements of ambition, artifice, pride, and power meet an unfortunate end.

Without doubt the most excellent of Promethean stories, however, is that written by Aeschylus 2,500 years ago. In *Prometheus Bound* we find in luminous, mythical outline many of the themes we have encountered in this essay, for an interesting feature of Aeschylus's treatment of the legend is that it emphasizes the importance of technology in Prometheus's crime against the gods. The fall of man is in Aeschylus's view closely linked to the introduction of science and the arts and crafts. Chained to a desolate rock for eternity, Prometheus describes his plight.

Prometheus   I caused mortals to cease foreseeing doom.
Chorus   What cure did you provide them with against that sickness?
Prometheus   I placed in them blind hopes.

Chorus That was a great gift you gave to men.
Prometheus Besides this, I gave them fire.
Chorus And do creatures of a day now possess bright-faced fire?
Prometheus Yes, and from it they shall learn many crafts.
Chorus These are the charges on which—
Prometheus Zeus tortures me and gives me no respite.[37]

The theft of fire, Aeschylus makes clear, was in its primary consequence the theft of all technical skills and inventions later given to mortals. "I hunted out the secret spring of fire," Prometheus exclaims, "that filled the narthex stem, which when revealed became the teacher of each craft to men, a great resource. This is the sin committed for which I stand accountant, and I pay nailed in my chains under the open sky."[38] As the brash protagonist recounts the specific items he has bestowed upon the human race, it becomes evident that Aeschylus's tale represents the movement of primitive man to civilized society. "They did not know of building houses with bricks to face the sun; know how to work in wood. They lived like swarming ants in holes in the ground, in the sunless caves of the earth."[39] The fire enabled mankind to develop agriculture, mathematics, astronomy, domesticated animals, carriages, and a host of valuable techniques. But Prometheus ends his proud description on a sorry note.

> It was I and none other who discovered ships, the sail-driven wagons that the sea buffets. Such were the contrivances that I discovered for men— alas for me! For I myself am *without contrivance to rid myself of my present affliction* [emphasis added].[40]

Prometheus's problem is something like our own. Modern people have filled the world with the most remarkable array of contrivances and innovations. If it now happens that these works cannot be fundamentally reconsidered and reconstructed, humankind faces a woefully permanent bondage to the power of its own inventions. But if it is still thinkable to dismantle, to learn and start again, there is a prospect of liberation. Perhaps means can be found to rid the human world of our self-made afflictions.

Notes

1. Mary Shelley, *Frankenstein, or The Modern Prometheus,* in *Three Gothic Novels,* ed. Peter Fairclough (Harmondsworth: Penguin Books, 1968), p. 295. The Penguin edition reprints the text of *Frankenstein* published in London in 1831 by Colburn and Bentley and contains Mary Shelley's final revisions. Compare

to the first edition in *Frankenstein, or The Modern Prometheus* (the 1818 text), ed. James Rieger (New York: Bobbs-Merrill, 1974). Mary Shelley's life and writings are discussed in Eileen Bigland, *Mary Shelley* (London: Cassell, 1959); Elizabeth Nitchie, *Mary Shelley, Author of "Frankenstein"* (New Brunswick: Rutgers University Press, 1953); Margaret (Carter) Leighton, *Shelley's Mary: The Life of Mary Godwin Shelley* (New York: Farrar, Straus, & Giroux, 1973). Those interested in a quick, provocative account of the story behind the novel can turn to Samuel Rosenberg's essay, "Frankenstein, or Daddy's Little Monster," in *The Confessions of a Trivialist* (Baltimore: Penguin Books, 1972).

2. Shelley, *Frankenstein*, p. 307.

3. Ibid., p. 312.

4. Ibid., pp. 318–319.

5. Ibid., p. 319.

6. Ibid., p. 364.

7. Ibid., p. 363.

8. Ibid., p. 364.

9. Ibid., p. 366.

10. Ibid, p. 413.

11. Ibid, p. 435.

12. Ibid., pp. 435–436.

13. Ibid., p. 436.

14. Ibid., p. 439.

15. See Rosenberg, "Frankenstein." There is now a growing literature on the historical background of the novel. See N. H. Brailsford, *Shelley, Godwin and Their Circle* (New York: Henry Holt and Co., 1913); Christopher Small, *Mary Shelley's Frankenstein—Tracing the Myth* (Pittsburgh: University of Pittsburgh Press, 1973); Radu Florescu, *In Search of Frankenstein* (Boston: New York Graphic Society, 1975); Ellen Moers, "Female Gothic: The Monster's Mother," *New York Review of Books*, March 21, 1974.

16. See Mario Praz's introduction to *Three Gothic Novels*, pp. 25–31.

17. Percy Bysshe Shelley, *Prometheus Unbound*, in *The Selected Poetry and Prose of Shelley*, ed. Harold Bloom (New York: New American Library, 1966), p. 121.

18. Ibid., p. 124.

19. Spiro T. Agnew, "Address by the Vice President of the United States to the Printing Industries of America Convention," transcript, New York, July 12, 1972. See also Mr. Agnew's philosophy of progress in "Address by the Vice President of the United States at the Alaska Republican Luncheon," transcript, Fairbanks, Alaska, July 24, 1972.

20. The French philosophes, for example, saw progress as the development of education and moral virtue, as well as the growth of science and technology. They appreciated the contributions of the past to this development, rejecting the now common view that only the latest thing counts. In the *Encyclopedia*, Diderot summarizes this outlook: "The aim of an *encyclopedia* is to collect all the knowledge scattered over the face of the earth, to present its general outlines and structure to the men with whom we live, and to transmit this to those

who will come after us, so that the work of past centuries may be useful to the following centuries, that our children, by becoming more educated, may at the same time become more virtuous and happier, and that we may not die without having deserved well of the human race." *The Encyclopedia, Selections,* ed. and trans. Stephen J. Gendzier (New York: Harper & Row, 1967), p. 92.

21. Representative works in this genre are: *Technology: Processes of Assessment and Choice,* Report of the National Academy of Sciences, Committee on Science and Astronautics, U.S. House of Representatives (Washington, D.C.: Government Printing Office, 1969); Herman Kahn and Anthony J. Wiener, *The Year 2000: A Framework for Speculation on the Next Thirty-Three Years* (New York: Macmillan, 1967); *Harvard University Program on Technology and Society, 1964–1972: A Final Review* (Cambridge, Mass.: Harvard University Press, 1972). Other prominent voices in the conversation include the followers of R. Buckminster Fuller, for example, John McHale, *The Future of the Future* (New York: George Braziller, Inc., 1969), and the ubiquitous environmentalists, for example, Barry Commoner, *The Closing Circle: Nature, Man and Technology* (New York: Alfred A. Knopf, 1971).

22. See *The Impacts of Snow Enhancement: Technology of Winter Orographic Snowpack Augmentation in the Upper Colorado River Basin,* comp. Leo W. Weisbecker (Norman, Okla.: University of Oklahoma Press, 1974).

23. Under the National Environmental Policy Act of 1970, all government agencies doing work likely to have an impact on the environment are required to prepare a detailed study covering the likely effects of their projects on the environment. These impact statements are supposed to consider alternative plans and to show why the one chosen is preferable. The law has given birth to a small new industry of impact statement writing, much of which operates under the influence of the science of public relations.

24. For the distinction between "Technologies and Supporting Systems," see *Technology,* pp. 15–18.

25. See, for example, Charles Reich's *The Greening of America* (New York: Random House, 1970). One began to wonder about the viability of Reich's "Consciousness III" when one noticed how well it fit with the operation of the modern corporation.

26. Stanley Cavell, "The Avoidance of Love," in his *Must We Mean What We Say* (New York: Charles Scribner's Sons, 1969), pp. 337–353; Nadezhda Mandelstam, *Hope Abandoned: A Memoir,* trans. Max Hayward (London: Collins and Harvill Press, 1974), p. 182.

27. See Hannah Arendt, *The Human Condition* (Chicago: University of Chicago Press, 1958), and chap. 6 of *On Revolution* (New York: The Viking Press, 1963); Sheldon Wolin, *Politics and Vision* (Boston: Little, Brown, 1960); Carole Pateman, *Participation and Democratic Theory* (Cambridge: Cambridge University Press, 1970).

28. See Rosebeth Moss Kanter, *Commitment and Community: Communes and Utopias in Sociological Perspective* (Cambridge, Mass.: Harvard University Press, 1972); Sam Dolgoff, ed., *The Anarchist Collectives: Workers' Self-Management in the Spanish Revolution, 1936–1938* (New York: Free Life Editions, 1974); Albert

Soboul, *The Sans-Culottes: The Popular Movement and Revolutionary Government, 1793–1794*, trans. Remy Inglish Hall (New York: Doubleday & Company, Anchor Books, 1972); Stewart Edwards, *The Parish Commune 1871* (Chicago: Quadrangle Books, 1971).

29. See Terrence E. Cook and Patrick M. Morgan, eds., *Participatory Democracy* (San Francisco: Canfield Press, 1971).

30. Paul Goodman, *People or Personnel and Like a Conquered Province* (New York: Random House, Vintage Books, 1968), pp. 297–316; Murray Bookchin, *Post-Scarcity Anarchism* (Berkeley: Ramparts Books, 1971); Herbert Marcuse, *An Essay on Liberation* (Boston: Beacon Press, 1969); Jacques Ellul, *Autopsy of Revolution*, trans. Patricia Wolf (New York: Alfred A. Knopf, 1971), chap. 5.

31. Erich Fromm finds in this tendency the foremost principle of action in the technological society: "something ought to be done because it is technically possible to do so." *The Revolution of Hope* (New York: Bantam Books, 1968), p. 33. In some understandings this principle or the motive it expresses is taken to be "the technological imperative" itself. I have not adopted that definition here, preferring to employ the term in the context presented in chapters 2 and 6, Winner, *Autonomous Technology* (Cambridge, Mass.: MIT Press, 1977). The phenomenon Fromm and others have noticed is perhaps best called "technomania."

32. Gertrude Stein, *Look At Me Now and Here I Am, Writings and Lectures 1909–45*, ed. Patricia Meyerowitz (Baltimore: Penguin Books, 1971), p. 58.

33. See E. F. Schumacher, *Small is Beautiful: Economics As If People Mattered* (New York: Harper & Row, 1973); Ivan Illich, *Tools for Conviviality* (New York: Harper & Row, 1973); Wilson Clark, *Energy for Survival: The Alternative to Extinction* (Garden City: Doubleday & Company, Anchor Books, 1975). See also *The Journal of the New Alchemists* (Woods Hole, Mass.: The New Alchemy Institute, 1973, 1974).

34. One peculiar response of thinkers now worried about the technological society is to pretend in effect that one *already* lives in the future. A sophisticated technology has been directed toward more intelligent ends and given a more humane structure. Through the proper selection of new devices, the problems of the old order have been surmounted. But in these future fantasies, of which "postindustrialism" is now the most popular, there is almost no attempt to stipulate what will have happened to the technologies we will supposedly have "gone beyond." Since there are new technologies of information processing, for example, we are somehow entitled to assume that the world of industrial technology has vanished.

35. Glenn T. Seaborg and Roger Corliss, *Man and Atom* (New York: E. P. Dutton, 1972), p. 265.

36. See Malcolm I. Thomis, *The Luddites: Machine-Breaking in Regency England* (New York: Schocken Books, 1970); George Rudé, *The Crowd in History* (New York: John Wiley & Sons, 1964), pp. 79–92. I owe the formulation of the Luddite criteria to Larry Spence.

37. Aeschylus, *Prometheus Bound*, in *Aeschylus II*, ed. David Grene and Richard Lattimore and trans. David Grene (New York: Washington Square Press, 1967), pp. 148–149.

38. Ibid., p. 144.
39. Ibid., p. 56.
40. Ibid.

Study Questions

1. What is the main point of Winner's article? Why do you think it might be correct? Why do you think it might be wrong? Explain and defend your answers.
2. What assumptions does Winner make about technology? about values? Do you think these assumptions are correct? Explain and defend your answers.
3. If society accepts Winner's main conclusions, what consequences might follow? Would these consequences be desirable? Explain and defend your answers.
4. What are the points of similarity between the story of Frankenstein and technology as a whole?
5. Is there any way to mitigate "Frankenstein's Problem" today?
6. Is progress today associated with improvement for humankind or with technological achievement? Are these the same things? Why or why not?

Technology, Demography, and the Anachronism of Traditional Rights

Robert E. McGinn

ABSTRACT   *Theories of the influence of technology on modern Western soci-ety have failed to take into account the important role played by a widespread pattern of sociotechnical practice. The pattern in question involves the inter-play of technology, rights, and numbers. This paper argues that in the context of an ever more potent technological arsenal and an ever increasing number of individuals who have access to its elements and believe themselves entitled to use them in maximalist ways, adherence to the traditional notion of individ-ual human rights is anachronistic and increasingly problematic for the quality of life in contemporary society. To combat this situation, I criticise the idea of human rights as timelessly valid and offer a contextualised account of these constructs, one able to take on board the implications of their maximalist exer-cise in a populous technological society. I conclude by illustrating the strug-gles being waged over the adaptation of human rights to techno-demographic reality in two areas of contemporary Western society: urban planning and medicine.*

Introduction

Critics of the influence of technology on society debit the unhappy outcomes they decry to different causal accounts. Some target *specific*

Originally published as "Technology, Demography, and the Anachronism of Traditional Rights," *Journal of Applied Philosophy*, Vol. 11, No. 1, 1994. Copyright © 1994 Society for Applied Philosophy, Blackwell Publishers, 108 Cowley Road, Oxford, OX4 1JF, UK and 3 Cambridge Center, Cambridge, MA 02142, USA.

characteristics or purposes of technologies that they hold are inherently objectionable. For example, certain critics believe biotechnologies such as human *in vitro* fertilisation and the genetic engineering of transgenic animals to be morally wrong, regardless of who controls or uses them, and attribute what they see as the negative social consequences of these innovations to their defining characteristics or informing purposes. Others, eschewing the technological determinism implicit in such a viewpoint, find fault with the *social contexts* of technological developments and hold these contexts—more precisely, those who shape and control them—responsible for such unhappy social outcomes as result. For example, some critics have blamed the tragic medical consequences of silicone gel breast implants on a profit-driven rush to market these devices and on lax government regulation. Still other critics focus on *users* of technologies, pointing to problematic aspects of the use and operation of the technics and technical systems at their disposal. For example, some attributed the fatal crash of a DC-10 in Chicago in 1979 and the rash of reports in the mid 1980s of spontaneous acceleration of Audi automobiles upon braking to the alleged carelessness of maintenance workers and consumers.

Questions of the validity and relative value of these theory-laden approaches aside, this paper identifies and analyses an important source of problematic technology-related influence on society of a quite different nature. Neither wholly technical, nor wholly social, nor wholly individual in nature, the source discussed below combines technical, social, and individual elements.

The source in question is a recurrent pattern of sociotechnical practice characteristic of contemporary Western societies. The pattern poses a challenge to professionals in fields as diverse as medicine, city planning, environmental management, and engineering. While not intrinsically problematic—indeed, the pattern sometimes yields beneficial consequences—the pattern is *potentially* problematic. Its manifestations frequently dilute or jeopardise the quality of life in societies in which they unfold. Unless appropriate changes are forthcoming, the pattern's effects promise to be even more destructive in the future. In what follows, I shall describe and clarify the general pattern, explore its sources of strength, elaborate a conceptual/theoretical change that will be necessary to bring the pattern under control and mitigate its negative effects, and survey some conflicts over recent efforts to do just that in two social arenas: urban planning and medicine.

The Pattern: Nature and Meaning

The pattern in question involves the interplay of technology, rights, and numbers. It may be characterised thus:

'technological maximality,' unfolding under the auspices of 'traditional rights' supposedly held and exercised by a large and increasing number of parties, is apt to dilute or diminish contemporary societal quality of life.

Let us begin by defining the three key expressions in this formulation. First, in speaking of an item of technology or a technology-related phenomenon as exhibiting 'technological maximality'(TM), I mean the *quality of embodying in one or more of its aspects or dimensions the greatest scale or highest degree previously attained or currently possible in that aspect or dimension.*

Thus understood, TM can be manifested in various forms. Some hinge on the characteristics of technological products and systems, while others have to do with aspects of their production, diffusion, use, or operation. Making material artifacts (technics) and sociotechnical systems of hitherto unequalled or unsurpassed scale or performance might be viewed as paradigmatic forms of TM. However, the TM concept is also intended to encompass maximalist phenomena having to do with processes as well as products. Examples of technological maximality of process include producing or diffusing as many units as possible of a technic in a given time interval or domain, and using a technic or system as intensively or extensively as possible in a given domain or situation. It is important to recognise that technological maximality can obtain even where no large-scale or super-powerful technics or technical systems are involved. Technological maximality can be reflected in *how* humans interact with and use their technics and systems as much as in technic and system characteristics proper. TM, one might say, has adverbial as well as substantive modes. In sum, technology can be maximalist in one or more of the following nine senses:

- product size or scale
- product performance (power, speed, efficiency, scope, etc.)
- speed of production of a technic or system
- volume of production of a technic or system
- speed of diffusion of a technic or system
- domain of diffusion of a technic or system
- intensity of use or operation of a technic or system
- domain of use or operation of a technic or system
- duration of use or operation of a technic or system

Secondly, 'traditional rights' are entitlements of individuals as traditionally conceived in modern Western societies. For example, in the traditional Western conception individual rights have been viewed as timelessly valid and morally inviolable. Traditional individual rights often interpreted in this absolutist way include the right to life as well as liberty, property, and procreative rights.

Thirdly, the 'large and increasing number of parties' factor refers to the presence in most kinds of context in contemporary Western societies of many, indeed also a growing number of, parties—usually individual humans—each of whom supposedly holds rights of the above sort and may exercise them in, among other ways, technologically maximalist behaviour.

Before proceeding, I want to stress that this paper is neither a critique of 'technological maximality' per se nor a celebration of E. F. Schumacher's 'small is beautiful' idea. For, like the above triadic pattern, TM (or, for that matter, technological minimality) per se is not inherently morally objectionable or problematic. Specimens of technological maximality such as the then unprecedentedly large medieval Gothic cathedrals and the mammoth Saturn V rockets of the kind that took Apollo XI toward the moon suffice to refute any such claim. Rather, it is the *conjunction* of the three above-mentioned factors in repeated patterns of sociotechnical practice—*large and increasing numbers of parties engaged in technologically maximalist practices as something that each party supposedly has a morally inviolable right to do*—that is apt to put societal quality of life at risk. With this in mind, in what follows we shall refer to the combustible mixture of these three interrelated factors as 'the troubling triad.'

The triadic pattern is surprisingly widespread. Consider the following examples:

1. the intensive, often protracted use of life-prolonging technologies or technological procedures in thousands of cases of terminally ill or irreversibly comatose patients, or in the case of those needing an organ transplant or other life-sustaining treatment, such uses supposedly being called for by the inviolable right to life;
2. the proliferation of mopeds, all-terrain, snowmobile, and other kinds of versatile transport vehicles in special or fragile environmental areas, such use supposedly being sanctioned by rider mobility rights; and
3. the erection of growing numbers of high-rise buildings in city centres, as supposedly permitted by owner or developer property rights.

As suggested by these examples, our pattern of sociotechnical practice unfolds in diverse spheres of human activity. Problematic phenomena exemplifying the pattern in other arenas include the infestation of American national parks by tens of thousands of small tourist aircraft overflights per year; the depletion of ocean fishing areas through the use of hundreds of enormous, mechanically operated, nylon monofilament nets; and the decimation of old-growth forests in the northwestern U.S. through the use of myriad potent chain saws. The untoward effects exacted by the unfolding of our triadic pattern include steep financial and psychological tolls, the depletion and degradation of environmental resources, and the dilution and disappearance of urban amenities. In short, the costs of the ongoing operation of the triadic pattern are substantial and increasing.

To this point, the pattern identified above makes reference to a number of individual agents, each of whom engages in or is involved with a specimen of technologically maximal behaviour, e.g., having life-prolonging technologies applied intensively to herself or himself, using a technic 'extensively'—meaning either 'in a spatially widespread manner' or 'frequently'—in a fragile, limited, or distinctive domain; or erecting a megastructure.

However, as characterised above, our pattern obscures the fact that TM can be present in *aggregative* as well as non-aggregative situations. Each of a large number of individuals, acting under the auspices of a right construed in traditional fashion, can engage in behaviour that while not technologically maximal in itself becomes so when aggregated over all relevant agents. Of course, aggregating over a number of cases each of which is *already* technologically maximal compounds the maximality in question, and probably also its effects on society. We may say, therefore, that there is *individual* TM (where the individual behaviour in question is technologically maximal) and *aggregative* TM, the latter having two subspecies: *simple-aggregative*, where the individual behaviour is *not* technologically maximal, and *compound-aggregative*, where the individual behaviour is *already* technologically maximal.

One reason why simple-aggregative TM is troubling is that individual agents may have putative rights to engage in specimens of non-technologically-maximal behaviour that, taken individually, seem innocuous or of negligible import. However, contemporary environments or contexts do not automatically become larger or more robust in proportion to technic performance improvements, increasing costs of contemporary technics and systems, or the increasing number of those with access to or affected by these items. Hence, the aggregation of individually permissible behaviour over all relevant agents with access to technics can result in substantial harm to societal quality of life. One can therefore speak of 'public harms of aggregation.' For example, the failure of each of a large number of people to recycle their garbage is technological behaviour that when aggregated provides an instance of problematic technological maximality of use. Aggregating the individually innocuous effects of a large number of people driving motor vehicles that emit pollutants yields the same story: individually innocuous behaviour can, when aggregated over a large group, yield a significant, noxious outcome.

The Pattern: Sources of Strength

How is the power of the pattern under discussion to be accounted for? Put differently, why does the troubling triad come under so little critical scrutiny when it has such untoward effects on individual and societal quality of life? In the case of simple-aggregative TM, the reason is that the effects of the behaviour of the individual agent are negligibly problematic. It is difficult to induce a person to restrict her or his behaviour when it is not perceivably linkable to the doing of significant harm to some recognised protectable individual or societal interest.

More generally, the strength of the pattern derives from the effects of factors of various sorts that lend impetus to its constituent elements. Let us examine each element in the pattern in turn.

Technological Maximality

The modern drive to achieve increases in efficiency and economies of scale and thereby reap enhanced profits is unquestionably an important factor that fuels various modes of technological maximality. One thinks in this connection of maximalist technics such as the Boeing 747 and the Alaska pipeline as well as the diffusion speed and scope modes of TM for personal technics like the VCR and CD player.

However, economic considerations do not tell the entire causal story. Cultural phenomena also play an important role and help explain the low level of resistance to our pattern. Technological maximality is encouraged by the 'technological fix' mentality deeply entrenched in Western countries. Should anything go awry as a result of some technological maximalist practice, one can always, it is assumed, concoct a technological fix to remedy or at least patch up the situation in time. Moreover, there is much individual and group prestige to be garnered in modern Western societies by producing, possessing, or using the biggest, fastest, or more potent technic or technological project; more generally, by being, technologically speaking, 'the-firstest-with-the-mostest.' Further, influential sectors of Western opinion gauge societal progress and even a society's level of civilisation by the degree to which it attains and practises certain forms of technological maximality. Small-scale, appropriate technology may be fine for developing countries but resorting to it would be seen as culturally retrogressive for a technologically 'advanced' society such as the U.S.

Technological maximality is often associated with construction projects. In 1985, American developer Donald Trump announced what proved to be abortive plans for a 150-storey, 1,800-foot-tall Television City on the West Side of Manhattan, a megastructure that he revealingly called 'the world's greatest building.'[1] The demise in contemporary Western society of shared qualitative standards for making comparative value judgments has created a vacuum often filled by primarily quantitative standards of value. This, in turn, has fuelled technological maximality as a route to invidious distinction. If a building is quantitatively 'the greatest,' it must surely be qualitatively 'the best,' a convenient confusion of quality and quantity.

TM in the sense of virtually unrestricted technic use throughout special environments is greatly encouraged by modern Western cultural attitudes toward nature. Unlike in many traditional societies, land and space are typically perceived as homogeneous in character. No domains of land or space are sacred areas, hence possibly off limits to certain technological activity. On the contrary, in the contemporary U.S., nature is often regarded more as a playground for technology-intensive human activity. Dune buggy riders were incensed when environmentalists sued to force the National Park Service to ban off-road vehicles from the fragile dunes at Cape Cod National Seashore. The leader of the Massachusetts Beach Buggy Association lamented that 'it seems like every year they come up with more ways to deprive people of recreational activities,'[2] a comment that comes close to suggesting that rider rights have been violated.

The U.S. has no monopoly on TM. For example, France has a long tradition of technological maximality. The country's fascination with 'grands travaux,' large-scale technical projects conceived by politicians, public engineers, or civil servants, is several centuries old.[3] Encompassing classic projects such as Napoleon's Arc de Triomphe, Hausmann's transformation of central Paris, and Eiffel's Tower, the maximalist trend has also been manifested in the nation-wide SNCF electric railroad system and the Anglo-French Concorde supersonic transport airplane. More recent specimens suggesting that TM is alive and well in France include audacious undertakings such as the Channel Tunnel, ever more potent nuclear power stations, the T.G.V. *(très grand vitesse)* train, and the Mitterand government's plan for building the world's largest library, dubbed by critics the 'T.G.B.' *(très grand bibliothèque).*[4] Such projects are not pursued solely or primarily for economic motives but for reasons of national prestige and grandeur, certification of governmental power and competence, as symbols of cultural superiority, and as monuments to individual politicians.

Another cultural factor that fosters certain modes of TM is the relatively democratic consumer culture established in the U.S. and other Western countries in the twentieth century. For the American people, innovative technics should not be reserved for the competitive advantage and enjoyment of the privileged few. Rather, based on experience with technics such as the automobile, the phone, and the television, it is believed and expected that such items should and will become available to the great mass of the American people. This expectation, cultivated by corporate advertising in order to ensure sufficient demand for what industry has the capacity to produce, in turn greatly facilitates technological maximality of production and diffusion.

Traditional Rights

Many modern Western societies are founded on belief in what were once called the 'rights of man,' a term that succeeded the earlier phrase 'natural rights.' Building on Locke's thought about natural rights, the Bill of Rights enacted by the British Parliament in 1689 provided for rights to, among other things, life, liberty, and property. The U.S. Declaration of Independence of 1776 declared that 'all men . . . are endowed by their Creator with certain inalienable rights; that among these are life, liberty, and the pursuit of happiness.' The French 'Declaration des droits de l'homme et du citoyen' of 1789 asserts that 'the purpose of all political association is the conservation of the natural and inalienable rights of man: these rights are liberty, property, security and resistance to oppression.' In the 1940s Eleanor Roosevelt promoted use of the current expression 'human rights' when she determined through her work in the United Nations 'that the rights of men were not understood in some parts of the world to include the rights of women.'[5] Although later articles of the 1948 U.N. Universal Declaration of Human Rights make reference to novel 'economic and social rights' that are more clearly reflections of a particular stage of societal development, the document's Preamble refers to inalienable human rights and its early articles are couched in the language of 'the old natural rights tradition.'[6] Thus, in the dominant modern Western conception, individual rights are immutable, morally inviolable, and, for many, God-given.

What has this development to do with technological maximality? Things are declared as rights in a society under particular historical circumstances. When the declaration that something is a fundamental right in a society is supported by that society's dominant political-economic forces, it is safe to assume that recognition of and respect for that right is congruent with and adaptive in relation to prevailing social conditions. However, given the millennial history of the perceived close relationship between morality and religion, to get citizens of a society to take a declared 'right of man' seriously, it has often seemed prudent to represent rights thus designated as having some kind of transcendental seal of approval: e.g., God's blessing, correspondence with the alleged inherent fabric of the universe, or reference to them in some putatively sacred document. The right in question is thereby imbued with an immutable character, as if, although originating in specific historical circumstances, the right was nevertheless timelessly valid. Such a conception of rights can support even technologically maximal exercises of particular rights of this sort.

However, the specific sets of social-historical circumstances that gave birth to such rights eventually changed, whereas, on the whole, the perceived nature of the rights in question has not. Continuing to affirm the same things as categorical rights can become dysfunctional under new, downstream social-historical conditions; in particular, when the technics and systems to which citizens and society have access have changed radically. Endowing traditional rights with a quasi-sacred status to elicit respect for them has made it more difficult to delimit or retire them as rights further down the historical road, e.g., in the present context of rampant technological maximality. In essence, the cultural strategy used to legitimate traditional rights has bestowed on them considerable intellectual inertia, something which has proved difficult to alter even though technological and demographic changes have radically transformed the context in which those rights are exercised and take effect.

A recent example of how continued affirmation of traditional rights in the context of unprecdedented technological maximality can impede or disrupt societal functioning is that of the automated telephone dialer. These devices can systematically call and leave prerecorded messages at every number in a telephone exchange, including listed and unlisted numbers, cellular telephones, pagers, corporate switchboards, and unattended answering machines. By one estimate, at least 20,000 such machines are likely to be at work each day dialing some 20 million numbers around the United States. As a consequence of the potency and number of autodialers, significant communications breakdowns have already occurred.[7]

When Oregon legislators banned the commercial use of autodialers, two small-business owners who used the devices in telemarketing brought suit to invalidate the legislation on the grounds that, among other things, it violated their right to free speech. One issue here is whether U.S. society should leave its traditional robust right to free speech intact when threats to or violations of other important protectable interests, e.g., privacy, emergency preparedness, and efficient organisational operation, result from exercise of this right in revolutionary technological contexts such as those created by the intensive commercial use of autodialers and fax machines for 'junk calls' and 'junk mail.' Significantly, the American Civil Liberties Union, which supported the plaintiffs in the Oregon case, argued for preserving the traditional free speech right unabridged.

In November 1991 Congress passed the Telephone Consumer Protection Act that banned the use of autodialers for calling homes, except for emergency notification or if a party had explicitly agreed to receive such calls. However, the decision to ban turned on the annoying personal experiences of Congressional representatives and their constituents with unsolicited sales calls, not on any principled confrontation with the tension between traditional rights, technological maximality, and increasing numbers.[8] Not surprisingly, in 1993 a U.S. District Court Judge blocked enforcement of the law, ruling that it violated the constitutional right to free speech.[9]

Increasing Numbers

The positive attitude in the U.S. toward an increasing national population was adaptive in the early years of the Republic when more people were needed to settle the country and fuel economic growth. Today, even while evidencing concern over rapid population increases in less developed countries, the U.S. retains strongly pronatalist tax policies and evidences residues of the long-standing belief that when it comes to population 'more is better.' The 'land of unlimited opportunity' myth, belief that America has an unlimited capacity to absorb population increases without undermining its quality of life, and conviction that intergenerational fairness requires that just as America opened its doors widely to earlier generations of impoverished or persecuted peoples so also should it continue to do so today; these and other beliefs militate against taking the difficult steps that might decrease or further slow the rate of increase of the American population, hence of the number of rights claimants.

The Pattern as Self-Reinforcing

Not only do powerful cultural factors foster each of the three elements of the pattern, it is also self-reinforcing. For example, reproductive freedom, derived from the right of freedom or liberty, is sacrosanct in contemporary Western societies. This belief aids and abets the increasing numbers factor, something that in turn fuels technological maximality (e.g., in technic and system size and production and diffusion rates) to support the resultant growing population. Put differently, the increasing numbers factor intensifies the interaction of the elements of the troubling triad. Under such circumstances the latter can undergo a kind of chain reaction: increases in any of its elements tend to evoke increases in one or both of the other two, and so forth. The rights to life, liberty, property and the pursuit of happiness have traditionally been construed as 'negative rights,' i.e., as entitlements *not to be done to* in certain ways: not to be physically attacked, constrained, deprived of one's property, etc. But, in the context of new technologies, some such rights have also taken on a positive facet: entitlement of the individual *to be done to* in certain ways, e.g., to be provided with access to various kinds of life-sustaining medical technologies and to be provided with certain kinds of information in possession of another party. A positive-faceted right to life encourages further technological maximality in both development and use, something that in turn increases the number of rights holders.

Toward a Contextualised Theory of Human Rights

A society that generates an ever more potent technological arsenal and, in the name of democratic consumerism, makes its elements available in ever larger numbers to a growing citizenry whose members believe they have inviolable rights to make, access, and use those items in individually or aggregatively technologically maximalist ways, risks and may even invite progressive impairment of its quality of life. Substantial changes will be necessary if this scenario is to be avoided, especially in the U.S.

What changes might help avoid this outcome?

Decrease, stabilise, or at least substantially cut the rate of increase in the number of rights holders.

To think that any such possibility could be achieved in the foreseeable future is utopian at this juncture in Western cultural history. In spite of projections about the environmental consequences of a doubled or trebled world population, no politician of standing has raised the question of population limitation as a desirable goal for the U.S. or any other Western society. For this possibility to be realisable, it would seem that certain traditional rights, viz., those relating to reproductive behaviour and mobility, would have to be significantly reined in, a most unlikely prospect.

Put a tighter leash on individual technologically maximal behaviour. As with the previous possibility, this option too would seem to require abridging certain traditional exercise rights or changing the underlying, quasi-categorical traditional conception of individual rights to a more conditional one. Alternatively, if one could demonstrate that untrammelled operation of the pattern is producing effects that undermine various intangible individual or societal interests, this might furnish a reason for leash-tightening. However, for various reasons, such demonstrations, even if feasible, are rarely socially persuasive.[10]

This situation suggests that one thing that may be crucial to avoiding the above scenario is elaboration and diffusion of *a new theory of moral rights*. While detailed elaboration and defence of such a theory is not feasible here, an acceptable theory should at least include accounts of the basis, function, status, and grounds for limitation of individual rights. Brief remarks on these components follow.

Basis

Western intellectual development has reached a stage in which individual moral rights can be given a more empirical, naturalistic basis. It should be acknowledged that the epistemological plausibility of rights talk need not, indeed should not, depend upon untestable beliefs in the existence and largesse of a deity interested in protecting the vital interests of individual human beings by endowing them with inalienable rights. Human rights can be plausibly anchored in basic human needs, i.e., universal features of human 'wiring' that must be satisfied to an adequate degree if the individual is to survive or thrive.[11] The notion then would be that something qualifies as an individual human right if and only if its protection is vital to the fulfilment of one or more underlying basic human needs. This *bottom-up* approach has the virtue of making discourse about moral rights more empirically grounded than traditional top-down theological or metaphysical approaches.

Function

In the new theory I propose, moral rights have a mundane though important function: to serve as conceptual spotlights that focus attention on aspects of human life that are essential to individual survival or thrival. The reason why such searchlights are needed is that such aspects of human life are ever at risk of being neglected because of political or social inequalities, socially conditioned preoccupation with ephemera, or the tendency of human agents to overlook or discount the interests of parties outside of their respective immediate geographical and temporal circles.

Status

Joel Feinberg has distinguished three degrees of absoluteness for individual moral rights:[12]

1. a right can be absolute in the sense of 'bounded exceptionlessness,' i.e., binding without exception in a finite, bounded domain, as with, e.g., the right to freedom of speech;
2. a right can be absolute in the (higher) sense of an 'ideal directive,' i.e., always deserving of respectful, favourable consideration, even when, after all things have been considered, it is concluded that the right must regrettably be overridden, as with, e.g., the right to privacy, and
3. a right can be absolute in the (still higher) sense of 'unbounded exceptionlessness' and 'non-conflictability,' i.e., binding without exception in an unbounded domain and not intrinsically susceptible to conflict with itself or another right, in the way that, e.g., the right to free speech is conflictable, as exemplified in the hectoring of a speaker. The right not to be subjected to gratuitous torture is a plausible candidate for a right that is absolute in this third sense.

In the theory we propose, individual moral rights will not be absolute in the third, highest degree, only in the first or second degree, depending on technological and demographic circumstances and on the effects on societal quality of life of aggregated maximalist exercise of the right in question.

Grounds for Decreasing the Absoluteness of Individual Rights

There are at least six kinds of circumstantial grounds that may justify restriction or limitation of an individual moral right because of the bearing of its technologically maximal exercise on societal quality of life:

1. if the very existence of society is called into question by the exercise of a putative right, e.g., exercise of the right to self-defence by the acquisition of the capability of making and using weapons or other technologies of mass destruction;
2. if continued effective social functioning is threatened by the exer-

cise of a right, e.g., the disruption of telecommunication by the operation of automatic phone dialers operated under the auspices of the right of free speech;

3. if some natural resource vital to society is threatened through the exercise of a right, e.g., the reduction of fishing areas or forests to non-sustainable conditions by technologically maximal harvesting practices;

4. if a seriously debilitating financial cost is imposed on society by the widespread or frequent exercise of a right, as with mushrooming public health care payments for private kidney dialysis treatment in the name of the right to life;

5. if some phenomenon of significant aesthetic, cultural, historical, or spiritual value to a people is jeopardised by the exercise of a right, e.g., the destruction of a recognised architectural landmark by affixing its façade to a newly built, incongruous, megastructure under the aegis of a private property right; and

6. if some highly valued social amenity would be seriously damaged or eliminated through the exercise of a right. For example, between 1981 and 1989 convivial public space at the Federal Plaza in downtown Manhattan was effectively eliminated by the installation of an enormous sculpture (Richard Serra's 120 ft long by 14 ft high 'Tilted Arc'). The artist unsuccessfully sued the government attempting to halt removal of the work as a violation of his First Amendment right of free speech, while many of his supporters cited the right to free artistic expression.[13]

In the case of simple-aggregative TM, the only option to acquiescence is to demonstrate the significant harm done to a protectable societal interest by the aggregated act and attempt to effect an ethical revaluation of putatively harmless individual behaviour; in other words, to lower the threshold of individual wrongdoing to reflect the manifest wrong effected by aggregation. With such a revaluation, the individual would have no right to act as he or she once did because of the newly declared immorality of the individual act. This process may be underway vis-à-vis the individual's disposal of home refuse without separation for recyling.

In sum, we need a *contextualised theory of human rights*. An acceptable theory of rights in contemporary technological society must be able to take on board the implications of their exercise in a context in which a rapidly changing, potent technological arsenal is diffused throughout a populous, materialistic, democratic society. Use of such a technological arsenal by a large and growing number of rights holders has considerable potential for diluting or diminishing societal quality of life.

Indeed, insistence on untrammelled, entitled use of potent or pervasive technics by a large number of individuals can be self-defeating, e.g., by yielding a state of social affairs incompatible with other social goals whose realisation the group also highly values.

At a deeper level, what is called into question here is the viability of modern Western individualism. Can, say, contemporary U.S. society afford to continue to promote technology-based individualism in the context of the diffusion and use of multiple potent technics by a large and ever growing population? Or is the traditional concept of individualism itself in need of revision or retirement? The ideology of individualism in all areas of life may have been a viable one in the early modern era, one with a less potent and diverse technological arsenal and a less populous society. But can contemporary Western societies have their ideological cake and eat it too? Can individualism continue to be celebrated and promoted even as a greater and greater number of citizens have access to powerful technics and systems that they, however technologically unsocialised, believe themselves entitled to use in maximalist ways?

Recent Struggles to Adapt Individual Rights to Technological Maximality and Increasing Numbers

In recent years, struggles to adapt individual rights to the realities of technological maximality in populous democratic societies have been waged incessantly on several professional fronts. Let us briefly discuss some pertinent developments in two such fields: urban planning and medicine.

Urban Planning

Two urban planning concerns involving our pattern, over which there were protracted struggles in the 1980s, are building construction and the unrestricted movement of cars. In 1986, the city of San Francisco, California, became the first large city in U.S. history to impose significant limits on the proliferation of downtown high-rise buildings. After several unsuccessful previous efforts, a citizen initiative was finally approved that established a building height limit and a cap on the amount of new high-rise floor space that can be added to the downtown area each year. The majority of San Francisco voters came to believe that the aggregate effects of the continued exercise of essentially unrestricted individual property rights by land owners and developers in technologically maximal ways—entitled erection of numerous high-rises—was undermining the quality of city life. In 1990, voters of Seattle, Washington, reached the same conclusion and approved a similar citizen initiative.

As for cars, the 1980s saw the adoption in a few Western countries of substantial limits on their use in cities. For example, to combat air pollution and enhance the quality of urban social life, citizens of Milan and Florence voted overwhelmingly in the mid 1980s to impose limits on the use of cars. In Milan, they are prohibited from entering the *centro storico* between 7:30 a.m. and 6:30 p.m., while in Florence much of the *centro storico* has been turned into a pedestrians-only zone. In California, the cities of Berkeley and Palo Alto installed barriers to prevent drivers from traversing residential streets in the course of crosstown travel. Revealingly, in a debate in the California State Senate over legislation authorising Berkeley to keep its barriers, one senator argued that 'We should be *entitled* to use all roadways . . . Certain individuals think they're too good to have other people drive down their streets' (emphasis added).[14] The phenomenon combated by the road barriers is a clear instance of aggregative TM of use unfolding under the auspices of traditional mobility rights exercised by large numbers of cardrivers. The senator's mind reading notwithstanding, it would seem that citizens, perceiving this pattern as jeopardising the safety of children and diluting the neighbourhood's residential character (read: quality of social life), prevailed on authorities to diminish the long-established domain of driver mobility rights.

Efforts to restrict individual property and mobility rights in urban settings in light of the quality-of-life consequences of their aggregated, technologically maximalist exercise have initiated a high-stakes struggle that promises to grow in importance and be vigorously contested for the foreseeable future.

Medicine

An important issue in the area of medicine that involves our pattern is the ongoing tension between the right to life and the widespread intensive use of life-prolongation technology. Following World War II, the change in the locus of dying from the home to the technology-intensive hospital enabled the full arsenal of modern medical technology to be mobilised in service of the right to life. However, the quality of the prolonged life was often so abysmal that efforts to pull back from application of technologically maximal life-extending medical care eventually surfaced.

The Karen Ann Quinlin case (1975–1985) was a landmark in the United States. The Quinlins asked their comatose daughter's doctor to disconnect her respirator. He refused, as did the New Jersey Court of Appeals.[15] The latter argued, significantly for our purposes, that 'the right to life and the preservation of it are "interests of the highest order." ' In other words, in the Appeals Court's view, respecting the traditional individual right to life was held to require ongoing provision of technologically maximal medical care. The New Jersey Supreme Court eventually found for the Quinlins, not by revoking this idea but by finding that a patient's privacy interest grows in proportion to the invasiveness of the medical care to which the patient is subjected, and that that interest can be exercised in a proxy vein by the patient's parents.[16]

The equally celebrated Nancy Cruzan case (1983–1990) was essentially an extension of Quinlin, except that the technological means of life extension that Nancy's parents sought to terminate were her food and hydration tubes. Many who opposed the Cruzans believed that removal of these tubes was tantamount to killing their comatose daughter, i.e., to violating her right to life. In their view, respect for Nancy's right to life required continued application of these technological means without limitation of time, regardless of the quality of life being sustained. The Missouri Supreme Court concluded that the state's interest in the preservation of life is 'unqualified,' i.e., that the right to life is inviolable. The Court held that in the absence of 'clear and convincing evidence' that a patient would not want to be kept alive by machines in the state into which he or she had fallen, i.e., would not wish to exercise her or his right to life under such circumstances, the perceived absoluteness of the right to life drove continued application of the life-prolonging technology.[17]

The 1990 case of Helga Wanglie, seemingly commonplace at the outset, took on revolutionary potential. Hospitalised after fracturing her hip, Mrs Wanglie suffered a respiratory attack that cut off oxygen to her brain. By the time she could be resuscitated, the patient had incurred severe brain damage and lapsed into a vegetative state believed irreversible by hospital doctors. Despite this prognosis and after extensive consultation with the doctors, Mrs Wanglie's family refused to authorise disconnection of the respirator that prolonged her life, asserting that the patient 'want[ed] everything done.' According to Mr Wanglie, 'she told me, "Only He who gave life has the right to take life." '[18]

Unprecedentedly, believing that further medical care was inappropriate, the hospital brought suit in court to obtain authorisation to disconnect the patient's respirator against her family's wishes. Predictably, this suit was unsuccessful, but Mrs Wanglie died shortly thereafter.[19,20] Had the suit succeeded, it would have marked a significant departure from traditional thinking and practice concerning the right to life. Care would have been terminated not at the behest of patient or guardians, something increasingly familiar in recent years, but rather as the result of a conclusion by a care-providing institution that further treatment was 'futile.' Projected quality of patient life would have taken precedence over the patient's inviolable right to life as asserted by guardians and the absoluteness of the right to life would have been diminished. Consensus that further treatment, however intensive or extensive, offered no reasonable chance of restoring cognitive functioning would have been established as a sufficient condition for mandatory cessation of care.

There are thousands of adults and children in the U.S. and other Western societies whose lives of grim quality are sustained by technological maximality in the name of the right to life, understood by many as categorically binding.[21] The financial and psychological tolls exacted by this specimen of compound-aggregative TM are enormous and will continue to grow until the right to life—its nature and limits—is adapted to the individual and aggregative implications of the technologies used on its behalf.

The troubling triadic pattern should be of concern to many kinds of professional practitioners, not just public officials. Professionals such as urban designers, environmental managers, engineers, and physicians are increasingly confronted in their respective practices with problematic consequences of the continued operation of the troubling triad. Each such individual must decide whether to conduct her or his professional practice—processing building permits, managing natural resource use, designing technics and sociotechnical systems, and treating patients—on the basis of traditional individualistic conceptions of rights unmodified by contemporary technological capabilities and demographic realities, or to alter the concepts and constraints informing her or his practice to reflect extant forms of technological maximality. The fundamental reason why the triadic pattern should be of concern to practising professionals is that failure to combat it is tantamount to acquiescing in the increasingly serious individual and societal harms apt to result from its predictable repeated manifestations. Professionals have an important role to play in raising societal consciousness about the costs of continuing to rely on anachronistic concepts of individual rights in contemporary technological societies. To date, doctors have made some progress in this effort but other professional groups have not even begun to rise to the challenge.

Conclusion

In the coming years U.S. citizens and other Westerners will face some critical choices. If we persist in gratifying our seemingly insatiable appetite for technological maximality, carried out under the auspices of anachronistic conceptions of rights claimed by ever increasing numbers of people, we shall pay an increasingly steep price in the form of a diminishing societal quality of life. Consciousness-raising, through education and responsible activism, though maddeningly slow, seems the most viable route to developing the societal ability to make discriminating choices about technological practices and their aggregated effects. However accomplished, developing that ability is essential if we are to secure a future of quality for our children and theirs. Taming the troubling triadic pattern would be an excellent place to begin this quest. The technodemographic anachronism of selected traditional rights should be recognised and a new, naturalistic, non-absolutist theory of human rights should be elaborated, one that stands in dynamic relationship to evolving technological capabilities and demographic trends. Whether or not such a new theory of rights emerges, becomes embodied in law, and alters the contours of professional practice in the next few decades will be critically important to society in the twenty-first century and beyond.

Notes

1. *New York Times*, 19 November, 1985, p. 1.
2. *Newsweek*, 25 July, 1983, p. 22.
3. See for example: Cecil O. Smith Jr (1990) The Longest Run: Public Engineers and Planning in France, *American Historical Review*, 95, No. 3, pp. 657–692.
4. *New York Times*, Section II, 22 December, 1991, p. 36.
5. Maurice Cranston (1983) Are There Any Human Rights?, *Daedalus*, 112, No. 4, p. 1.
6. Maurice Cranston (1973) *What Are Human Rights?* (New York, Taplinger), pp. 53–54.
7. *New York Times*, 30 October, 1991, p. A1.
8. *New York Times*, 28 November, 1991, pp. D1 and D3.
9. *New York Times*, 23 May, 1993, I, p. 26.
10. Robert E. McGinn (1979) In Defense of Intangibles: the Responsibility-Feasibility Dilemma in Modern Technological Innovation, *Science, Technology, and Human Values*, No. 29, pp. 4–10.
11. See for example: David Braybrooke (1968) Let Needs Diminish That Preferences May Prosper, in *Studies in Moral Philosophy* (Oxford, Blackwell), pp. 86–107, and the same author's (1987) *Meeting Needs* (Princeton, Princeton University Press) for careful analysis of the concept of basic human needs. For discussion of the testability of claims that something is a bona fide basic human need, see Amatai Etzioni (1968), Basic Human Needs, Alienation, and Inauthenticity, *American Sociological Review*, 33, pp. 870–885. On the relation-

ship between human rights and human needs, see also C. B. Macpherson, quoted in D. D. Raphael (1967), *Political Theory and the Rights of Man* (London, Macmillan), p. 14.

12. Joel Feinberg (1973), *Social Philosophy* (Englewood Cliffs, NJ: Prentice-Hall), pp. 85–88.

13. See for example: J. Hitt, (ed.) The Storm in the Plaza, *Harper's Magazine*, July 1985, pp. 27–33.

14. *San Francisco Chronicle*, 2 July, 1983, p. 6.

15. *In re Quinlin*, 137 N.J. super 227 (1975).

16. *In re Quinlin*, 70 N.J. 10,335 A. 2d 647 (1976).

17. After the U.S. Supreme Court decision upholding the Missouri Supreme Court was handed down, three of the patient's friends provided new evidence of her expressed wish to be spared existence in a technologically sustained vegetative state. This led, in a lower court rehearing, to a judgment permitting parental exercise of the patient's recognised privacy interest through the withdrawal of her food and hydration tubes. Nancy Cruzan expired twelve days after this decision was announced. See *New York Times*, 15 December, 1990, A1 and A9, and 27 December, 1990, A1 and A13.

18. *New York Times*, 10 January, 1991, A16.

19. Ibid., 2 July, 1991, A12.

20. Ibid., 6 July, 1991, I, 8.

21. U.S. Congress, Office of Technology Assessment (1987) *Technology-Dependent Children: Hospital v. Home Care Sustaining Technologies and the Elderly* (Washington, D.C., U.S. Government Printing Office).

Study Questions

1. What is the main point of McGinn's article? Why do you think it might be correct? Why do you think it might be wrong? Explain and defend your answers.
2. What assumptions does McGinn make about technology? about ethics and human rights? Do you think these assumptions are correct? Explain and defend your answers.
3. If society accepts McGinn's main conclusions, what consequences might follow? Would these consequences be desirable? Explain and defend your answers.
4. What is the "troubling triad" about which McGinn speaks? Explain.
5. What makes the pattern of technological maximality so powerful? Explain.
6. From the point of view of ethical or moral theories, what is the difference between individual practices and their cumulative effects?
7. What dangers are associated with McGinn's proposal to contextualize human rights?

3.4

Economic Evaluations of Technology

Kristin Shrader-Frechette

Historians tell us that some forms of technology assessment may date back more than 4000 years. In ancient Mesopotamia, for example, technical projects were submitted to the scrutiny of priests.[1] Despite such apparently venerable traditions and the growing popularity of the term "technology assessment," however, there is little agreement about the approach to which the term refers. In the U.S. Office of Technology Assessment (OTA), for instance, the methods and skills used "depend on the technology under study, the requesting client, . . . the time for and setting of the project, . . . and the resources available for a study."[2] Although nearly all the reports employ cost-benefit analysis, beyond this central method of welfare economics, there are "no agreed upon techniques for carrying out a technology assessment."[3] The means for evaluating noneconomic (e.g., biological, chemical, legal, or political) aspects of diverse technological impacts appear to vary widely.

One of the main reasons for the absence of a uniformly accepted methodology is that, despite the fact that "science . . . underlies technology and provides the fundamental knowledge for technological application,"[4] a technology ordinarily is evaluated as successful, acceptable, or desirable for quite different reasons than is a scientific theory. Scientific theories are usually assessed on the basis of fairly well-established criteria for their *truth,* e.g., simplicity, explanatory value, or predictive power. Technological inventions, however, are judged

Originally published as "Technology Assessment as Applied Philosophy of Science," *Science, Technology, & Human Values,* Vol. 6, No. 33, Fall 1980. Copyright © 1980 by the Massachusetts Institute of Technology and the President and Fellows of Harvard College.

primarily in the context of the *purpose* for which they are used (e.g., nuclear power)[5] and their success must be evaluated the way one judges the correctness of a particular route for a journey. As the Cheshire Cat responded, when Alice asked which way she ought to go: "that depends a good deal on where you want to get to."[6]

The Role of Applied Philosophy of Science in Assessment

Because assessors utilize the data, models, and theories of the natural and social sciences in order to analyze the effects of technological applications, philosophers of science have a clear role to play in technology assessment. According to Emilio Daddario, first director of the OTA, "assessment work will be directed by the performers and managers of applied science. The value framework against which technological consequences are judged will be erected by the social sciences, the arts, and humanities. . . ."[7]

A philosopher of science can contribute to technology assessment and hence ultimately to the guidance of technology-related public policy on at least two levels: that of conceptual and methodological analysis, and that of concrete applications, where the concepts and methods are employed in specific contexts.[8] At the level of *conceptual analysis,* for example, the central notions of philosophy of science (e.g., "prediction") or of some particular science (e.g., "Pareto optimality" in economics) are scrutinized critically in order to provide clarification, to evaluate claims of logical connection, or to explore the foundations of scientific argument. These inquiries may then be used at the second level, to support or question forms of practical argument which appeal to conceptual warrants for their justification.

At the level of *concrete applications,* the more theoretical questions of the earlier analysis make contact with issues of fact and public policy regarding technology. The goal at this stage is to employ the methods of philosophy of science in contexts calling for practical responses. For example, on the basis of (1) a conceptual analysis of "prediction," (2) a factual investigation of nuclear reactor properties, and (3) an evaluation of mathematical methods used in WASH-1400, a philosopher of science might challenge the well-known argument that a core melt can be predicted to occur once in every 17 thousand reactor years.[9] Another specific claim, made by an assessment team on which I served, is that only nuclear- and coal-generated energy will be able to help meet electricity demand by the year 2000.[10] A philosopher of science could investigate this thesis by providing: (1) a methodological analysis of the principle of "exclusion,"[11] (2) a factual investigation of the short-term energy-generating potential of all sources of electricity (including solar, geothermal, etc.), and (3) an evaluation of the various assumptions employed in models used to predict future energy demand.

As these examples indicate, the philosopher of science may be helpful at several tasks. He or she may help ascertain the extent to which a given scientific (e.g., econometric) model adequately depicts the technological situation being analyzed, or may undertake an analysis of the concepts, methods, and assumptions employed by the scientists and engineers working on an assessment team.[12] Both tasks could help to encourage the practice (almost never followed in technology assessment) of stating explicitly, even highlighting, the limitations of a particular study or model.[13]

A Case Study in Applied Philosophy of Science: Level One, Conceptual Analysis

To illustrate how philosophers of science may aid in technology assessment, I will undertake a brief case study in welfare economics. It is widely held (and probably true) that technology assessors "base their analyses most on engineering cost estimates,"[14] and that economics is therefore "the final test of public policy."[15] Many economists believe themselves to be the most rigorous and successful of all the social scientists;[16] and, indeed, the science of economics "has always been concerned with problems of wealth and welfare—which are conspicuously problems of value." As Sidney Hook remarks, "the great philosophers have concerned themselves with economic questions to a point where the history of philosophic thought and the history of economic thought are to some extent overlapping."[17] [One thinks immediately, in this regard, of Adam Smith, John Stuart Mill, and John Maynard Keynes, all of whom were philosophers.] Given that economics is probably the single most important science employed in technology assessments, and that cost-benefit analysis is the central method used in them, it makes sense to analyze the scientific concepts regarded as "the most basic" and "the most celebrated" in economics,[18] i.e., the "Pareto Optimum" and "Pareto Improvement," from which the entire "rationale of cost-benefit analysis derives."[19] Investigation of these concepts should provide a framework for understanding how the vast majority of technology assessors and welfare economists calculate costs and benefits.[20]

Generally speaking, the Pareto Optimum is that "position from which it is not possible, by any reallocation of factors, to make anyone better off without making at least one person worse off."[21] More specifically, a *"potential* (potential because gains *can* be, but may not be distributed in this way). Pareto Improvement" is "measured in principle as the algebraic sum of all CVs" (all compensating variations), where a CV for a given individual is

> the sum of money which, if received or paid after the economic change in question, would make the individual no better or worse off than before the change. If, for example, the price of a loaf of bread falls by 10 cents, the CV is the maximum sum a man would pay in order to be allowed to buy bread at this lower price. *Per contra,* if the loaf rises by 10 cents the CV is the minimum sum the man must receive if he is to continue to feel as well off as he was before the rise in price.[22]

When the CVs of the gainers (a positive sum) are added to the CVs of the losers (a negative sum), and the result is positive, then the gainers can compensate the losers; consequently, the economic change causing the gains and losses may be said to result in an excess of benefits over costs, where benefits and costs are understood in terms of compensating variations.[23] "An amount of money calculated at the given set of prices will suffice to measure the CV."[24]

There appear to be at least three basic notions packed into the concept of "compensating variation": (1) that it is a measure, when CVs are summed to equal a (potential) Pareto Improvement, of how gains can be distributed so as to make everyone in the community "better off";[25] (2) that the criterion for whether one is "better off" is how he "feels" subjectively;[26] and (3) that feelings of being better off are measured by a "sum of money," judged by the individual and "calculated at the given set of prices" on the market.[27] Some significant, often unrecognized, *assumptions* underlie these three notions.

For one thing, using CV sums to measure everyone's being "better off" presupposes that gains and losses, for all individuals in all situations, can be computed numerically. This is something notoriously difficult to do.[28] Pareto himself pointed out that, for only 700 commodities and merely 100 persons, his theory required solving not less than 70,699 equations.[29] Hence the concept of compensating variation appears applicable only *in principle*. And, as one economic theorist has pointed out, neither Pareto nor his followers have even proved that, in principle, a solution for a CV could be computed "in a concrete case of application."[30]

Another important assumption built into this method is that it is acceptable to employ an economic change to improve the community well-being, even though distributional effects are ignored. Mishan expresses this point by noting that the use of CV sums according to Pareto theory "ignores the resulting change in the distribution of incomes."[31] Suppose, for example, that an economic change made a given set of individuals better off by a total of $10x$ dollars, at the expense of another set of individuals made worse off by a total of x dollars. This "desirable" change would produce an excess gain of $9x$ dollars for the community as a whole. Nevertheless, according to the CV concept, even redistributive measures directed at compensating the second set of individuals (by at least x dollars) would result in disproportionate benefits for the two sets of persons. As the example suggests, a likely consequence of accepting this assumption is not only that distributional inequities will occur, but also that those in the community who are made worse off, or less well off, are to be found primarily among lower-income groups.[32] This is probably one of the reasons why Biderman cautioned that the use of quantitative economic criteria of well-being ought not to be assumed to lead to greater equity and even-handedness in policy. Rather, because it allows economists to eschew evaluation of distributional effects, employment of the CV concept is likely to "reflect the dominant ideological orientations of the most powerful and articulate groups affected by the phenomena measured."[33]

The notion, identified previously, that the CV criterion for whether one is "better off" is how he "feels" subjectively as measured in quantitative terms,[34] also has a particular ideological or normative component. Obviously built into this criterion is the goal of maximizing individual well-being, which involves the assumption that welfare is defined in terms of egoistic hedonism. Arrow admits as much in discussing compensating variations and Pareto Optimality.[35] What is important, I think, is that, whether this egoistic and hedonistic presupposition is right or wrong, it is normative to the core, a fact apparently ignored or denied by many economists (e.g., Friedman, Mises) who claim their discipline is value-free and objective.[36] One author has gone so far as to assert: "my charge against most economists is that they are ready to exclude other [than CV-based or Pareto-based] normative viewpoints as unscientific, while permitting this one to crawl under the fence."[37]

Also implicit in the CV concept is the assumption that individual costs and benefits are understood by reference to what people prefer, as measured by their compensating variations. The difficulty, however, is that persons often prefer things that don't actually increase their well-being (e.g., smoking, a particular marriage partner).[38] The discrepancy is important, since it seems rational to maximize benefits and minimize costs only if they are truly connected with human welfare. If a person's best interests are assumed to be identical with what he prefers, however, then at least four undesirable consequences follow. First, the *quality* of the choices is ignored.[39] Second, there is no recognition of the classical Platonic-Aristotelian doctrine that there is a difference between what makes men good or secures justice and what fulfills their wants, or, in other words, between needs and wants.[40] Third, the question of a distinction between *utility* and *morality* is begged.[41] And fourth, group welfare is assumed to be merely the aggregate of individual preferences (the sum of all CVs). Clearly, however, public well-being is not simply the aggregate of individual preference. An individual's personal welfare might be maximized if, for example, he alone made a choice to use a disproportionate share of natural resources; *if* it were conceived of as the aggregate of CVs in which everyone made this choice, the public welfare would not be optimized. Moreover, in a rapidly changing situation, leaders must act on the basis of likely future events, not merely on the aggregate of individual preferences. It is very difficult to determine which *present* public policy is likely to maximize *future* benefits, particularly if that *social* policy is defined as merely the aggregate of *individual* costs and benefits (CVs).[42] Because of these four consequences, it is not clear that compensating variations actually measure well-being.[43] What seems to be going on here is that, because of the way the concept of CV is defined, welfare economists are employing a stipulative definition of "value" by virtue of their technical understanding of "well-being." Provided the definition is correct, there is not necessarily a problem. What appears likely, however, is that it is not correct and that this technical *economic-science sense* of the term will be forgotten and then exported to other contexts (e.g., used in a technology assessment) where it will be used, in an *ordinary-language sense,* as a misleading attempt to clarify specific ethical issues.

A final difficulty with the assumption that the individual is best able to determine his own CV is that it ignores the fact that wealthy and poor individuals are not able to make equally desirable judgments based on monetary criteria. Consider a case in which an individual determines the maximum sum of money (the CV) he would pay in order to be allowed to buy an electrostatic air filter for his furnace, after its cost has decreased by x dollars. Obviously, a rich man would be able to pay a much higher CV than would a poor man. It has been shown statistically that, as income increases, people are willing to pay a much higher CV for environmental quality,[44] as well as for medical care, improved life expectancy, transportation safety equipment, home repairs, and job safety.[45]

As these statistics suggest, there are clear discriminatory effects of measuring CV on the basis of what a person "is willing to pay, or to receive, for the estimated change of risk."[46] Nevertheless, subjective determination of CV (either through the questionnaire approach or observation)[47] is touted by cost-benefit analysts as "the only economically justifiable" means of applying the concept.[48] This interpretation of CV, however, presents several other difficulties. For example, the consequences of discrimination on the basis of wealth might be contrary to Rawls's notion of justice as fairness and, in fact, contrary to the egalitarian ethical framework regarded as part of democratic institutions.[49] Also, the cost-benefit model of economic transactions apparently does not maximize individual freedom as much as has been alleged by market proponents.[50] A person is not free to pay a particular price in order to obtain some benefit—e.g., reduced health risk—if his income is less than that of other people.

Employment of the subjectively determined CV points up not merely problems with discrimination and lack of equal freedom and economic opportunity, but also difficulties with one of the classical simplifying assumptions employed in econometric and technological modeling, i.e., aggregation. Affirming the adequacy of computing costs and benefits according to subjectively measured CVs likewise affirms the adequacy of aggregating and of comparing costs and benefits determined according to quite different criteria.[51] The problem with this assumption generally is that it suggests wrongly that given data (e.g., CVs) are homogeneous when in fact they are not. And, to the extent that they are not, conclusions regarding aggregates may be either misleading or false.[52] For all these reasons, it would be well for economists to discuss the degree of error possibly arising from using subjective valuations of compensating variations.

Another notion built into the CV concept is (3) that one's feelings of being better off are measured by a "sum of money" judged by the individual and "calculated at the given set of prices" on the market.[53] Probably the most basic presupposition implicit in (3) is that market prices are measures of the value of goods.[54] Oscar Wilde's famous remark, however, that the cynic knows the price of everything and the value of nothing, suggests that value includes something price does not. Price takes account neither of "the intrinsic service [a thing] . . . is capable of yielding by its right use," nor of the ability to satisfy desire, nor of the "capacity for satisfying wholesome human wants." Hobson and others believed that price was a function of the intensity of human wants, not of the intensity of correct or desirable human wants. But if price is not based on authentically good wants, then price might not always be based on value, but on what Hobson calls "illth."[55] Moreover, the value of goods is not *causally* determined by economic exchange, as Anderson noted, any more than the amount of water in a vessel is causally determined by measuring it. All these considerations suggest that using "price" as a measure of "value" gives it a misleading ontological status.[56]

There are also some clear reasons, *economically* speaking, why market prices diverge from authentic values: (a) the distorting effects of monopoly;[57] (b) the failure to compute effects of externalities;[58] (c) speculative instabilities in the market,[59] and (d) the absence of monetary-term values for natural resources (e.g., air) and "public goods," even though these items obviously have great utility to those who use them.[60] All these ill effects, visited on market-based economic statistics such as prices, are widely known and discussed among both classical and nonclassical economists. What may not be recognized, however, is that if (and this is usual) market prices are used to determine CVs, then use of the CV concept automatically entails a *normative bias* in favor of the status quo. This is because whenever one attempts to bring the economy toward a Pareto Optimum by using the existing set of prices, they themselves are a function of "the existing income-distribution."[61] But if this distribution is neither egalitarian nor socially just, then to a similar extent, the optimization achieved through calculation of CVs also will not be.[62]

The discrepancy between market price and value illustrates a common difficulty with economic statistics. The problem is that the economist-observer selects what to include (e.g., price) in a particular concept (e.g., compensating variation), since complex phenomena (e.g., value) are never exhaustively describable.[63] As a consequence, however, the concept is so narrow as to be useless for many practical applications. Since the terms of definition (e.g., price) of the concept (e.g., CV) are irrelevant to the purpose for which it was introduced (e.g., for calculating "value"), the concept is one of "specious accuracy."[64]

Of course, the problem with assuming that price equals value when computing CVs is not merely that price purports to measure value, but does not; rather, it is that we are likely to treat a surrogate concept (price) as though it were identical with one that we wish to represent (value). In our dalliance with the surrogate, we forget the inferences we have made and we create a false aura of "hardness" about our data.[65] Logically speaking, however, perhaps the most serious difficulty is that the critical question is simply begged. If human values are defined by the cost-benefit theorist in terms merely of choice (measured quantitatively by price), then to claim that the uncontrolled market maximizes human welfare (measured in summed CVs) is to state a tautology. This judgment about the way costs and benefits should be allocated allegedly is rationally evaluated by criteria that antecedently presuppose acceptance of the market system. What is required is to *test* the thesis that market allocations constitute maximum welfare, or social justice, and not merely to assume it.[66]

If most of these points—about the assumptions built into, and the consequences following from, the three notions included in the concept of "compensating variation"—are reasonably correct, then several conclusions may be drawn. With respect to *validity*, it is clear that the CV concept might provide a measure of welfare, but only as defined within a cooperative, egoistic, hedonistic framework. Likewise, a CV might be a valid criterion of an individual's well-being, so long as his preferences are indicators of his authentic welfare. The validity of such a measure, however, is undercut by the fact that the CV is often a function of the individual's wealth. So far as *coverage* is concerned, the concept of summed CVs is somewhat adequate in providing a utilitarian estimate of societal well-being, but inadequate inasmuch as it takes no account of distributional inequities. The *comprehensibility* of the concept is also limited, because market prices (ordinarily used to measure CVs) do not take account of factors such as spillovers, the authentic costs of natural resources and public goods, the distorting effects of market instabilities and monopolies, and the whole range of phenomena referred to as "values." The *experimental utility* of compensating variation is perhaps its weakest point. Not only is it difficult to compute CVs for nonmarket items, but also the later summing of CVs (according to Pareto theory) is in practice impossible because of its complexity.

The constraints on the validity, coverage, comprehensibility, and experimental utility of the CV concept point to the fact that it cannot be said to be "value-free" or "objective," a claim Friedman and others have made on behalf of the whole of economics.[67] If this is the case, then it would be well to follow the suggestions of persons such as Marc Tool and Gunnar Myrdal, who argue that explicitly stated value premises and error estimates be included in economic inquiry.[68]

Level Two: Applications of Conceptual Analysis to Technology Assessment

As the preceding analysis suggests, using the concept of compensating variation to assess technological impacts is likely to result in an assessment in which

1. distributional effects of costs and benefits are ignored;
2. subjective feelings are often used as criteria for assessing well-being;

3. the individual is assumed to be the best judge both of his and of the public welfare; and
4. implicit acceptance of the market system is built into the calculations.

In this section I will examine how each of these four notions applies to several representative technology assessments. I will argue that each of them leads to a number of undesirable consequences:

1) to a) employment of value-laden interpretations of Pareto Optima and "compensating variations";
   b) advocacy of policy sanctioning violations of minority rights in order to serve the majority;
   c) neglect of negative effects of technological decisions on public policy and foreign affairs;
2) to a) support of assessment conclusions in which actual societal well-being often is not maximized;
   b) confusion of needs for a given technology with demands for it;
3) to a) CV-based discrimination against the poor;
   b) inconsistent valuations on the same items (e.g., human life) in similar situations;
4) to a) calculation of "value" in terms of "market price";
   b) failure to include assessment of the social costs of technology;
   c) support of assessment conclusions based on incomplete analysis of all costs and benefits;
   d) misrepresentation of the relative costs and benefits of various, often competing, technologies; and
   e) apparent assessment bias in favor of technology, industry, and maintaining the status quo.

As a first example, consider a recent OTA study described as an analysis of "the costs and potential economic, social, and environmental impacts of coal slurry pipelines."[69] Although they did not say so explicitly, the authors apparently employed the standard cost-benefit concept of Pareto Optimality obtained by summing CVs, in order to arrive at their conclusion that "slurry pipelines can, according to this analysis, transport coal more economically than other modes [of transport, e.g., railway]."[70] The methodologically puzzling aspect of this conclusion, however, is that although allegedly *all* costs and benefits of the proposed technology were considered, at least one key social cost was apparently not calculated, that of the distributional inequities caused by use of the pipelines. Clearly, one set of persons (those in the Western United States) would be negatively affected by slurry use of scarce water resources, while a quite different set of people (those in Midwestern and Eastern regions) would be positively affected by receiving Western coal. Ignoring both these facts, the assessors merely asserted that it was legal to impose disproportionate costs and benefits on different persons: "Constitutional power is adequate to do that ["to make water available for use in a coal slurry"], whether the source of the power is the inability of a state to thwart Federal policy . . . or the power of Congress"; the Federal government does not have to respect private rights regarding "allocation of navigable waters."[71]

Failure to include costs associated with disparate distributional effects of technological impacts also occurs in numerous other assessments, e.g., in studies of pesticides,[72] and in the only allegedly complete evaluation of nuclear fission technology.[73] Although these omissions are not surprising, they do suggest that the coal slurry, pesticide, and fission studies are not as value-free as their proponents claim. For one thing, all these reports sanction employment of utilitarian, rather than egalitarian, framework for calculating and evaluating costs and benefits.[74] This, in turn, means that the normative bias of the assessments is toward possible public policy violations of minority rights in order to serve the alleged good either of industry or of the majority.[75]

On a very practical level, the failure to include costs of distributional inequities in technology assessments also can have wide-reaching, often unsuspected, implications for foreign affairs. The authors of the recent Harvard Business School study of energy technologies point out that one interesting consequence of the failure to calculate the costs of distributional inequities arising from U.S. use of one-third of all oil consumed daily in the world, is increased pressure on the international oil market. Yet such pressure is not in the best interests of the U.S., because it wants Japan and the European nations to give up the breeder reactor. However, these countries clearly will be unable to forgo this option so long as the U.S. constricts the oil market, and so long as the foreign-policy consequences of its actions are not calculated as part of the cost of employing petroleum-based technologies.[76]

Besides failure to include costs of distributional inequities, other aspects of the economists' concept of "compensating variation" also apparently play a key role in how costs and benefits are calculated in technology assessments. One key assumption built into the concept is that the individual is the best judge of what policies will maximize his benefits and minimize his costs. Such a presupposition, however, does not always lead to actual optimization of individual welfare, as was noted earlier. To illustrate this point, consider the OTA assessment of private automobile technology. In this study, the authors noted the costs of air pollution and point out that, given the continuation of current trends (which they implicitly sanction), half of the U.S. population will be exposed to extremely hazardous levels of air pollution in the year 2000 as a result of employment of private automobiles.[77] They also discuss the rising costs of petroleum, highway congestion, owning and operating a car, and accidents.[78] Next, the costs and benefits are calculated, presumably as a sum of CVs, in order to determine what policy (e.g., promotion of private auto transport or mass transit) will maximize well-being. In the name of U.S. citizens, the assessors employ the notion that each individual's *feelings* are to be taken as the criterion of what actually maximizes his well-being. They claim implicitly that, on this basis, continued use of private auto transport is more desirable than increased employment of mass transit because of the high cost that persons place on more use of public transportation. Although no statistical studies were done, they maintain: "Americans have come to regard personal mobility as an inalienable right, and the automobile is viewed as the principal means to achieve this end."[79]

Because of the very information cited in the technology assessment, however, there is strong reason to believe that this subjective estimate of value—allegedly placed by individuals on personal, private-auto mobility—might not be accurate. This value might not outweigh the *costs*, in terms of pollution, resource depletion, congestion, accidents, etc. Nevertheless, so long as each individual's *feelings* (and one might question whether these feelings are the case) are taken as the criterion for the value placed on certain costs and benefits, then the quality of preferences is ignored, and authentic well-being cannot be distinguished from the satisfaction of wants.[80]

Of course, these problems should not be taken to mean that "big brother" should set the prices on the costs and benefits associated with private employment of autos. My point is, rather, that any technology assessor should note the *methodological difficulties* inherent in assuming that preference (real or apparent) is automatically a correct measure of well-being, and should not uncritically employ this notion built into standard economic methodology. Moreover, it is clear that there are some instances in which, problems of paternalism aside, individual preferences are not naïvely taken as indicators of authentic well-being. For a technology assessment of microwave ovens or CT scanners, for example,[81] it could be argued that persons' subjective feelings should not be used as measures of the costs of being subjected to unnecessary doses of ionizing radiation, especially if the individuals polled were ignorant of the hazards. Rather the technology assessor ought to use, for example, the standard BEIR report to obtain dose-response estimates of radiation-induced injury; then he could use these figures as a basis for computing the health costs associated with the injuries in question. Particularly in the case of technology assessments, it appears dangerous to use uncritically (i.e., without *caveats*) the CV notion that an individual's feelings are the best criterion for valuing risks affecting him, since many technology-related costs (e.g., pollution and its effects) are often not calculable in detail by the layman. Such caveats are almost wholly absent in contemporary technology assessments.

Another undesirable consequence (discussed earlier), following from acceptance of the CV notion that the individual is best able to affix prices to the costs and benefits affecting him, is that wealthy and poor individuals do not enjoy the same freedom to price their own costs and benefits as they would like. As a result, individual CVs are aggregated (for purposes of obtaining a Pareto Optimum), even though the conditions under which they were calculated are so different as to raise the question of whether they represent a homogenous parameter on the basis of which different technologies might be assessed. In this regard, the subjective value placed on the life of a coal miner is an especially interesting case in point. First, coal miners tend to place a lower dollar value on their own lives than do members of other occupational groups.[82] And second, the prices affixed by others to the coal miners' lives are of critical importance, because these values are used in calculating relative costs and benefits of alternative technologies for generating electricity.[83] Any discrepancies in valuation could easily affect the outcome of the comparative assessment. For example, if a high-technology mode of generating electricity (e.g., nuclear power) tended to employ more skilled workers, and if they lived under fewer economic constraints than persons less skilled, those in the former group would probably place a higher price on their lives. As a consequence, subjectively valued costs of occupation-related deaths and injuries for this technology might be calculated to be far greater, for equal numbers of employees, than would similar costs for a low-technology means of generating electricity (e.g., coal). This suggests not only that problems of equity (e.g., why should one's wealth determine the value of his life?) arise in comparative assessments of alternative technologies, but also that subjective notions of valuation (based on the definition of CV) could lead to questionable conclusions as to which technology in fact maximizes benefits and minimizes costs.[84]

The most serious problems besetting the CV concept appear to arise, however, not from subjective valuation, but from employment of the market system to calculate costs and benefits. As a consequence, the CV concept becomes enmeshed in all the problems associated with the assumption that market price equals value. Although the U.S. "has applied various assessment mechanisms," it has "relied primarily on the 'market' system for guiding and shaping the nature of new technological applications."[85] A recent assessment of railroads, for example, notes that the costs ascribed to various risks of the technology (e.g., derailments) are judged as acceptable or not "through traditional marketplace operations."[86] This means that, although one of the five major "concerns" of the OTA study of railroad technology was "data collection,"[87] no social costs, or externalities, of implementation of the technology were included in the cost-benefit calculations, since the market system excludes consideration of social costs. This failure to calculate externalities (e.g., approximately 4,000 major evacuations per year because of accidents involving hazardous materials)[88] in computing the costs of railroad technology is a significant methodological problem for a number of reasons. Between 1.04 and 2.5 million carloads of hazardous materials are shipped annually, and roughly 65% of *all* tank cars loaded with liquefied petroleum gas, sulfuric acid, anhydrous ammonia, and liquid caustic soda are involved annually in the release of hazardous materials.[89] Moreover, approximately twice as many serious accidents per train mile occur on U.S. railroads as on Canadian railroads. The U.S. has developed no accident or risk data to estimate either the social costs of these accidents or how to avoid them; the Canadians have done both.[90] In the light of these facts, it is surprising that the most comprehensive and recent U.S. government assessment of railroad technology does *not:* (1) calculate social costs; (2) note the probable effects of failing to do so; (3) recommend that they be computed in the future; or (4) acknowledge the methodological difficulties in computing rail technology costs only on the basis of market considerations. Despite these shortcomings, the report concludes that no new laws or regulations are needed to solve the problem of the high accident rate (which has doubled, per train mile, in the last ten years) of U.S. railroads.[91]

What is disturbing here is not only that social costs are ignored, but also that the assessors seem not to see that their omission presents a great methodological difficulty with potentially serious effects on both public policy and human well-being. Most importantly, if the complete costs (including externalities) of employing railroad technology, according to existing U.S. laws and regulations, are not *known*, how could the authors conclude that no new laws or regulations were needed to solve the accident problem? This suggests that the methodology implicitly contributes to a bias toward both the status quo (no new laws or regulations) and the railroad industry, as opposed to the public.

The same pattern of ignoring externalities occurs in other technology assessments—for example, in recent OTA studies of private auto transport,[92] pesticide technology,[93] and coal slurry technology.[94] In the few assessments in which spillovers or social costs are considered, there appears to be less bias in favor of the status quo, fewer doubtful economic conclusions regarding costs and benefits, and less prejudice toward acceptance of the costs of the technology in question. An OTA study of the direct use of coal, for example, analyzed the relevant economic, technological, and social costs and benefits, including the aesthetic impacts,[95] and then drew a conclusion favorable to direct use of coal. The conclusion, however, was couched in the context of warnings about the admitted uncertainties involved in "external costs, institutional and social constraints, and other nonmarket factors associated with coal use."[96] Much of the same methodological caution, also apparently contributing to less obviously biased conclusions, is likewise exercised in a recent OTA analysis of solar technology.[97]

In addition to neglect of externalities, another consequence of using (market-based) CVs in technology assessments is that the calculations do not include the distorting effects of monopolies, subsidies, and market imperfections. In one famous analysis (WASH-1224) of the relative costs and benefits of using coal (as compared to nuclear) technology for generation of electricity, the effects of more than $100 billion in government subsidies of nuclear technology were ignored in the computations. As a consequence of this omission, the report concluded that nuclear-generated electricity was more cost-effective than coal-generated power. If mere market prices had not been used as the basis of calculations, and if the prices of both means of generating electricity had included the nonmarket costs of government subsidies, then the cost-benefit analysis would have reached the opposite conclusion, i.e., that coal-generated electricity is more cost-effective than nuclear-generated power.[98]

Besides threatening the validity of one's conclusions regarding the economics of various technologies, market-based methodological assumptions also contribute toward an apparent normative bias in policy formation, as the railroad technology study illustrated (see above). The disastrous consequences of this normative bias are perhaps best exemplified by U.S. cost-benefit calculations which have ignored implicit subsidies to petroleum-based technologies. Stobaugh and Yergin, authors of the famous Harvard Business School assessment of various energy technologies, explain why this is so. They point out that in early 1979 the average market value of U.S.-produced oil, because it was kept down by government controls (maintaining prices far below replacement costs), was $9 a barrel, while the world market price for oil delivered to the U.S. was $15 per barrel. If all market distortions, externalities, subsidies, and effects of monopoly are taken into account, say Stobaugh and Yergin, the *real* cost of oil in early 1979 was approximately $35 per barrel.[99]

Misleading econometric assumptions, such as the authenticity of market price, have "helped to create some of the impasses and stalemates in U.S. energy policy."[100] These include excessive promotion of the use of petroleum (between 1973 and 1979, U.S. oil imports doubled);[101] a weakening of the dollar and increased likelihood of economic collapse;[102] and a failure to provide incentives for conservation and for use of solar technology.[103] Hence, because of both the need to follow sound economic methodology and the practical, social-political consequences of applications of this scientific methodology, it makes sense for assessors of technology either to admit explicitly (which they almost never do) the methodological problems surrounding their use of the CV concept, or to examine alternative methods.[104] If this were done, they would be more likely to avoid invalid conclusions based on incomplete cost data and to note explicitly the cost-benefit *assumptions* on which their conclusions were contingent. They also might thus avoid both begging the questions they were attempting to analyze and drawing apparently biased (pro technology, pro the status quo) conclusions on the basis of methodologically questionable premises.

Potentials and Limits of Applying Philosophy of Science in Technology Assessment

If these reflections on the applications of conceptual and methodological analysis to technology assessment suggest one central proposition, it is an unsurprising one: doing cost-benefit analysis unavoidably involves making value judgments. These judgments occur implicitly, by virtue of factors such as employment of a market-based concept of compensating variation, as well as explicitly, as a result of sanctioning policy consequences following from scientifically questionable methodological principles.[105] These judgments are significant since, contrary to what philosophers of science might suspect, most economists apparently think they "have limited themselves to what they believed was a purely objective position."[106]

If it is the case that economists cannot escape normative conclusions, then one of the potential values of applied philosophy of science is that it might help economists avoid, not the tendency to make such judgments, but rather the tendency to be unaware that they are doing so. Moreover, philosophers of science are often in a position (theoretically, at least) to be more objective about a given technology than are the scientists and engineers "in the employ of bureaus or companies that are interested in furthering technology"; yet these are the people who usually perform assessments under government contract.[107]

The limits to the work of a philosopher of science in such applied areas, however, are numerous. Some of these constraints arise from the disanalogies between methodological analysis of science and methodological analysis of technology assessments. Doing straightforward philosophy of science, for example, usually does not require much expertise in ethics and, in fact, has been explicitly defined as *excluding* ethical evaluation of scientific conclusions and their applications.[108] Philosophical analyses of technological studies, on the other hand, are frequently bound up with such evaluations; this is in part because technology irrevocably involves issues of application (and, therefore, questions about human actions) and issues of policy (and, therefore, questions of oughtness). Since the philosopher of science often is not trained in ethical analysis, he may be limited in this area of technology assessment.

Philosophy of science also proceeds according to relatively clearer criteria for the analysis of science (e.g., simplicity, predictive power) than the criteria used for evaluating technology. This appears to be in part a result of the fact that questions of technological applications and policy appear to have many more facets (legal, political, social) than issues of scientific theory *per se.* Conversely, a technology assessment often is more limited, in a practical sense, than is a methodological analysis of science; the OTA philosophy regarding the scope of their technology evaluations is that the reports should consider all those and only those factors relevant to the specific type of assessment request issuing from agencies, U.S. senators, or congressmen.[109] Philosophers of science are usually not forced to operate within such investigative limits, unless they, too, are working under particular funding constraints.

Issues in the philosophy of science are also not as highly politicized as are those in technology assessment, probably because the philosophical issues do not often involve marketplace consequences of industrial gain and loss. Because of this politicization, doing applied philosophy of science is (at least in theory) more problematic and professionally hazardous than doing typical theoretical work in the discipline. My own experiences on technology assessment teams have been confirmed by several fellow scientists and philosophers of science: If one's conclusions appear to threaten particular vested interests, then he or she can sometimes expect political, professional, personal, and financial pressures to be brought to bear. This situation is not unique, of course, to the *philosopher* who does technology assessment. Nevertheless, it does indicate the existence of problems not normally encountered in doing theoretical philosophy of science. If the risks are higher, however, perhaps the rewards are too. No small pleasure often accompanies the attempt to give birth to one's "Ralph Nader" suspicions through the respected avenues of the Socratic gadfly.

Notes

1. L.H. Mayo, "The Management of Technology Assessment," in R.G. Kasper, ed., *Technology Assessment: Understanding the Social Consequences of Technological Applications* (New York: Praeger, 1972), p. 73.

2. U.S. Congress, Office of Technology Assessment (OTA), *Annual Report to the Congress for 1978* (Washington, DC: U.S. Government Printing Office, 1978), p. 74.

3. U.S. Congress, OTA, *Technology Assessment in Business and Government* (Washington, DC: U.S. Government Printing Office, 1977), p. 13. This same point is made in numerous analyses; for example, in U.S. Congress, OTA, *Annual Report to the Congress for 1976* (Washington, DC: U.S. Government Printing Office, 1976), pp. 66–67. See William Fisher, Assistant Secretary for Energy and Minerals, U.S. Department of the Interior, in U.S. Congress, *Technology Assessment Activities in the Industrial, Academic, and Governmental Communities*, Hearings Before the Technology Assessment Board of the Office of Technology Assessment, 94th Congress, Second Session, June 8–10 and 14, 1976 (Washington, DC: U.S. Government Printing Office, 1976), p. 27. See also R. G. Kasper, *op. cit.* (Note 1), p. 4; and J.R. Ravetz, *Scientific Knowledge and Its Social Problems* (Oxford: Clarendon Press, 1971), pp. 369–70. For information on the use of cost-benefit analysis in technology assessment, see Notes 14, 15, 20, 69, 70, 72, 73, 78, 82, 83, 92–96, 98 below.

4. R.C. Dorf, *Technology, Society, and Man* (San Francisco, CA: Boyd and Fraser, 1974), p. 20. See Don Idhe, *Technics and Praxis* (Boston, MA: D. Reidel, 1979), pp. xvii–xxvi, 109ff., for a critique of the "paradigm" that technology is applied science.

5. Ravetz, *op. cit.* (Note 3), pp. 361, 426–32, makes this same point about purpose being central to the evaluation of a technology. Because of its focus on purpose (versus truth), technology is not simply "applied science." A study of the origins of 84 British technological innovations revealed, for example, that "demand pull," for some industrial or governmental purpose, occurred much more frequently than "discovery push" from theoretical science [J. Languish, "The Changing Relationship Between Science and Technology," 250 *Nature* 5468 (August 23, 1974): 614]. See M.D. Reagan, *Science and the Federal Patron* (New York: Oxford University Press, 1969), p. 9.

6. Even though it has provided no generally accepted criteria for which way technology assessments "ought to go," OTA has been guided by the Technology Assessment Act of 1972. The act authorized it to provide "competent, unbiased information concerning the physical, biological, economic, social, and political effects" of applications of technology.

U.S. Congress, OTA, *Annual Report 1978, op. cit.* (Note 2), p. 113. Technology assessments are widely held to be (a) "unbiased"; (b) outside the realm of "policy" or value judgments [U.S. Congress, OTA, *Technology Assessment Activities, op. cit.* (Note 3), pp. 200, 220; U.S. Congress, OTA, *Annual Report 1976, op. cit.* (Note 3), p. 66]; and (c) "nonpartisan" and "objective" [U.S. Congress, OTA, *Annual Report to the Congress for 1977* (Washington, DC: U.S. Government Printing Office, 1977), p. 4; U.S. Congress, OTA, *Annual Report 1978, op. cit.* (Note 2), p. 7; U.S. Congress, OTA, *Technology Assessment in Business, op. cit.* (Note 3), p. 9].

I will say more later about this alleged neutrality and objectivity. For further information on technology assessment, see R.A. Bauer, *Second-Order Consequences: A Methodological Essay on the Impact of Technology* (Cambridge, MA: The MIT Press, 1969), pp. vii–ix. See also R.A. Carpenter, "Technology Assessment and the Congress," in Kasper, *op. cit.* (Note 1), Robert Coburn, "Technology Assessment, Human Good, and Freedom," in K.E. Goodpaster and K.M. Sayre, eds., *Ethics and the Problems of the 21st Century* (Notre Dame, IN: University of Notre Dame Press, 1979), p. 106. See also U.S. Congress, OTA, *Technology Assessment Activities, op. cit.* (Note 3), pp. 27–28; Dorf, *op. cit.* (Note 4), pp. 1ff., 241–61, 289–95, 307, 387; Ravetz, *op. cit.* (Note 3), pp. 369–401.

7. E.Q. Daddario, "Foreword," in Bauer, *op. cit.* (Note 6), p. vi.

8. These two levels follow the three stages outlined by Goodpaster and Sayre, *op. cit.* (Note 6), pp. vii–viii. They describe the areas in which moral philosophers may make a contribution to social action. There are several reasons why the methodological analyses characteristic of philosophy of science are particularly needed in technology assessment.

For one thing, "the conceptual impoverishment undergone by scientific theory when used as a means for practical ends can be frightful" [Mario Bunge, "Towards a Philosophy of Technology," in A.C. Michalos, ed., *Philosophical Problems of Science and Technology* (Boston, MA: Allyn and Bacon, 1974), p. 32]. This means that, because technological studies are more guilty of "oversimplification" and "superficiality" (*ibid.*, p. 32), the methodological analyses of them are needed all the more.

Both governmental and industrial policymakers admit that the biggest problem with such studies is "the lack of conceptual thinking on the technology assessment function, including alternative notions of adequacy of assessment" [L.H. Mayo, "The Management of Technology Assessment," in Kasper, *op. cit.* (Note 1), p. 99]. See also U.S. Congress, OTA, *Technology Assessment Activities, op. cit.* (Note 3), p. 229.

The difficulty, they say, is not what "research must be done, but how to do it" in terms of "methodological requirements" [U.S. Congress, OTA, *A Review of the US Environmental Protection Agency Environmental Research Outlook: FY 1976 through 1980* (Washington, DC: U.S. Government Printing Office, 1976), pp. 8, 53, 97–98].

Also, there is a great tendency to perform the assessments "as though they are engineering studies" [E.F. Huddle, "The Social Function of Technology Assessment," in Kasper, *op. cit.* (Note 1), p. 170].

Assessment teams composed only of engineers and natural scientists (and not also social scientists and philosophers of science) often attempt either to deal with issues outside their areas of professional expertise, or to ignore them. As a consequence, technological problems are frequently defined in terms "too narrow to do justice to the complexity of real situations" [Dorothy Nelkin, *The Politics of Housing Innovation* (Ithaca, NY: Cornell University Press, 1971), p. 88].

Hence it is not surprising that one government spokesperson commented: "the experimental skills of scientists and engineers do not necessarily coincide with those [conceptual, methodological] skills necessary for identifying and analyzing the consequences of new or existing technologies" [C.H. Danhof, "Assessment Information Systems," in Kasper, *op. cit.* (Note 1), p. 23]. This is, in part, because "the higher the policy level [discussion of technology], the less strictly technical are the factors that have to be taken into account" [Reagan, *op. cit.* (Note 5), p. 97].

9. See K.S. Shrader-Frechette, *Nuclear Power and Public Policy: Social and Ethical Problems of Fission Technology* (Boston, MA: D. Reidel, 1980), pp. 83–85.

10. See J.J. Stukel and B.R. Kennan, *Ohio River Basin Energy Study*, I-A (Washington, DC: U.S. Environmental Protection Agency, November 1977), pp. 12–15.

As Koreisha and Stobaugh have pointed out, use of "exclusion" in technology assessments often takes the form of excluding analyses of social-science-related parameters and concentrating on purely technical ones. An excellent example of use of the principle is found in the MIT Energy Self-Sufficiency Study. Here, say Koreisha and Stobaugh, the authors admitted that politics was a crucial factor in assessing feasible energy technologies, and that problems such as nuclear safety and environmental protection were obviously affected by politics. After this admission, however, the group simply stated: "Such issues, though both appropriate and important to the debate . . . are beyond the scope of this report" [Sergio Koreisha and Robert Stobaugh, in "Appendix: Limits to Models," in Robert Stobaugh and Daniel Yergin, eds., *Energy Future: Report of the Energy Project at the Harvard Business School* (New York: Random House, 1979), pp. 263–64].

11. When appeal is made to "exclusion," one attempts to justify his assumption that the influence of any factor not included in the model is unimportant in affecting the conclusions. For discussion of exclusion, see Koreisha and Stobaugh, *ibid.,* pp. 237–38.

12. In this regard, see Shrader-Frechette, "The Ohio River Basin Energy Study: Methodological and Ethical Problems," in J.J. Stukel and B.R. Keenan, *Ohio River Basin Energy Study*, IV (Washington, DC: U.S. Environmental Protection Agency, 1978), pp. 50ff.

13. One way to determine the limiting assumptions of a particular model, for example, is by testing the validity of the predictions. This might be done by discovering what happens when specific changes are made in key parameters, and then by determining whether the results are plausible or not. An implausible change could indicate faulty assumptions or incorrect logical relationships in the model. See Koreisha and Stobaugh, "Appendix," in Stobaugh and Yergin, *op. cit.* (Note 10), pp. 234–65.

14. Koreisha and Stobaugh, *ibid.*, p. 234. See also E.F. Schumacher, *Small Is Beautiful: Economics As If People Mattered* (New York: Harper & Row, 1973), p. 38.

15. John Kenneth Galbraith, *The New Industrial State* (Boston, MA: Houghton Mifflin, 1967), p. 408. See also E.J. Mishan, *Welfare Economics* (New York: Random House, 1969), p. 5; and Ravetz, *op. cit.* (Note 3), p. 396.

16. Theodore Roszak, "Introduction," to Schumacher, *op. cit.* (Note 14). Carl Kaysen says that "economics has clearly been the most successful of the social sciences" ["The Business Corporation as a Creator of Values," in Sidney Hook, ed., *Human Values and Economic Policy* (New York: New York University Press, 1967), p. 209].

17. Hook, *ibid.*, p. ix. See also Kenneth Boulding, *Economics as a Science* (New York: McGraw-Hill, 1970), p. 117; and A.L. Macfie, "Welfare in Economic Theory," 3 *The Philosophical Quarterly* 10 (January 1953): 59.

18. Oskar Morgenstern, *On the Accuracy of Economic Observations* (Princeton, NJ: Princeton University Press, 1963), p. 50.

19. E.J. Mishan, *Cost-Benefit Analysis* (New York: Praeger, 1976), p. 445. See also M.W. Jones-Lee, *The Value of Life: An Economic Analysis* (Chicago, IL: University of Chicago Press, 1976), p. 1. Together with allocation theory, cost-benefit analysis comprises what is generally known as "welfare economics," a "study which endeavors to formulate propositions by which we may rank on the scale of better or worse, alternative economic situations open to society" [Mishan, *Cost-Benefit Analysis*, p. 382]. See also pp. 386–89 and Mishan, *op. cit.* (Note 15), pp. 13 and 15.

20. By virtue of the fact that, at least since 1939, they have "grant[ed] a monopoly" to the Pareto notion, most welfare economists and cost-benefit theorists rank alternative economic situations by means of appeal to this concept [S.S. Alexander, "Human Values and Economists' Values," in Hook, *op. cit.* (Note 16), p. 108, and Mishan, *op. cit.* (Note 15), p. 22]. Since most technology assessments are based upon cost-benefit analyses (see Note 3), they also employ the Pareto concept. Its use, however, raises an interesting epistemological issue. (1) Can welfare economists be said *not* to be using the notions of Pareto Optimum and "compensating variation," since they do not include *all* cost-benefit parameters in their calculations? or (2) may they be said to employ modified versions of these two concepts, since they are not practically usable as defined in economic theory? Whether either (1) or (2), or neither, is the case will not substantially affect the discussion here. Although most economists would probably agree with point (2), the purpose of examining the Pareto concept is to uncover the methodological notions built into it, since, by whatever name these notions are called, they are clearly a central part of most cost-benefit analyses and technology assessments. My goal is to evaluate the desirability of using these methodological notions, and not necessarily to determine their status as Pareto-based or not.

21. Mishan, *op. cit.* (Note 15), p. 22; see also pp. 22–30.

22. Mishan, *op. cit.* (Note 19), p. 391. See also Jones-Lee, *op. cit.* (Note 19), pp. 6–14 and Mishan, *op. cit.* (Note 15), pp. 227–30.

23. Mishan, *op. cit.* (Note 19), p. 391; see also pp. 390–402. Jones-Lee, *op. cit.* (Note 19), p. 5.

24. Mishan, *op. cit.* (Note 15), p. 113; see also pp. 107–113.

25. Mishan, *op. cit.* (Note 19), p. 390; see also Notes 21–23.

26. See Note 22. Mishan [*op. cit.* (Note 19), p. 309] specifically affirms this aspect of compensating variation.

27. See Notes 22 and 24.

28. Jones-Lee [*op. cit.* (Note19), p. 3] and Robert Coburn [in Goodpaster and Sayre, *op. cit.* (Note 6), p. 109] also point out the same difficulty.

29. Morgenstern, *op. cit.* (Note 18), pp. 100–101.

30. *Ibid.,* p. 101.

31. Mishan, *op. cit.* (Note 19), p. 392.

32. Mishan makes a similar point about the poor [*Ibid.,* p. 393].

33. A.D. Biderman, "Social Indicators and Goals," in R.A. Bauer, ed., *Social Indicators* (Cambridge, MA: MIT Press, 1966), pp. 131–32.

34. In the preceding example of an economic change causing one set of persons to be better off by a total of $10x$ dollars, at the expense of another set who were made worse off by a total of x dollars, a member of the second group might not "feel" as well off as a member of the first, particularly if he knew what the relative gains and losses of the two sets were. This suggests that Pareto-induced distributional inequities probably increase in proportion to the ignorance of members of the disadvantaged set regarding their relative gains and losses.

35. Cited in V.C. Walsh, "Axiomatic Choice Theory and Values," in Hook, *op. cit.* (Note 16), p. 197. One of the great classics in the last 25 years of debate on welfare economics is K. J. Arrow's *Social Choice and Individual Values.*

36. See Milton Friedman, "Value Judgments in Economics," in Hook, *op. cit.* (Note 16), pp. 85–88, and E.C. Pasour, "Benevolence and the Market," 24, *Modern Age* 2 (Spring 1980): 168–70. For criticisms of this claim, see Michael Freeden, "Introduction," in J.A. Hobson, *Confessions of an Economic Heretic* (Sussex, England: Harvester Press, 1976), p. vi. See also Boulding, *op. cit.* (Note 17), p. 119.

37. S.S. Alexander, "Human Values and Economists' Values," in Hook, *op. cit.* (Note 16), p. 108.

38. Coburn in Goodpaster and Sayre, *op. cit.* (Note 6), pp. 109–10, makes this same point.

39. Gail Kennedy ["Social Choice and Policy Formation" in Hook, *op. cit.* (Note 16), p. 142] makes a similar observation, as does also John Ladd ["The Use of Mechanical Models for the Solution of Ethical Problems" in Hook, *op. cit.,* pp. 167–68].

40. This inadequacy is mentioned by M.A. Lutz and K. Lux, *The Challenge of Humanistic Economics* (London: Benjamin/Cummings, 1979), p. 4. They say it is characteristic of "conventional, mainstream economics." John Ladd [in Hook, *op. cit.* (Note 16), p. 168] argues the same position as Lutz and Lux.

41. R.B. Brandt and John Ladd also make this point [See their essays in Hook, *op. cit.* (Note 16), p. 37 and p. 159, respectively].

42. A similar observation is made by Gail Kennedy, "Social Choice and Policy Formation," in Hook, *op. cit.* (Note 16), p. 148.

43. Economists have answered this charge by appeal to the notion of "rational preference *ordering*," as a substitute for relying merely on an individual's feelings. This approach, nevertheless, has epistemological drawbacks. How could one obtain an ordering except through a questionnaire? And if one is unclear in the beginning about how to maximize his welfare, then how could a questionnaire remove these difficulties? This "solution" seems to serve only to transfer the lack of clarity from the relationship between "preference" and "welfare" to the relationship between "rational ordering" and "welfare." Just as one's preferences are no infallible guide to maximizing one's well-being, so also what is "rational" according to the economists' sense of CV is not necessarily "rational" in the best, or even ordinary, sense of the term. [A related point is made by Coburn in Goodpaster and Sayre, *op. cit.* (Note 6), p. 111; and by R.B. Brandt, "Personal Values and the Justification of Institutions," in Hook, *op. cit.* (Note 16), p. 31.]

44. B.A. Emmett, *et al.*, "The Distribution of Environmental Quality: Some Canadian Evidence," in D.F. Burkhardt and W.H. Ittelson, eds. *Environmental Assessment of Socioeconomic Systems* (New York: Plenum, 1978), pp. 367–71, 374.

45. P.S. Albin, "Economic Values and the Value of Human Life," in Hook, *op. cit.* (Note 16), p. 97; Jones-Lee, *op. cit.* (Note 19), pp. 20–55.

46. Mishan, *op. cit.* (Note 19), p. 318.

47. Jones-Lee, *op. cit.* (Note 19), p. 39; see also p. 72. Mishan, *op. cit.* (Note 19), pp. 319–20.

48. Mishan, *op. cit.* (Note 19), p. 318 and pp. 319–20. Jones-Lee, *op. cit.* (Note 19), p. 72. Some theorists say this problem of subjectivity could be solved by using an average value, computed from the CVs of both wealthy and poor individuals. This solution, however, is also arguably discriminatory. Obviously, a corporate executive, making $300,000 per year, for example, could argue that if his CV were lowered through the averaging process, then the new value would not represent an accurate economic compensation for him. This is probably why nearly all cost-benefit theorists insist on the notion of subjective (rather than average) determination of an individual's compensating variation.

49. See M.C. Tool, *The Discretionary Economy: A Normative Theory of Political Economy* (Santa Monica, CA: Goodyear, 1979), esp. pp. 208, 320–24, 334. See also John Rawls, *A Theory of Justice* (Cambridge, MA: Harvard University Press, 1973), pp. 14–5, 100–114, 342–50; and Shrader-Frechette, *op.cit.* (Note 9), pp. 31–35 and 122–23.

50. See Pasour, *op. cit.* (Note 36), pp. 168–78.

51. See Note 52.

52. For a discussion of "aggregation," see Koreisha and Stobaugh in Stobaugh and Yergin, *op. cit.* (Note 10), pp. 238–65.

53. See Note 27. As Mishan [*op. cit.* (Note 19), pp. 391–92] puts it, "the value of goods and 'bads' to men, either as consumers or producers (factory owners), is not worked out from scratch. Market prices can be deemed to provide these values in the first instance, following which they can be corrected for 'market failure'."

54. Although Aristotle distinguished between the "fair price" (value) and the market price, in the last century the distinction has been abandoned, and economics has moved from a normative to a positive (in Comte's sense) emphasis. As one theorist put it, to the extent that there is a valid distinction between the price and the value, to that degree contemporary economics "could well use a modern Aristotle" [Adolf Lowe, "The Normative Roots of Economic Values," in Hook, *op. cit.* (Note 16), pp. 180 and 171–73].

55. Hobson, *op. cit.* (Note 36), pp. 39–40, and B.M. Anderson, *Social Value: A Study in Economic Theory Critical and Constructive* (New York: A.M. Kelley, 1966), pp. 26, 31, 162.

56. See K.E. Boulding, "The Basis of Value Judgments in Economics," in Hook, *op.cit.* (Note 16), pp. 67–69. See also Anderson, *op. cit.* (Note 55), p. 24.

57. See the Boulding essay, *ibid.* Morgenstern [*op. cit.* (Note 18), p. 19] ties the "errors of economic statistics," such as price, in part to the fact of the prevalence of monopolies. In an economy characterized by monopoly, statistics regarding price are not trustworthy because of "secret rebates granted to different customers." Moreover, "sales prices constitute some of the most closely guarded secrets in many businesses." For both these reasons it is likely that price does not equal value, and that actual price does not equal official market price.

58. See Dorf, *op. cit.* (Note 4), pp. 223–40, and H.R. Bowen, Chairman, National Commission on Technology, Automation, and Economic Progress, *Applying Technology to Unmet Needs* (Washington, DC: U.S. Government Printing Office, 1966), pp. v–138. See also Boulding in Hook, *op. cit.* (Note 16), pp. 67–68, and Mishan, *op.cit.* (Note 19), pp. 393–94.

Externalities (also known as "spillovers," "diseconomies," or "disamenities") are social benefits or costs (e.g., the cost of factory pollution to homeowners nearby) which are "external" to cost-benefit calculation, and hence do not enter the calculation of the market price. For this reason, says Mishan [*The Costs of Economic Growth* (New York: Praeger, 1967), p. 53], "one can no longer take it for granted that the market price of a good is an index of its marginal price to society." Another way of making this same point [*ibid.*, p. 57] is to say that diseconomies cause social marginal costs of some goods to exceed their corresponding private marginal costs; this means that the social *value* of some goods is significantly less than the (private) market *price*.

59. See Boulding, in Hook, *op. cit.* (Note 16), pp. 67–68, and Schumacher, *op. cit.* (Note 14), pp. 38–49.

60. There are no monetary-term values for natural resources because the "cost" of using natural resources is measured in terms of low entropy and is subject to the limitations imposed by natural laws (e.g., the finite nature of nonrenewable resources). For this reason, the price mechanism is unable to offset any shortages of land, energy, or materials. To assume otherwise is to commit the fallacy of "entropy bootlegging." [Nicholas Georgescu-Roegen, *Energy and Economic Myths: Institutional and Analytical Economic Essays* (New York: Pergamon Press, 1976), pp. xv, 10, 14–15]. See also Schumacher, *op. cit.* (Note 14), pp. 41–49; Emmett, in Burkhardt and Ittelson, *op. cit.* (Note 44), p. 363; and Bauer, *op. cit.* (Note 6), p. 54.

61. Mishan, *op. cit.* (Note 58), p. 49. See also Tool, *op. cit.* (Note 49), pp. 280–85.

62. See Pasour, *op. cit.* (Note 36), pp. 168–78.

63. Morgenstern, *op. cit.* (Note 18), p. 26.

64. This particular notion of "specious accuracy" has been developed by Morgenstern, *ibid.*, p. 62.

65. This insight is discussed by R.A. Bauer, "Detection and Anticipation of Impact: The Nature of the Task,"in Bauer, *op. cit.* (Note 33), p. 46. See also the Biderman essay in the same collection (p. 97).

66. Tool, *op. cit.* (Note 49), p. 334, and Hook, "Basic Values and Economic Policy," in Hook, *op. cit.* (Note 16), p. 247.

67. See Pasour, *op. cit.* (Note 36), p. 168, and Friedman, "Value Judgments in Economics," in Hook, *op. cit.* (Note 16), pp. 85–88.

68. See Tool, *op. cit.* (Note 49), p. xvi. One eminent expert in methodology believes that "economics has a long way to go before it will be ready" to admit the existence of these value premises and error components [Morgenstern, *op. cit.* (Note 18), pp. 60–61, vii, and 7].

69. U.S. Congress, OTA, *A Technology Assessment of Coal Slurry Pipelines* (Washington, DC: U.S. Government Printing Office, 1978), p. v.

70. *Ibid.*, p. 15.

71. *Ibid.*, pp. 131–32.

72. For discussion of this point, see Shrader-Frechette, *Environmental Ethics* (Pacific Grove, CA: Boxwood Press, 1980), Ch. 11.

73. For an analysis of distributional inequities in cost-benefit studies of nuclear technology, see Shrader-Frechette, *op. cit.* (Note 9).

74. For discussion of the utilitarian versus egalitarian frameworks for calculating and assessing costs and benefits of technology, see Shrader-Frechette, *op. cit.* (Note 9). See also John Stuart Mill, *Utilitarianism, Liberty, and Representative Government* (New York: Dutton, 1910), pp. 6–24; Jeremy Bentham, *The Utilitarians: An Introduction to the Principles of Morals and Legislation* (Garden City, NY: Doubleday, 1961), pp. 17–22; J.J.C. Smart, "An Outline of a System of Utilitarian Ethics," in *Utilitarianism: For and Against*, J.J.C. Smart and B. Williams, eds. (Cambridge, England: Cambridge University Press, 1973), pp. 3–74; Rawls, *op. cit.* (Note 49), pp. 14–15; Charles Fried, *An Anatomy of Values* (Cambridge, MA: Harvard University Press, 1970), pp. 42–43; Charles Fried, *Right and Wrong* (Cambridge, MA: Harvard University Press, 1978), pp. 116–17, 126–27; Alan Donagan, *The Theory of Morality* (Chicago, IL: University of Chicago Press, 1977), pp. 221–39; and Alasdair MacIntyre, "Utilitarianism and Cost-Benefit Analysis: An Essay on the Relevance of Moral Philosophy to Bureaucratic Theory," in *Values in the Electric Power Industry*, K.M. Sayre, ed. (Notre Dame, IN: University of Notre Dame Press, 1977), pp. 217–37.

75. See Note 74, esp. Shrader-Frechette, *op. cit.* (Note 9), pp. 31–35.

76. Daniel Yergin, "Conservation: The Key Energy Source," in Stobaugh and Yergin, *op. cit.* (Note 10), p. 137.

77. U.S. Congress, OTA, *Technology Assessment of Changes in the Future Use and Characteristics of the Automobile Transportation System*, 2 volumes (Washington, DC: U.S. Government Printing Office, 1979). See Volume I, p. 16.

78. *Ibid.*, Volume I, p. 31, for example.

79. *Ibid.*, Volume II, p. 25.

80. Lutz and Lux [*op. cit.* (Note 40), p. 4] note: "Conventional, mainstream economics has not been able to adequately deal with values because it has not seen economics in terms of needs. . . . what *has* been its focus? The answer is *wants.*"

81. See Congress, OTA, *Policy Implications of the Computed Tomography (CT) Scanner* (Washington, DC: U.S. Government Printing Office, 1978).

82. See Jones-Lee [*op. cit.* (Note 19), p. 39) for a table on values ascribed to human life.

83. See, for example, U.S. Atomic Energy Commission, *Comparative Risk-Cost-Benefit Study of Alternative Sources of Electrical Energy*, WASH-1224, (Washington, DC: U.S. Government Printing Office, 1974).

84. See Jones-Lee, *op. cit.* (Note 19), pp. 150ff.; and Mishan, *op. cit.* (Note 19), pp. 403–406.

85. Mayo in Kasper, *op. cit.* (Note 1), p. 78.

86. U.S. Congress, OTA, *An Evaluation of Railroad Safety* (Washington, DC: U.S. Government Printing Office, 1978), p. 37.

87. *Ibid.*, p. 156.

88. *Ibid.*, pp. 14 and 141.

89. *Ibid.*

90. U.S. Congress, OTA, *Railroad Safety–U.S.–Canadian Comparison* (Washington, DC: U.S. Government Printing Office, 1979), pp. vii–viii.

91. U.S. Congress, OTA, *op.cit.* (Note 86), p. xi.

92. Certain social costs, or externalities, of the technology were mentioned, pollution, noise, community disruption, death and injury, for example [U.S. Congress, OTA, *op. cit.* (Note 77), Volume II, pp. 75, 295, esp. p. 251; and Volume I, pp. 15–16]. According to Bowen [*op. cit.* (Note 58), pp. v–138] the most frequently used cost for air pollution, in terms of damage to property [but *excluding* health costs, absence from work, etc., which "constitute the most significant economic loss of all"), is $65 per capita per year. This is an annual cost of more than $12 billion in the U.S., and it still excludes the most important monetary losses.

Calculation of the value of these parameters was neither made nor included in the overall cost-benefit assessment of the technology, however. In spite of this failure, the authors of the report concluded both that extensive use of mass transit was not cost-effective, and that considerations of Americans' desire for personal mobility, via the auto, outweighed the costs incurred by a massive system of private transport. [U.S. Congress, OTA, *op. cit.* (Note 77), Volume II, p. 228]. This suggests that, because the cost-benefit analysis presupposed the methodological validity of employing only market costs, the assessment is biased in favor of the status quo, i.e., continuation of present trends in the use of private autos. The authors would have done better, I think, either to have included imperfect estimates of the value of externalities within their considerations, or to have omitted them and included careful statements as to the limits on their conclusions that were imposed by these omissions.

93. Authors of one famous study of the costs and benefits of employing pesticides drew their conclusions on the basis of calculating only three (market) parameters: (1) the value of the average corn and soybean crop per pound of herbicide and insecticide; (2) the average cost per pound of herbicide and insecticide; and (3) the increased market price of corn and soybeans to consumers, as a result of reduced yield, if no chemicals had been used [U.S. Congress, OTA, *Pest Management Strategies*, Volume 2 (Washington, DC: U.S. Government Printing Office, 1979), pp. 68–71, 79–81]. On the sole basis of benefit (1) and costs (2) and (3), the authors calculated the aggregate CV for the decrease in consumers' welfare, which they said would amount to $3.5 billion annually (pp. 79–81). Obviously, however, carcinogenic, mutagenic, and teratogenic effects of pesticide use, as well as occupational costs and environmental damages, were ignored in the cost-benefit calculations.

94. The report included an analysis, based on market parameters, for "the costs that pipelines will have to pay for water usage, such as pumping, transportation, and purchase price," and for the price of "the water's value in alternative uses" [U.S. Congress, OTA, *op. cit.* (Note 69), p. 84]. Omitted in the calculations were the real costs of *any* use of the scarce Western water supply, the price of resource depletion (especially for future water users), the benefits of water conservation, and the costs of water pollution through slurry use. Nevertheless the authors of the technology assessment concluded that slurry transportation was a more cost-effective means of coal transport than was the use of railways [see Note 70].

95. U.S. Congress, OTA, *The Direct Use of Coal* (Washington, DC: U.S. Government Printing Office, 1979), pp. 316–34.

96. *Ibid.*, p. 372. The authors cautioned that they were "not so clear" about the "validity of the . . . analysis" because of the methodological problems they encountered in estimating nonmarket costs such as externalities (p. 372).

97. See U.S. Congress, OTA, *Application of Solar Technology to Today's Energy Needs*, Volume 1 (Washington, DC: U.S. Government Printing Office, 1975), p. 3. Some of the external benefits considered in this assessment include the labor intensiveness of various energy technologies, energy self-sufficiency, environmental impacts, and conservation of fossil fuels (pp. 12 and 22–23). Some of the external costs, on the other hand, are ill effects of onsite technology, regulatory barriers, and inadequate financial incentives (p. 3). As a consequence of consideration of all these social benefits and costs, the economic methodology of this assessment appears not to force one into a bias that is pro-technology or pro the status quo.

98. For a complete analysis of this example, including actual cost parameters involved, see Shrader-Frechette, *op. cit.* (Note 9), pp. 108–134.

99. Stobaugh and Yergin, "The End of Easy Oil" in Stobaugh and Yergin, *op. cit.* (Note 10), pp. 9 and 11.

100. *Ibid.*, p. 11.

101. *Ibid.*, p. 4.

102. *Ibid.*, p. 6.

103. See Note 100; also "Conclusion: Toward a Balanced Energy Program" in

Stobaugh and Yergin, *ibid.*, p. 227. Also in same collection, see the essay by M.A. Maidique, "Solar America," p. 211.

104. For one such alternative, see Lutz and Lux, *op. cit.* (Note 40), pp. 304–307; see also pp. 297–304.

105. A.L. Macfie ["Welfare in Economic Theory," 3 *The Philosophical Quarterly* 10 (January 1953): 59] asserts: "welfare judgments about the real world are value judgments."

106. Lutz and Lux, *op. cit.* (Note 40), p. 3.

107. C.H. Danhof, "Assessment Information Systems," in Kasper, *op. cit.* (Note 1), p. 22. For illustrations of how "expert bias" operates in assessment of chemical and nuclear technologies, see Robert van den Bosch, *The Pesticide Conspiracy* (Garden City, NY: Doubleday, 1978), and J.W. Gofman and A.R. Tamplin, *Poisoned Power* [Emmaus, PA: Rodale Press, 1971).

108. May Brodbeck ["The Nature and Function of the Philosophy of Science," in H. Feigl and M. Brodbeck eds., *Readings in the Philosophy of Science* (New York: Appleton-Century-Crofts, 1953], pp. 3–6 considers four meanings of the term "philosophy of science" and concludes (probably rightly, I think) that it refers neither to ethical nor to socio-psychological evaluation of science, nor to philosophy of nature. It refers, instead, to "logical analysis of science."

109. In its 1976 annual report [U.S. Congress, OTA, *op. cit.* (Note 3), p. 63], the OTA warned: "the method employed, the personnel involved, and the skills tapped depend on the technology being assessed, the client for whom the assessment is undertaken, the nature of the issues at stake, and the time available for, and the setting of, the project." Using even more explicit remarks, a number of persons note the "political" basis for the range of types of assessments, the differences in methodology, and the "dichotomy" between the government sponsor's constituency and the society in general [Herbert Fox, Chairman, *Technology Assessment: State of the Field,* Second Report of the Technology Assessment Panel of the Engineers Joint Council (New York: Engineers Joint Council, 1976), p. 4; and U.S. Congress, OTA, *Technology Assessment Activities, op. cit.* (Note 3), p. 227]. In fact, if there is "no *explicit* congressional mandate," key issues will not be raised in government-sponsored technology assessments. [U.S. Congress, OTA, *op. cit.* (Note 8), p. 91].

Study Questions

1. What is the main point of Shrader-Frechette's article? Why do you think it might be correct? Why do you think it might be wrong? Explain and defend your answers.
2. What assumptions does Shrader-Frechette make about technology? about ethics? Do you think these assumptions are correct? Explain and defend your answers.
3. If society accepts Shrader-Frechette's main conclusions, what consequences might follow? Would these consequences be desirable? Explain and defend your answers.

4. What is Pareto Optimum? compensating variation? specious accuracy? What are the main assumptions built into the notion of compensating variation?
5. Use the Shrader-Frechette article to give examples of problematic policy consequences that follow from uncritical use of economic methods in technology assessment.
6. What is the point of the coal slurry example and the railroad-technology example? Explain.

3.5

Sociological versus Metascientific Views of Technological Risk Assessment

Deborah G. Mayo

In this chapter I shall discuss what seems to me to be a systematic ambiguity running through the large and complex risk-assessment literature. The ambiguity concerns the question of separability: can (and ought) risk assessment be separated from the policy values of risk management? Roughly, risk assessment is the process of estimating the risk associated with a practice or substance, and risk management is the process of deciding what to do about such risks. The separability question asks whether the empirical, scientific, and technical questions in estimating the risks either can or should be separated (conceptually or institutionally) from the social, political, and ethical questions of how the risks should be managed. For example, is it possible (advisable) for risk-estimation methods to be separated from social or policy values? Can (should) risk analysts work independently of policymakers (or at least of policy pressures)? The preponderant answer to the variants of the separability question in recent risk-research literature is no. Such denials of either the possibility or desirability of separation may be termed *nonseparatist* positions. What needs to be recognized, however, is that advocating a nonseparatist position masks radically

Originally published as "Sociological Versus Metascientific Views of Risk Assessment," Chapter 12 in *Acceptable Evidence: Science and Values in Risk Management*. Edited by Deborah G. Mayo and Rachelle D. Hollander, New York: Oxford University Press, 1991.

different views about the nature of risk-assessment controversies and of how best to improve risk assessment.

These nonseparatist views, I suggest, may be divided into two broad camps (although individuals in each camp differ in degree), which I label the *sociological* view and the *metascientific* view. The difference between the two may be found in what each finds to be problematic about any attempt to separate assessment and management. Whereas the former (sociological) view argues against separatist attempts on the grounds that they give too small a role to societal (and other non-scientific) values, the latter (metascientific) view does so on the grounds that they give too small a role to scientific and methodological understanding. . . . The problem I am raising results because nonsepa-ratists of the metascientific stripe are lumped along with nonseparat-ists of the sociological stripe.

Although they are not often put forward as such, I believe that each of the two camps of nonseparatists corresponds to a different underly-ing philosophy of scientific knowledge and scientific rationality. I grant that the distinction I am seeking is neither clear-cut nor uniform, which is doubtless part of the reason it has largely gone unrecognized. However, failing to see this key difference between those who argue against separatism has, in my opinion, so obscured the debates over how to improve risk assessment that even an oversimplified partition-ing of nonseparatists seems necessary.

Processes of Risk Assessment

The term *risk assessment* may be used more broadly[1] than I intend here, to include risk perceptions (or even risk management). To avoid a fur-ther ambiguity that often infects discussions of risk-assessment contro-versies, I shall delineate at the outset what I understand risk assessment to cover. It then will be possible to ask, without begging key questions, whether the components of risk assessment do or do not include these broader aspects. Following the characterization spelled out in a 1983 report by the National Academy of Science (p. 18), I understand risk assessment to include the following four steps:

1. *Hazard Identification:* Characterizing the nature and strength of the causal evidence as to whether exposure to an agent can increase the incidence of a health condition (cancer, birth defect, etc.) in humans, lab animals, or other test systems.
2. *Dose Response Assessment:* Estimating the incidence of an effect as a function of exposure in various populations of interest, extrapo-

lating from high to low dose and from animals to humans. It should describe and justify the methods of extrapolation used and should characterize the statistical and biological uncertainties.

3. *Exposure Assessment:* Measuring or estimating the extent of human exposure to an agent that exists or would exist under specified circumstances in various subgroups.

4. *Risk Characterization:* Estimating the incidence of health effect under the various conditions in the specified subgroups. This combines the previous steps and includes a description of the uncertainties at each step.

The NAS–NRC Report and Separability

Far from being of merely conceptual interest, the issue of separability is fundamental to debates directly affecting actual practices of government regulation of hazards. An extremely rich source of material in which to trace recent arguments concerning separability is the period from the early 1980s to the present. I shall focus on the practices and the philosophy of the Environmental Protection Agency (EPA) during 1981 and 1982, referring to it as the "Gorsuch EPA," as Anne Gorsuch (later Anne Burford) was the EPA administrator at the time.

The problem that arose in case after case[2] (often resulting in the Gorsuch EPA's being brought to trial) was that many widespread procedures for risk assessment were being repudiated in favor of "scientific" assessments (and in some cases reassessments) that fairly blatantly reflected the antiregulatory policy favored by the affected industry and by the Reagan administration generally. As an outgrowth of such concerns, Congress requested a study of these problems and some proposals for reforms. This resulted in a report by the National Academy of Science–National Research Council (NAS–NRC) in 1983. The concern, as the report states, is this:

> With a scientific base that is still evolving, with large uncertainties to be addressed in each decision, and with the presence of great external pressures, some see a danger that the scientific interpretations in risk assessments will be distorted by policy considerations, and they seek new institutional safeguards against such distortion. (p. 14)

To avoid such distortion, the proposed reforms suggest that risk assessment be separated from risk management.

Among the institutional reforms suggested, the NAS–NRC report focuses on two: reorganization, to ensure that risk assessments are protected from inappropriate policy influences, and the development and use of uniform guidelines for carrying out risk assessment. Although the report did not recommend institutional separation, it did believe it important to strive to enable risk analysts to work independently of policy pressures and to distinguish risk assessment from risk management conceptually:

> We recommend that regulatory agencies take steps to establish and maintain a clear conceptual distinction between assessment of risks and consideration of risk management alternatives; that is, the scientific findings and policy judgments embodied in risk assessments should be explicitly distinguished from the political, economic, and technical considerations that influence the design and choice of regulatory strategies. (p. 7)

To implement its recommendations the NAS–NRC proposed setting out uniform inference guidelines for interpreting scientific and technical information relevant to risk assessment. The use of such guidelines, it believes, "will aid in maintaining the distinction between risk assessment and risk management" (p. 7). It also recommends an independent Board on Risk Assessment Methods whose main function would be to assess critically the evolving scientific basis of risk assessment and to make explicit the underlying assumptions and policy ramifications of the inference options (p. 8).

The NAS–NRC report, then, was the basis for implementing separation at regulatory agencies such as the EPA (Environmental Protection Agency) and OSHA (Occupational Safety and Health Administration). William Ruckelshaus, who replaced Gorsuch after this controversy erupted, made it clear in his 1983 statement that he was relying on the NAS–NRC report in stressing the importance of separating risk assessment and risk management. Even in the face of conflicting political pressures, Ruckelshaus declared that "risk assessment at EPA must be based only on scientific evidence and scientific consensus. Nothing will erode public confidence faster than the suspicion that policy considerations have been allowed to influence the assessment of risk" (Ruckelshaus 1983, pp. 1027–28).

The attempt to improve risk assessment through such separation has been challenged both in principle and in regard to the form it has actually taken. It is often argued in work in the social studies of science that any attempt to enhance the neutrality of risk assessment by separating risk assessment and risk management is wrongheaded because it presupposes a view of risk assessment as a matter of objective, impartial, empirical fact, while viewing risk-management policy as invested with social values, subjective, emotional, or aesthetic feelings not adjudicable in a "rational" manner. What is being contested is the view underlying Ruckelshaus's remark that "although there is an objective way to assess risk, there is, of course, no purely objective way to manage it" (Ruckelshaus 1983, p. 1028). Not surprisingly, challenges to the objectivity and rationality of scientific risk assessment, and the corresponding arguments against separatism, are closely connected with the more general challenge from philosophers, historians, and sociologists to a certain image of scientific rationality and objectivity. The image being questioned is the naive positivist or what I shall call the *old image* of scientific rationality, in which science is thought to be value free.

Overview

I shall begin by considering how the challenge to the plausibility of separating assessment from policy relates to a more general philosophical challenge to what may be called the old image of scientific rationality, and I shall outline the argument that leads the sociological view to reject separability. Then I shall compare the implications of the meta-scientific view, which also rejects separability, with those of the sociological view and two other positions. My conclusion is that what matters most is not whether or not a view espouses separatism but whether it adheres (implicitly or explicitly) to the old image of science or, alternatively, sets the stage for a new or postpositive image, in which scientific scrutiny—though neither algorithmic nor value free—can nevertheless appraise objectively the adequacy of risk assessments. By "appraising objectively" I mean determining (at least approximately) what the data do and do not say about the actual extent of a given risk. Last I shall illustrate these issues and arguments by considering a controversial ruling during the Gorsuch EPA pertaining to the significance of the risks associated with formaldehyde.

The Old Image of Scientific Rationality and the Sociological Challenge to Separability

The Old (or Naive Positivist) Image of Scientific Rationality

According to the old image of science, stemming from the positivist tradition that prevailed from the 1930s to 1960s, the rationality of science rests on objective rules for appraising hypotheses and adjudicating between competing hypotheses. Demonstrating the rationality of science required articulating objective, value-free rules to assess hypotheses.

Philosophers, however, failed to articulate such rules satisfactorily, and the entire view that science follows impartial algorithms came under challenge by Kuhn (1962) and others. Actual scientific debates often last several decades and are not adjudicated simply by empirical rules; "extrascientific" values enter as well. This has been taken tos how the untenability of the old image of scientific rationality. But some philosophers, presumably holding the old image ideal to be the only type of rationality worth having, have taken this as grounds for abandoning altogether the view that science is characterized by rational methods. Instead they hold that such extrascientific values— metaphysical beliefs, goals, subjective interests and the like—play a greater role than does empirical evidence in evaluating hypotheses and adjudicating scientific disputes. According to extreme versions of this view (e.g., Feyerabend) beliefs about the world are not constrained at all by what we learn from empirical evidence in science. Sociology of knowledge has further strengthened this "antipositivist" sentiment by revealing how social contexts have influenced scientific theory appraisal in specific instances.[3]

Based on the old image, if scientists disagree in the face of the same empirical data—one group concluding that a substance is carcinogenic, say, and another group concluding it is not—there is a violation of scientific rationality. Such disagreement seems to illustrate, as Hamlin (1986) notes, that "experts on one side, or even both sides, are falling so much under the sway of 'interests' that they violate central norms of impartiality, emotional neutrality, universality . . . and the like . . . their behavior has been seen as the prostituting of science, as the selling of credibility to the highest bidder" (p. 486). Were it possible to avoid or somehow to neutralize these interests, adjudication of the scientific disagreement would be forthcoming. According to this old image perspective, it makes sense to seek reforms by means of new alignments among experts and policymakers, for example, the type of independent board recommended in the NAS–NRC report.

But this perspective has been questioned in both philosophy of science generally and the social studies of science. As Hamlin continues:

> A growing body of literature in the social studies of science takes issue with this formulation. It suggests that it is meaningless to think about a disinterested science; especially in such policy-relevant areas as ecology and public health, it will be impossible to impose any unarbitrary separation of scientific from social issues, and unreasonable to expect that "un-

biased" assessors will have no interest of their own to represent. In this view there can be no disinterested parties, but simply parties with different, more-or-less conflicting or compatible interests. (1986, p. 487)

The Sociological View

Such questioning of the old image is taken as the basis for variants of what I shall call the sociological view. An example of such a view is put forward by Brian Wynne, who claims:

> The debates over the safety of many environmental pollutants are structurally conditioned by the fact that underlying the overt technical discourse is a symbolic discourse in which those who have previously committed themselves to a particular scientific point of view, with particular policy implications, are attempting to defend their long-term credibility. (1982, p. 133)

The meaning of such "symbolic discourse" (which he compares with poetry, art, and religion) is "socially malleable," in contrast with the "cold steel of hard empirical fact" assumed in the old or positivistic view to be characteristic of science. Wynne still employs the term "scientific rationality," but now it is seen as "intrinsically conditioned by social commitments" (p. 139).

Another type of sociological view stems from a variant of this criticism of the old image. The criticism begins with the recognition that facts are not only necessarily theory laden but also fail unequivocally to pick out a best hypothesis, that is, facts underdetermine hypotheses. Thus, if decisions about hypotheses are reached despite these knowledge gaps, they must be the result of "subjective judgments" influenced, if not wholly determined by, social and ethical values. This is one of the tacks employed by Douglas and Wildavsky (1982) in their sociocultural theory of risk. They maintain that "the risk assessors offer an objective analysis. We know that it is not objective so long as they are dealing with uncertainties and operating on big guesses. They slide their personal bias into the calculations unobserved" (p. 80). The knowledge gaps, they continue, must be filled with educated guesses, and the remainder of their book is devoted to showing "that the kinds of guesses about natural existence depend very largely on the kinds of moral education of the people doing the guessing" (p. 80).

Douglas and Wildavsky go even further: In their view, not only do such social values enter to fill knowledge gaps in reaching risk assessments, but the very methods, models, and interpretations are themselves social constructions. They are led to an extreme position that reduces risk assessment to cultural or moral judgment—a "social reductionism." As S. O. Funtowicz and J. R. Ravetz put it, social reductionism may be seen as "a methodological position which assumes that every debate over technological risks is really a conflict among contradictory 'ways of life' and that the awareness of this would be enough to settle the question" (1985, p. 223).[4]

For our discussion, we can place under the heading of the sociological view of risk assessment two views (both of which come in degrees):

1. Social factors necessarily enter into risk assessment because there are knowledge gaps, and there is nothing to fill them except social and moral judgments (social relativism).
2. Risk assessment is entirely the outcome of socially determined methods and judgments, which are social constructs (social reductionism).

What matters for my argument is that both of these views lead to similar consequences for risk assessment: Risk-assessment judgments (at least those containing uncertainty) are policy judgments, and risk-assessment disagreements largely reflect disagreements about policy (including moral, social, economic, or other nonscientific[5]) values. Thus, risk-assessment judgments and the adjudication of disagreements necessitate going beyond empirical evidence, scientific criteria, and analytic methods to extrascientific (social and policy) considerations. In the sociological view, the primary issue is not estimating objective physical risks, but instead social and politically conditioned attitudes toward physical risk. Wynne adds: "Determining objective physical risks will still be valid of course, but the lingering tendency to start from this scientific vantage point and add social perceptions as qualifications to the 'objective' physical picture must be completely reversed" (1982, p. 138).

The result is that the methods of science are given little, if any, role in an unbiased adjudication of disagreements over risk assessments. How could they, after all, if each view is as biased as the next? According to Douglas and Wildavsky, "everyone, expert and layman alike, is biased. No one has a social theory above the battle" (1982, p. 80). If a judgment is biased as long as it is the output of a human or of a method that humans articulate, then, of course, all judgments are biased. But then for a risk assessment to be biased becomes trivially true: By definition there is no way to criticize a risk assessment as biased. There would be no sense in criticizing a risk assessment as a misinterpretation of data, in the sense of incorrectly asserting what the data indicate about the actual extent of a given risk. Because in the sociological view, interpreting scientific results is necessarily colored by social and political contexts, such criticism would be simply a criticism of the social and moral views of the assessor. The claim is not just that science cannot tell us which risks to find acceptable—that much is not controversial. The claim is that science cannot tell us the extent of a given harm because, given the uncertainties, such an assessment is always a matter of one's ethical and policy values. No wonder, then, that Douglas and Wildavsky conclude: "Science and risk assessment cannot tell us what we need to know about threats of danger since they explicitly try to exclude moral ideas about the good life" (p. 81). Can this conclusion be avoided?

The Sociological Argument Recapitulated: Premise (P)

Let us recapitulate the link between rejecting the old image of rationality and the view that risk-assessments always reflect (to varying degrees) prior policy or social commitments. The first part of the argument is that a strict demarcation between risk assessment and risk management (or of "facts" and "values") is plausible only if the old image of science can be maintained—that is, only if there are methods for a strictly factual appraisal of hypotheses—in this case hypotheses about the actual risks caused by a given substance. The old image is unattainable; therefore strict separability is unattainable. Let us accept the argument thus far as sound. But how does this lead to the further claims (of the sociological view) that the risk estimate is largely or solely determined by prior social policy positions, that disagreements about estimates reflect different extrascientific values, and that there can be no unbiased scientific court of appeals? The additional premise required is something like this:

P: If strict separability is unattainable, then empirical, technical, and scientific methods cannot provide unbiased risk assessments or adjudicate objectively between conflicting assessments.

Again, by an objective adjudication I mean one constrained by what is actually the case about a risk, regardless of one's policy preferences.

However, to assume premise (P) is tantamount to assuming that the old image of scientific rationality is the only one possible or worth having. Although it is not typically recognized, workers in social studies of science are led to this sociological view not as a consequence of repudiating the old image of scientific rationality but by taking it all too seriously. Laudan (1989) makes an analogous point concerning the postpositivists Kuhn, Feyerabend, and Quine, as does Shapere (1986) with regard to work in social studies of science. For what supporters of the sociological view must be arguing is that unless risk assessment can be accomplished by (value-free) logical rules or algorithms, it is outside the domain of science proper (falling instead into the domain of extrascientific policy), and we are led to dethrone science as an adjudicator in assessment disputes. That is, their grounds for denying that risk assessment consists of applying scientific methods separable from policy stem from implicitly holding to the old image philosophy in which scientific methods must be dictated by neutral, logical rules.

So prevalent is the view underlying premise (P) that it seems to be assumed with little argument. Indeed, in the first issue of *Risk Analysis,* as Funtowicz and Ravetz (1985) also observe, Kaplan and Garrick (1981) assume from the outset that risk is radically relative and that there is no difference between risk and perceived risk. This view is a corollary of premise (P).

If premise (P) were true, then anyone rejecting the separability of risk assessment and policy (i.e., any nonseparatist) must embrace (P)'s consequent and repudiate the adjudicating power of science. And because (P) is so widely assumed, every contribution to the literature that takes a nonseparatist line tends to be regarded as giving yet further support for this repudiation. But this is a mistake. There is another view that, like the sociological view, challenges the tenability of separating risk assessment and policy, yet, unlike the sociological view, denies premise (P). This second view also argues against a rigid separation of risk assessment as cold hard facts divorced from risk management; it too objects to the old image. Where this second view differs from the sociological view is in upholding the ability of scientific methods and criteria to criticize objectively those risk-assessment judgments involving knowledge gaps. For reasons to be explained later, I shall refer to this second, critical view as the *metascientific* view.

The Metascientific Challenge to Separatism and the Denial of Premise (P)

My understanding of the term *metascience* should be distinguished from the term *transscience.* The latter is defined by Weinberg (1972) as being outside science proper, but metascience is within science, at least if science is understood in a genuinely postpositivist (new image) manner, which I shall be discussing. The term "metascience" seems appropriate because the critical scrutiny of the uncertainties in risk assessment involves a critical reflection that might be seen as one level removed from the process of risk assessment itself—much as we term metalogic the scrutiny of the philosophical foundations of logic.

The metascientific view need not actually be put forward in terms of an argument in favor of any general philosophy of science. More commonly, it is simply what underlies repudiations of the actual form that attempts at separation have taken, as in the period since the NAS–NRC report. A major criticism is that such attempts hamper communication between risk assessors and risk managers, thereby hindering rather than helping ensure adequate assessments (e.g., Jasanoff 1991 and Silbergeld 1991). The result is that managers are left to use assessments without knowing the underlying methodological assumptions and their associated uncertainties. The metascientific suggestion is that the problem can be ameliorated or at least improved by a better understanding of the uncertainties underlying choices of risk-assessment estimates. In striking contrast with the sociological view, these calls for nonseparatism are calls for greater scientific and methodological understanding, albeit an understanding that is based on a critical (or metascientific) scrutiny of the uncertainties involved.

Thus, although both the sociological and the metascientific views are nonseparatist—they both hold that risk assessment involves both science and policy—each sees this nonseparability as having very different consequences for the role of science in risk assessment. The proponent of the sociological view argues against separatist attempts because he considers it impossible to perform adequately the science of assessment without being involved in the policy and ethical values of risk management. The metascientist argues against separatist attempts because he considers it impossible to perform risk management adequately without understanding the science underlying the assessments. To clarify the distinction between the two positions, we need to begin by asking where the policy input into risk assessment lies according to the metascientific view. Although in the sociological view the answer is essentially "everywhere," the metascientific view is concerned with the entry of policy considerations only in specific places, places that the NAS–NRC report labels *risk-assessment policy (RAP)* judgments.

Risk-Assessment Policy (RAP)

Risk-assessment policy refers to the various judgments and decisions, sometimes called *inference options,* that are required to carry out risk-assessment estimates. Because these judgments include choices with no unequivocal scientific answers and because these choices have policy implications, they are intertwined with policy. The NAS–NRC report (1983, pp. 29–33) offers a useful delineation of more than fifty junctures at which, owing to uncertainty, an inferential or analytical choice must be made in the course of making risk assessments. Some of those relevant to the metascientific standpoint are the following:

1. Some RAP questions under different components of hazard identification:
   a. Epidemiologic data
      - What weight should be given to studies with different results? Should a study be weighted in accordance with its statistical power?
      - What weight should be given to different types of studies (prospective versus case control)?
      - What statistical significance level should be required for results to be considered positive?
   b. Animal-bioassay data
      - What degree of confirmation of positive results should be necessary?
      - Should negative results be disregarded or given less weight?
      - Should a study be weighed according to its statistical power?
      - How should the occurrence of rare tumors be treated? (Should it be considered evidence of carcinogenicity even if the finding is not statistically significant?)
      - What models should be used to extrapolate to humans?
2. Some RAP questions concerning dose response assessments
   a. Epidemiological data
      - What dose response models should be used to extrapolate from observed doses to relevant doses?
   b. Animal bioassay data
      - What mathematical models should be used to extrapolate from experimental doses to human exposures?
      - Should dose response relations be extrapolated according to best estimates or according to upper confidence limits? If the latter, what confidence limits should be used?

Although science and policy are intermingled in selecting among inference options at each stage of risk assessment (hazard identification, dose response assessment, exposure assessment, etc.), the policy entry here is intended to be distinguished from the broader social, ethical, and economic policy decisions in risk management. According to the NAS–NRC report: "At least some of the controversy surrounding regulatory actions has resulted from a blurring of the distinction between risk assessment policy and risk management policy" (1983, p. 3). To see how this blurring occurs and to consider whether it may be avoided, we need to be clear on how policy enters in RAP.[6]

How Policy Enters into Risk-Assessment Policy (RAP) Judgments

Policy enters into RAP judgments as follows: Insofar as there is more than one scientifically acceptable answer to these RAP questions, there will be latitude for choice among possible plausible responses. Each choice influences the risk assessment and so has a policy implication. In particular, the choice will influence the likelihood that a substance will be judged to pose a significant hazard to human health. The more an inference choice increases the likelihood that a substance will be judged a significant risk, the more *protective* or *conservative* it is. For example, deciding to use positive results from animal data as indicative of human risk is more conservative than, say, requiring positive human data. Although animal data provide imperfect predictions of human risk, deciding to use them makes it more likely that a human risk will be uncovered.

Because RAP options differ in their degree of protectiveness, it is possible to advance one's policy values—one's view of how protective an estimate should be required to be—by making suitable choices of these inference options. The criticism lodged against the Gorsuch EPA (and other agencies during the early Reagan administration) was that antiregulatory (i.e., less protective) RAP choices were systematically allowed to influence risk assessments made by agency scientists. The result was to politicize the agency. Science in the Gorsuch EPA was science in support of the policy of freedom from regulation (dubbed "regulatory relief"). Under the guise of demanding stringent scientific evidence, these less protective RAP choices made it extremely unlikely that a substance would be deemed a significant human risk. Indeed, as Silbergeld (1991) notes, scientists who did not toe the antiregulatory line were excluded from serving on the agency's scientific staff or as advisers. This resulted in the explicit effort by Ruckelshaus in 1983 to separate risk assessment from risk management (the new separatism). But these separatist reforms failed to have their intended effect, which is the basis of the objections from the metascience corner.

We can now make plain what the metascience view finds objectionable about attempts to separate risk assessment from management. The problem in a nutshell is that in all separatist models, the purely scientific components of risk assessment are to be performed by scientists, and the components in which policy enters are to be performed by policymakers. Once it is recognized that risk assessment involves policy in the form of RAP choices, it follows that such choices should be the policymaker's task. But this allows two undesirable consequences: It permits policymakers (1) to fall into all manner of misinterpretations of the assessment evidence and (2) to introduce, either consciously or unconsciously, the very same biases in assessments that led to the separatist "reforms" in the first place.

In order to avoid such hidden biases in assessments, it is necessary to recognize the specific policy influences and implications of specific RAP choices—for example, the implications for protectiveness of choosing a maximum likelihood estimate, in contrast with an upper 95 percent confidence interval. But understanding the protectiveness of the RAP choices requires understanding the evidential meaning of the different types of risk estimates. And this requires understanding the scientific uncertainties involved. If RAP judgments are made by nonscientist policymakers, they are likely to be made by persons divorced from the original issues and uncertainties underlying the different risk estimates. At the same time, the scientist is limited to presenting possible RAP choices but is involved neither in making them nor in bringing out the implications for protectiveness.[7] For example, if the scientific work ends after reporting two possible estimates that may be used, say, a maximum likelihood estimate and an upper 95 percent confidence-bound estimate, then the scientist will not be around to explain how far off each is likely to be from the actual risk and why. Under the Gorsuch EPA the risk assessor could choose either, supporting his or her choice as good science. Under the new separatist reforms, the risk assessor could also choose either, supporting his or her choice as selecting from among two scientifically plausible options. Neither case requires the assessor to articulate and defend the standard of protectiveness effectively being held by choosing a given option.

Recapitulation of Four Positions: Some Shared Consequences

As with any attempt to analyze positions that differ in degree and often exist without explicit articulation, my description of four positions will be somewhat oversimplified. However, I believe it will bring to light their key consequences and philosophical underpinnings. The four positions are those of (1) the Gorsuch EPA, (2) the separatist reform that it engendered (the new separatism), (3) the sociological view, and (4) the metascientific view.

Although risk assessment in the Gorsuch EPA was thought to be in need of the separatist reforms, it could be said to have been espousing a separatist framework (though it did not actually succeed in following one). In the Gorsuch EPA, as in the new separatism, the "scientific" aspects of risk assessment were seen as separable from the value-laden management decisions. But the line between these two was drawn in different places. The Gorsuch EPA—as its defenders' arguments indicate—reflected the view that the decisions that we would call RAP decisions were within the province of science. What was effectively a call for less protective standards was typically couched as a call for better and more rigorous science.

The challenges to the Gorsuch EPA (analogous to the challenges to the positivist image) essentially claim that the cutoff between science and policy needs to be moved: The judgments needed to generate and interpret data in order to arrive at risk assessments also permit the entry of policy values (hence the term risk-assessment policy or RAP judgments). Thus, in the new separatist reforms, the judgments that go beyond the "hard facts" of science enter the domain of policy judgments: RAP judgments are moved from the domain of science to the domain of policy. So science *qua* science cannot help in deciding among RAP options. But allowing RAP to be performed by policymakers without a critical scientific oversight, we said, permits the original problem to recur. The consequence is that RAP judgments are made without bringing out the different implications for the protectiveness of different choices.

Before illustrating how this occurs, it should be noted that this consequence has implications beyond the separatist views. The same consequence follows from seriously embracing the sociological view. The reason is this: If going beyond the hard facts introduces subjective policy judgments, then any judgments beyond pure science belong to the realm of policymaking and ought to be made by policymakers, not scientists. This would follow for both the new separatists and for adherents to the sociological view. If according to the sociological view, policy judgments about risk estimates are more like "symbolic discourse"—more like moral or aesthetic judgments—then the judgment selected is a matter of subjective preference. I am not saying that the new separatist reforms explicitly endorse this sociological view of RAP judgments, for they do not. Rather, I am saying that the operationalization of these separatist reforms in fact yields the same results willingly embraced by the sociological view.[8]

That the separatist view embodied in the reforms and the sociological view should have this shared consequence is not surprising if one remembers that both views hold to the old image of science; the former, explicitly, the latter, implicitly. What follows is that the real difference in views turns not on whether they espouse separatism or nonseparatism but on whether or not they are based on the old image of science.

The philosophical underpinnings of a metascientific approach, I claim, genuinely depart from this old image of science to a postpositivist or "new image" of science. In contrast with the old image of science, the metascientific view acknowledges the lack of value-free, universal, algorithmic methods for reaching and evaluating claims about the world (in our case, risk-assessment claims). But far from understanding this to preclude objectivity, an explicit recognition of how value judgments can influence the statistical risk assessments (in RAP) can—according to the metascientist—be used to interpret assessments more objectively. One way to recognize the policy influences and implications of RAP judgments is to evaluate their corresponding protectiveness for the case at hand.[9] This requires critical metascientific tools.

I shall now turn to the question of why a metascientific approach is more adequate than the alternatives just considered are. As is appropriate for appraising any postpositivist view, I shall judge its adequacy not by a priori arguments but by considering how well it would fare (in contrast with the other views outlined) in an actual risk-assessment controversy.

The Formaldehyde Controversy

The formaldehyde controversy at the EPA illustrates the problems that arose from the agency's politicization of science, and the attempted cure—the new separatism.

The RAP Controversy in Assessing Formaldehyde

In order to arrive at a risk assessment in the case of formaldehyde—that is, to determine whether and the extent to which formaldehyde increases the risk of cancer in humans—many of the key RAP judgments discussed earlier had to be made. This is an unusually instructive example of disagreements at each of these decision points and of policy controversies that erupted from such disagreements. My focus will be on the RAP option around which the bulk of the risk-assessment controversy centered: the importance of positive animal studies in the face of negative or inconclusive epidemiological studies. As is often the case with carcinogenic risk assessment, information from prospective randomized treatment-control experiments was available only on animals, that is, in the case of formaldehyde, rats. Epidemiological studies on humans, in contrast, allowed only a retrospective analysis of cancer rates in various occupations. On the basis of the statistically significant increases in (nasal) cancer among treated rats, the Chemical Industry Institute of Toxicology (CIIT) reached the assessment that formaldehyde is carcinogenic in laboratory rats and reported this to the Environmental Protection Agency in November 1980. A panel of eminent scientists convened by the National Toxicology Program confirmed this hazard assessment and concluded that "formaldehyde should be presumed to pose a risk of cancer to humans."[10] The lengthy document detailing the formaldehyde risk assessments was entitled the "Priority Review Level 1" (PRL-1) and dated February 1981.

On the basis of hazard assessments, the EPA staff reached the policy decision to designate formaldehyde as a priority chemical under the EPA provision known as 4(f). On May 20, 1982, the U.S. House of Representatives, Subcommittee on Investigations and Oversight of the Committee on Science and Technology, held a hearing on this matter. Its review, entitled *Formaldehyde: Review of the Scientific Basis of EPA's Carcinogenic Risk Assessment* (hereafter referred to as *Hearing*) concluded:

> EPA believes that formaldehyde has met the criteria for 4(f) for the following reasons. First, the results of a recently reported bioassay study demonstrate that formaldehyde is carcinogenic in rats. A National Toxicology Program (NTP) panel, which evaluated the 18-month data, concluded that formaldehyde should be presumed to pose a risk of cancer to humans. . . . Second, review of the available information on the use of formaldehyde and resulting human exposure suggests that large numbers of people are potentially exposed to harmful concentrations of formaldehyde. Accordingly, the Agency finds that there may be a reasonable basis to conclude that formaldehyde presents a significant risk of widespread harm to humans from cancer. (pp. 5–6)

This last sentence is important because statute 4(f) requires only that there may be a reasonable basis and not that there is a reasonable basis to conclude that a significant risk exists.[11] In itself it does not call for any regulation but is simply a call for closer scrutiny based on an indication that there may be a significant cancer risk.

Then there was a change in administration; the Reagan administration entered, and along with it a new EPA administrator, Anne Gorsuch, and some new staff. In fact, formaldehyde was the first 4(f) recommendation brought before the new administrator for signing. But instead of signing it she had members of the new EPA staff carry out a reassessment of the hazard data in the PRL-1. The new and revised version of the data became the Todhunter memorandum, named for John Todhunter, a new EPA assistant administrator. Some of the changes included blatant erasures of the highest-risk estimates that had been given in the PRL-1. The original document read, "For most of identified subpopulations, the estimated risks are equal to or greater than 1 in 10,000, however in some instances the risks are in the range of 1 in 10 . . . 1 in 100 and 1 in 1,000." The Todhunter memorandum cut out everything after "1 in 10,000" and reported, "For most of identified subpopulations, the estimated risks are equal to or greater than 1 in 10,000."[12] (At least one witness at the hearing that resulted testified that he quit rather than make these changes.) Among the other most important changes was deciding to de-emphasize the positive rat studies and to emphasize the negative epidemiological studies on humans. Todhunter concludes, for example, "There does not appear to be any relationship, based on the existing data base on humans, between exposure [to formaldehyde] and cancer. Real human risk could be considered to be low on such a basis" (*Hearing,* p. 260). This hazard reassessment was given as the basis of a changed policy. On September 11, 1981, the original EPA staff recommendation to designate formaldehyde as a 4(f) priority chemical was reversed, and the opposite policy choice was made. Whether or not this shift in hazard assessment was justified was the subject of an enormous controversy leading to the aforementioned congressional hearing on formaldehyde. As the title of the report of the hearing makes plain, it was intended as a "review of the scientific basis of the EPA's carcinogenic risk assessment." (It makes for fascinating reading.) A key question for the subcommittee, as its chairman, Senator Albert Gore, Jr., states, is, "To what extent has EPA and Dr. Todhunter departed from the long standing principles for carcinogenic risk assessment and given the wide acceptance of these principles is there an accepted scientific basis for such departure?" (*Hearing,* p. 3).

Science Politicized in the Gorsuch EPA

RAP judgments, in the Gorsuch EPA, overtly reflected the agency's predetermined policy—to hold a very stringent standard before deeming a risk to be significant. As Ashford, Ryan, and Caldart (1983) note in their insightful article on this case:

> EPA's formaldehyde deliberations powerfully illustrate the ease with which matters of policy may be confused with matters of science. The agency's technical analysis hides significant procedural deficiencies. Whether intentional or not, the result is an invidious one: the analysis purports to justify, in the name of science, a risk assessment policy far less protective of human health than the agency's prior policy. (p. 342)

How was the political agenda (antiregulation) masked as science? We can answer this by referring to our points concerning the RAP decisions. Each choice has implications for the protectiveness of the risk assessment (the chance that it will be considered a significant hazard to humans). With an EPA purged of scientists (save those leaning toward avoiding regulation), there was plenty of leeway for them to choose, consistently, the inference option most likely to have a less protective outcome. (In addition to giving less weight to the well-documented positive animal results and interpreting negative epidemiological studies as indicating low or no increased human risk, they endorsed other less protective choices in the formaldehyde assessment, such as holding to the existence of a threshold for carcinogenicity of formaldehyde, discounting benign tumors, and preferring maximum likelihood estimates over upper confidence level estimates.)

After all, if RAP judgments are viewed as part of the science of risk assessment—as they were in the Gorsuch EPA—then it is appropriate for agency scientists to choose among RAP options. And when the choice among options is not determined by "hard scientific fact," then each option may be presented as scientifically plausible. In his attempt to demonstrate that each of the RAP options in the formaldehyde dispute were scientifically plausible, Todhunter strove to document "expert scientific support" of the positions he favored. It was precisely this tendency, in this case and others, that led to the suspicion that "expert scientific support" was being enlisted to justify any position that the EPA favored. This suspicion of the EPA—and the need to avoid it—led to the formaldehyde hearings. The chairman, Senator Albert Gore, Jr., remarked in his opening statement: "We have witnessed . . . a belief, becoming widespread, that industry has special access to EPA and that EPA is becoming a captive to the industries it was established to regulate" (*Hearing*, p. 1).

One of the main reasons for this suspicion was that the new administration did not base its decision against a 4(f) designation on any new data beyond the PRL-1 document that had been the basis for the original, and opposite, recommendation to designate formaldehyde under 4(f)—though, as we mentioned, it did conceal some evidence supporting a 4(f) designation. Rather, the new administration proceeded to hold a series of secret meetings restricted only to certain scientists and lawyers from the Formaldehyde Institute, the Formaldehyde Trade Association, and the EPA staff. These meetings were claimed to consist solely of scientific discussions. In these "scientific" meetings the participants reinterpreted the data and came to a conclusion opposite to the one that had been reached and approved by numerous scientists and agencies. As one attorney with the Natural Resources Defense Council (Jacquelin Warren) testified:

> There are no new data to support the reversal, only a reinterpretation which has been advocated by and is quite favorable to the interests of the formaldehyde industry. Those new assumptions, as we have heard, de-

236 *Deborah G. Mayo*

part radically from accepted principles of cancer risk assessment. They lack a sound scientific basis and leave the public subject to cancer risks that . . . regulatory statutes were designed to protect against. In our view, this has been an effort to get the Government off the back of the formaldehyde industry. (*Hearing*, p. 188)

The usual peer review was absent in the reassessment they carried out, and Todhunter denied that any such review was called for.[13]

Todhunter's defense reveals most clearly how policy may be masked as science and why it might be expedient to hold a view of risk assessment that permits such masking. Todhunter and his supporters urged the view that each of the RAP options in the formaldehyde dispute was scientifically plausible and that the choices made—in support of not triggering 4(f)—reflected not an implicit bias in favor of an antiregulatory outcome (favorable to the Formaldehyde Institute) but, rather, legitimate scientific disagreement. Consider the following portion of an exchange between Robert Walker—who is defending Todhunter—and Senator Gore:

> *Mr. Walker:* . . . we cannot even come up with good enough science to decide whether or not we ought to ban this substance, so I think that there is plenty of room for some discussion about the science involved in all of this and that there can be a variety of interpretations about the science that is available.
>
> We have one scientist here who is disagreeing with other scientists. That is entirely acceptable within the science community, or I did not realize that science was walking in lock step on all of these things.
>
> *Mr. Gore:* . . . Dr. Todhunter has stated . . . that the epidemiological evidence indicates that there is no increased risk to human beings. There is not really much of an argument about that. There are two people who say that, Dr. Todhunter and the Formaldehyde Institute. The rest of science takes a contrary view.

Reluctant to accept this, Walker continues:

> *Mr. Walker:* . . . a study that the Du Pont Co. now has that indicates that they would agree with Dr. Todhunter's assessment, so there seems to be considerable disagreement within the scientific community. . . . It seems to me that what we are into here is the politics of science and whose decision base you are going to use. (*Hearing*, pp. 137–38)

Walker, like the others defending Todhunter, is led in this last sentence to claim that the choice of data and their interpretation are inevitably political. (Only Todhunter, it seems, sticks to the position that his interpretation is free of policy bias.) In the belief that such intermingling of science and policy is a perversion of good science and of how science is supposed to operate at agencies such as the EPA, the hearing and subsequent proposed reforms called for separation. As one scientist (Roy Albert) testifying at the formaldehyde hearings put it: "It is absolutely essential to the soundness of regulatory decisions that their underlying scientific assessments of health risks be shielded from the political forces that operate at the level of the regulatory offices" (*Hearing*, p. 36).

The sociological view, we saw, rejects such an ideal of neutrality, which assumes that science can be separated from policy. But in taking this rejection as grounds for the various sociological positions outlined earlier, I shall argue that the sociological view effectively strengthens arguments that could be and were offered in defense of the Gorsuch EPA.

The Sociological View and the Defense of the Gorsuch EPA

In the sociological view, one cannot hope to appeal to science to arbitrate between the judgment of scientists claiming that the evidence warrants a 4(f) ruling and those such as Todhunter (and representatives of industry) who claim that it does not. For in this view, the conflict is inevitably an *ad hoc* attempt to justify a previously held social policy: All views are biased. If it is true that the acceptability of an interpretation of data leading to risk assessments largely reflects policy values, then the main ground for criticizing an interpretation is that it leads to policies deemed unfavorable, such as triggering 4(f). For example, to offer a criticism of the original assessment that formaldehyde may pose a significant harm on the grounds that a 4(f) designation would be economically undesirable (as lodged by the lobby for the Formaldehyde Institute), becomes as legitimate a criticism of the assessment as is an argument against its evidential warrant. This view just endorses as inevitable the position that conflicts over hazard assessments are largely conflicts over competing policy values. But if disagreements over hazard assessments are primarily disagreements over policy values, there would seem to be little justification for the allegations of a number of scientists at the hearings that Todhunter's assessment was irresponsible or incompetent on objective scientific grounds—for example, on grounds that it misconstrues what the data say about the actual extent of cancer risk. Indeed, the existence of objective grounds for criticism is what the sociological view wishes to deny.

Unsurprisingly, but of interest to us, this was precisely the argument given by those defending Todhunter's assessment against such allegations of incompetence and invalidity. The committee's minority member, Walker, expressed vehement opposition to the hearings altogether because they were framed as dealing with a scientific issue, whereas in his opinion "what we are dealing with here is a regulatory issue" (*Hearing*, p. 73). If this automatically entails that disagreements simply reflect differences in subjective policy values, one can perhaps understand Walker's portraying the hearings as "an obvious result of the efforts of a few disloyal and disgruntled employees of the EPA who" simply because they disagree with the decision not to place formaldehyde under 4(f) "feel justified in waging guerrilla warfare against the Agency and those in positions of authority" (*Hearing*, p. 4).

One of the few scientists who defended Todhunter's assessment against charges of poor science, Sorell Schwartz, did so by denying that there was anything scientific in the Todhunter memo. If it is not science, it cannot be poor science. The exchange between him and the chairman of the subcommittee, Senator Gore, is interesting:

Mr. Gore: Do you accept the scientific judgments in the Todhunter February 10 memo? Do you think it is good science?

Dr. Schwartz: I have not read anything in the Todhunter memo—

Mr. Gore: Oh—

Dr. Schwartz: No, no. I have read the Todhunter memo. However, I have never read anything in the Todhunter memo which to me involves science. I think what Dr. Todhunter did was act as a nonscientist, and that is to make a determination whether the risk as presented by the data was significant enough to be unacceptable under 4(f).

Mr. Gore: That is a little odd. He is the chief scientist responsible for the EPA toxics program and the agency published this document and released it to the public as the scientific justification for the decision. It discusses scientific judgments throughout. How can you say that it is written as a nonscientist, as a nonscientific document? (*Hearing*, p. 239)

When pressed, Schwartz admitted to disagreeing with the memo's statements on level of risk. He declared, for example, "I do not favor the idea that negative epidemiologic data signifies negative human effect." But he defended the Todhunter memo on the grounds that it dealt with "not the level of risk but the acceptability of that risk" (p.

239). However, Schwartz's defense misconstrues 4(f). The 4(f) priority unambiguously states that it is not about risk acceptability—which explicitly takes benefits into consideration (*Hearing*, pp. 491, 773). Rather, it is to indicate only whether there may be a reasonable basis for concluding that a substance presents a significant risk of human harm. The few other defenses of the Todhunter memo were also based on equating evidential issues with those of policy preferences. This seems to illustrate the rather disturbing point of Roberts, Thomas, and Dowling (1984): "Too many of the participants have good reasons *not* to distinguish scientific evidence from policy preferences, *not* to analyze carefully the various sources of technical disagreement and *not* to accept responsibility for some decisions or judgments" (p. 120).

The sociological view, we have seen, offers no ammunition with which to fight the type of politicization of science characteristic of the Gorsuch EPA. Unwittingly or not, it offers a basis for defending the status quo of that period. Although it is unclear whether holders of the sociological view would take this as a weakness of their view, it is clear that those proposing the new separatist reforms were striving to alter the status quo and to help ensure that risk assessment was free of political bias. Thus, if it turns out that separatist reforms also permit politicization to go unchecked, the reforms will have failed to reach their goal.

The new separatism recognizes that deciding which way to interpret data involves policy in the form of RAP judgments. Thus it places such decisions under the policy rubric. For example, after giving the scientific report of the positive rat studies and the negative epidemiological studies in the formaldehyde case, the RAP decision as to how to weigh them falls under policy. In this way, the sociological view, based on a general challenge to scientific rationality, undermines both the importance of and the ability to discern the actual physical risks. The new separatist reforms do the same, not by means of any general philosophical arguments, but by allowing the RAP choices to be made by nonscientists who cannot or do not articulate the implications for protectiveness of the given inference option. This is the basis for the metascientific criticism of the new separatism.

The Metascientific (or Metastatistical) Approach

How would the metascientific approach deal with this case? The metascientific view, we said, denies premise (P): It denies that the inability to divorce science from policy entails the inability to adjudicate objectively among risk assessments. The metascientist grants that the judgments required to reach risk assessments may reflect policy values, conventions, and the like and that political interests may be adduced to explain particular choices of RAP judgments (as the formaldehyde case clearly shows). Nevertheless, the metascientist holds that the question whether a given risk assessment is warranted by the evidence is not a matter of social and political values; it is a matter of what the risk actually is. One may hold, in other words, that what counts as "good science" in a given context may in fact be negotiated, while still maintaining that whether evidence and inferences based on that evidence are acceptable are not negotiable. Instead, the acceptability of evidence is constrained by the extent to which it actually warrants specific risk inferences based on it. The question of whether a risk inference is warranted is a question of how well it reflects what is really the case about the causal effect of the risky substance or the practice in question. The latitude in choosing RAP options does not preclude the objective scrutiny needed to answer this question.

An analogy with a weighing instrument is useful. My interest in whether I have gained as little as one-tenth pound may be a matter of my subjective values. But whether a scale with a digital readout in whole pounds, say, is a good tool for finding this out is not a matter of my subjective values. Given the scale chosen, whether or not a gain is detected depends on how much I have actually gained! And understanding the scale enables one to determine (at least approximately) what a given reading indicates about how much this is. Analogously, a critical understanding of tools used for estimating risks enables an understanding of the actual extent of risk that is or is not indicated by a given piece of evidence. This in turn enables one to determine the protectiveness of a given RAP judgment. It allows one to answer the question: According to the standard being required, what extent of risk must be fairly clearly indicated before it is taken to point to a significant human risk? Answering this question requires critical metascientific tools. What sort of tools can accomplish this in the case of formaldehyde?

My focus is on the RAP judgment in interpreting negative epidemiological results. One of the main questions raised at the hearings was: Does a failure to find a positive result (i.e., a statistically significant increase in risk) indicate that there is little or no risk? Todhunter and the Formaldehyde Institute say yes—that the negative epidemiological results are evidence that there is no increased risk to human beings. Yet Todhunter's own epidemiologist on the staff responsible for this work, wrote: "Before leaving [the EPA], I would again like to emphasize that the available epidemiologic data from studies on formaldehyde exposure are inconclusive and not supportive of no association, as purported by the Formaldehyde Institute" (*Hearing*, p. 137). What is the nature of this dispute? Only by understanding the principles of the statistical reasoning involved were the critics able to tell. What was pivotal was understanding how to interpret negative statistical results, as will be generally true in judicial (or other) reviews of decisions not to regulate or prioritize. Lacking this understanding, the courts may be unable to decide correctly between conflicting "expert" assessments. In order to rectify this situation the metascientist urges that we clarify what the disagreement really amounts to, that we go back and unearth what negative results do and do not warrant. Even without seeking an algorithm for risk assessment in general, understanding the implications of RAP judgments does admit systematization. In our example, what is called for is a way to systematize the reasoning for criticizing certain uses of statistical instruments (which unfortunately lend themselves to *un*critical use). . . . In the formaldehyde case, the hypotheses tested are assertions about a parameter that I will call Δ, the increased cancer risk in the population of humans. . . . The test hypothesis H in the formaldehyde case asserts that formaldehyde does not cause an increase in a person's risk of dying from cancer of a given type. That is, it asserts that there is a zero increase in the risk rate: $\Delta = 0$. The alternative hypotheses assert that formaldehyde causes a positive increase: $\Delta > 0$.

. . . How to Tell the Truth (About Negative Results) with Metastatistics

Ideally, the policy question of what counts as a substantively important increase in cancer risk is answered at the start, so that the test may be specified in order to have appropriately high probabilities of detecting all and only those increases. Substantively important increases in the formaldehyde case are those increases in cancer risk deemed serious enough to trigger 4(f)—a policy judgment. However, regardless of how the test has been specified—whether based on policy or other values—knowledge of the test's error probabilities, I claim, allows interpreting objectively what the data do and do not indicate about the increased risk—that is, about Δ. The way in which they may be used to this end involves reasoning that is both obvious and familiar.

It can be shown by the analogy with a weighing instrument. The null hypothesis H may be that I have gained no weight since I weighed in last week, say at 125 pounds. To test H, suppose I use this method: I weigh myself on a digital readout scale that expresses weight in whole pound units, and reject H only if a difference is registered. Suppose the result turns out to be the same weight in pounds as last week (125).

Am I warranted in concluding I have not gained any weight, even as little say, as one-tenth pound? Because my method had very little chance of detecting such a small increase, even if I had gained it, this negative result would be poor grounds for thinking I had gained no more than one-tenth pound. (My scale is too insensitive.) On the other hand, this negative report is a good indication that I have not gained as much as a full pound. For had I gained a pound, then it is likely that my scale would have registered some gain. Moreover, my negative reading is an even better indication that I have not gained as much as five pounds or more. A simple principle emerges:

> A failure to observe a difference in weight only indicates that my actual weight gain is less than *x* if it is very probable that the scale would have registered a larger difference in weight, if in fact I have gained as much as *x* pounds.

This leads to a precisely analogous metastatistical principle for interpreting a negative statistical result, which I label rule (M):

> RULE (M): A failure to observe a statistically significant difference only indicates that the actual increase is less than Δ' if it is very probable that the test would have resulted in a more significant difference than was observed, were the actual increase as large as Δ'.

To apply rule (M), we need a way to calculate the probability that a test T would have resulted in a more significant difference than was observed, were the actual increase as large as Δ'. . . .

Let us apply rule (M) to one of the negative results that Walker and others cited in defense of the Todhunter interpretation (*Hearing*, pp. 137–38): the Du Pont study. In a mortality study of Du Pont workers, the relative risk of dying from cancer among those in the study exposed to formaldehyde was not statistically significantly greater than among those not so exposed: The null hypothesis *H* was not rejected.[14] Du Pont concluded that "the data suggested that cancer mortality rates in the company's formaldehyde exposed workers were no higher than the rates among nonexposed workers" (*Hearing*, p. 284). They are inferring, in other words, that the increased risk Δ equals zero. The error in such an interpretation is that the failure to reject the null hypothesis of zero-increased risk is not the same as having positive evidence that the increased risk is zero. For such negative results may be common (i.e., probable) even if the underlying increase in risk is greater than zero. In fact, the Du Pont study had a very small chance of rejecting null hypothesis *H* even if the actual increase in risk had exceeded zero by substantial amounts—that is, the "power" against these alternatives is low. Hence, failing to reject *H* does not rule out these increases. For example, the Du Pont study had only a 4 percent chance of rejecting *H* even if there were a twofold increase in cancer of the pharynx or of the larynx in those exposed to formaldehyde. (As expressed in the table below, the study had a *power* of only 4 percent to detect twofold increases in those cancers.) Thus, failing to reject *H* does not rule out twofold increases in these types of cancers. The situation was even worse with nasal cancers and not much better with the others. This is indicated in the following chart adapted from a review of the Du Pont study by the National Institute for Occupational Safety and Health (NIOSH) (*Hearing*, p. 548):

	Lung	*Pharynx*	*Larynx*
Number of cases	181	7	8
Power to detect odds ratio = 2*	37%	4%	4%
Least significant odds ratio detect-ablet	2.9	57.5	42.5

*Assumes α = .05 (1 tail)
tAssumes α = .05 (1 tail) and power $(1 - \beta)$ = .80

. . . Although failure to reject does not indicate that the increase is zero, it does permit an inference about the likely upper bound of the unknown increase Δ. That is, a failure to reject *H* provides a reason to say that the data provide good grounds for asserting that the increased risk is no greater than such and such (upper bound). To find plausible upper bounds one need only determine the increase (i.e., the value of Δ) that very probably would have resulted in rejecting *H* (or at least obtaining a larger increased risk than was observed). . . .

In the Du Pont study, the test had a fairly high probability (.8) of rejecting *H* if the risk of lung cancer were three times higher among exposed than unexposed workers. Hence, a failure to reject *H* does indicate the increased risk of lung cancer is not as high as threefold. The risk of cancer of the larynx would have to be forty-two times higher among exposed than unexposed workers in order for the test to have a fairly high probability (.8) of obtaining a statistically significant difference (rejecting *H*). A failure to reject *H*, then, does indicate that the actual increased risk is not as high as forty-two-fold. For cancer of the pharynx, there would have to be a fifty-seven-fold increase in cancer risk before the test could have a good chance of rejecting *H*. So, even ignoring some methodological difficulties with the study, its negative statistical results at most indicate that the various cancers are no more than three or forty-two or fifty-seven times as likely among workers exposed to formaldehyde. They clearly do not warrant the conclusion reached by Du Pont and others that the study supports the claim of no increase in (relative) cancer risk among formaldehyde workers. The study does not even support the claim of low increased risk, given what Todhunter himself claimed the EPA counted as low.

Relevance to Assessing the Acceptability of the Evidence

The main purpose of the formaldehyde hearing was not the policy question of how high an increased risk may be before categorizing it as significant enough to trigger 4(f). It was to decide whether the decision not to prioritize formaldehyde represented a shift in the principles being used to reach this categorization. Doing so required understanding whether the evidence was acceptable for the assessments reached. It is necessary to ascertain the implications for protectiveness of the RAP judgments made. Metastatistical reasoning of the sort in rule (M) is the basis for doing so.

Thus (M) may be used to ascertain the approximate bound that the negative result warrants ruling out. This may be used to determine whether the standard of protectiveness being employed is in accordance with what agencies or individuals deem tolerable. (M) also allows one to compare this calculated estimate with the upper-bound risk associated with a different substance. If data on the latter are found to indicate an increased risk no greater than the former, and yet the latter leads to a different risk-management decision, then there are likely to be specific differences in policy values effectively operating in the two cases. These indeed were found. For example, applying the reasoning embodied in (M) showed that the epidemiological data could not rule out there being a risk of cancer as great as or greater than 1 in 100,000. As one scientist (Warren) testified:

> By comparison, the Consumer Product Safety Commission banned urea formaldehyde foam insulation based on an estimated risk of 10^{-5} (one in 100,000). That the EPA would disparage as insufficient merely to justify a closer look at formaldehyde [via 4(f)], risks others consider sufficient to warrant a ban is troubling indeed. (*Hearing*, p. 198)

Because limited sensitivity (or power) is common in epidemiological studies, positive results are typically not thought to be required to indicate that a substance may pose a significant risk to humans,[15] particularly when, as with formaldehyde, the substance is shown to cause cancer in rats and mice at doses reasonably comparable to those to which people may be exposed. Using the reasoning in rule (M), a number of scientists concluded: "The EPA decision on formaldehyde may constitute an *ad hoc* revision of the principles for assessing carcinogenic risk that have been widely accepted by the scientific community for over a decade" (*Hearing*, p. 179).

Todhunter denies this alleged shift in the agency's policy values for managing carcinogens. For example, Todhunter does not disagree with the general range of risk that agencies have tended to deem of public concern. Nevertheless, he claims formaldehyde does not meet the criteria of section 4(f) for the following reasons: "There is a limited but suggestive epidemiological base which supports the notion that any human problems with formaldehyde carcinogenicity may be of low incidence or undetectable. . . . [The ranges of risks] are of from low priority to no concern" (*Hearing*, p. 253). But even according to Todhunter's own interpretation of this range, to interpret the (negative statistical) data as indicative of a low incidence is, as rule (M) makes clear, to misinterpret it. Granted, the increased hazard may be undetectable with the tests used, but to take that as grounds against a 4(f) designation is to require positive epidemiological results before even recommending that the EPA take a closer look at a substance—which is a shift in the existing policy. Todhunter gives as the second reason for the decision: "There is suggestive evidence that there may be human exposure situations—which may not present carcinogenic risk which is of significance." He is surely correct about this, as it is always true that "there may be human exposure situations which may not present carcinogenic risk" of significance, namely, a zero or extremely low exposure. Were one to take this reason seriously, it seems one could always argue against a 4(f) designation. But this is to play a logical trick on the 4(f) wording. For, as the statute clearly states, a reasonable basis to conclude that exposure presents a significant risk is sufficient to trigger 4(f). Yet Todhunter's second reason is tantamount to construing the 4(f) designation as appropriate only if there is no exposure that does not present a significant risk! Only then would it make sense to say that finding a single exposure that does not present a risk is sufficient *not* to trigger 4(f). It is no wonder that a number of scientists concluded—using considerations identified in rule (M)—that "in order to justify its failure to address formaldehyde under 4(f) . . . EPA has rewritten both the science and the law" (*Hearing*, p. 195).

Because of the criticisms of the science underlying the EPA risk assessment, in 1985 the EPA finally did place formaldehyde under the 4(f) category, and the NAS–NRC report guidelines emphasized the importance of using power consideration in interpreting negative results. Additional metastatistical rules can be formulated for other types of RAP judgments. The appropriateness of choosing different extrapolation models awaits further biological knowledge, but here too, definite progress has been made, at least for carcinogenic assessments. Despite this progress, it would be erroneous to suppose that we have gotten past the sort of problems that arose in the formaldehyde episode. One thing is certain: Progress in this direction will be hampered if nonseparatists of the metascientific stripe continue to be confused with those of the sociological stripe. For according to the latter view, such metascientific rules cannot help.

[For a much fuller discussion of rules such as rule (M), see Mayo 1996, especially the discussion of "metastatistical rules" e.g., the Rule of Acceptance and Rule of Rejection.]

Conclusion

A major theme running through interdisciplinary discussions on risk is that risk assessment cannot and should not be separated from societal and policy values. However, under the banner of favoring nonseparatism are two radically different views of risk assessment—something that the literature has not explicitly recognized. Typically, all such calls for nonseparation have been taken as evidence for holding, to some degree, what I have termed the sociological view. That is, the untenability of strictly separating the science of assessment from the social and policy values of risk management is typically taken as grounds for denying that scientific methods can provide unbiased risk assessments or can adjudicate objectively between conflicting assessments. But to lump all nonseparatist positions under the sociological view, I have argued, is a mistake. Other nonseparatists take a different position, which I have called the metascientific view. In contrast with the sociological view, its calls for nonseparation are accompanied by calls for greater scientific and methodological understanding—albeit an understanding that allows for a critical or metascientific scrutiny of the uncertainties involved. By implicitly holding the overly stringent, old image conception of scientific objectivity—one that is precluded by the need to make uncertain judgments without algorithms—adherents to the sociological view are led to deny the attainability or importance of objective assessments of physical risks. As such, what matters most is not whether or not a view espouses separatism, but whether it adheres (implicitly or explicitly) to the old image of science or, alternatively, sets the stage for a new or post-positive image, in which metascientific scrutiny—not algorithms—can appraise objectively the adequacy of risk assessments.

Our case study focused on a metascientific scrutiny of risk-assessment policy (RAP) judgments as they arise in statistical tests of the existence of increases in risk. According to the sociological view, statistical methods in the arena of RAP judgments are tools for manipulation rather than instruments for an unbiased adjudication of conflicting risk assessments. Conflicting risk assessments are seen as largely, if not solely, conflicts over policy values or over different "ways of life."

Ironically, this entails an extreme subjectivism or relativism that undercuts the *raison d'être* of the sociological view for many of its adherents: to hold risk assessors and managers accountable to various societal values. As the metascientific view stresses, only by means of a critical understanding of the uncertainties underlying RAP judgments is it possible to distinguish adequately what is warranted by the evidence and what is prejudged by policy values. Thus, to exclude or downplay critical scientific scrutiny from this arena is to forfeit a crucial tool for holding risk policymakers accountable to the degrees of protectiveness deemed acceptable by society. . . .

Acknowledgment

A portion of this research was carried out during the tenure of a National Endowment for the Humanities Fellowship for College Teachers; I gratefully acknowledge that support. For many useful comments on earlier drafts, I thank Marjorie Grene and Harlan Miller. I am grateful to Rachelle Hollander for numerous discussions that helped me clarify the ideas presented here.

Notes

1. It may also be used more narrowly to include only strictly quantitative risk estimates.

2. For a good account of the practices of the EPA and other agencies during the Reagan administration, see Lash, Gillman, and Sheridan (1984).

3. For an excellent delineation of these positions, see Shapere (1986).

4. They go on to give excellent criticisms of the arguments for social relativism and social reductionism.

5. These are nonscientific, it should be kept in mind, in the traditional old image of science.

6. Many of these fifty-one or so components are straightforwardly statistical. If only in order to ascertain the consequences of adopting one or another answer or choice, one uses statistical considerations.

7. The NAS–NRC report had intended reforms far more amenable to the metascientific view than what actually came to pass. For example, the report stresses:

> The importance of distinguishing between risk assessment and risk management does not imply that they should be isolated from each other; in practice they interact, and communication in both directions is desirable and should not be disrupted. (p. 6)

> AAlthough we conclude that the mixing of science and policy in risk assessment cannot be eliminated, we believe that most of the intrusions of policy can be identified and that a major contribution to the integrity of the risk assessment process would be the development of a procedure to ensure that the judgments made in risk assessments, and the underlying rationale for such judgments, be made explicit. (p. 49)

8. What is more, it has led to such sociological tactics as focusing on differences of opinions about risk rather than the original disagreements, which are not about opinions but evidential uncertainties.

9. Some options, such as choosing to count positive animal studies as indicative of human risks, will always be more protective than will requiring positive human results, say. But the protectiveness of other options will depend on the substance being considered.

10. Among those judging the CIIT assessment valid were the National Toxicology Program, the Mt. Sinai School of Medicine Environmental Cancer Information Unit, and the International Agency for Research on Cancer (IARC). Ashford, Ryan, and Caldart (1983) provide a good discussion in support of the validity of this assessment, in contrast with an assessment by the Consumer Product and Safety Commission.

11. Section 4(f) provides that "Upon the receipt of—

(1) any test data required . . . or

(2) any other information available to the Administrator, which indicates to the Administrator that there may be a reasonable basis to conclude that a chemical substance or mixture presents or will present a significant risk of serious or widespread harm to human beings from cancer, gene mutations, or birth defects, the Administrator shall, within the 180-day period beginning on the date of the receipt of such data or information, initiate appropriate action . . . to prevent or reduce to a sufficient extent such risk or publish in the Federal Register a finding that such risk is not unreasonable. (*Hearing*, pp. 476–77)

An excellent discussion of the implications of and requirements under 4(f) occurs in the Primer on 4(f), *Hearing*, pp. 475–523.

12. I have replaced the exponential notation in the original document with equivalent fractions here. For documentation of this and several other of Todhunter's revisions, see *Hearing*, pp. 349–65.

13. Todhunter claimed additional review was not needed because the data had been well reviewed already by the many scientists who concurred with the original PRL-1. The fact that his reassessment of these same data was now being used to support the opposite policy on 4(f) did not seem to faze him. . . .

14. In the Du Pont study, 481 cancer deaths among male employees between 1957 and 1979 constituted the cases. These were matched on relevant factors with controls who did not die of cancer. The statistic observed was the relative odds ratio, the ratio of the odds of having been exposed to formaldehyde among cases and controls. For simplicity, I refer here to the risk rather than the relative risk. . . .

15. As the formaldehyde options report notes:

Generally, even the largest and most expertly performed epidemiological studies can seldom detect increases in cancer risk of less than 10% (1 in 10). However, from the public health standpoint, exposures to carcinogens may be considered problems if they increase the risk of cancer by 1 case in 1000 exposed persons or less. (*Hearing*, p. 763)

For a discussion of a study performed to document the problem of insufficiently powerful tests, see Freiman et al. (1978).

References

Ashford, N. A., Ryan, C. W., and Caldart, C. C. (1983). "A Hard Look at Federal Regulation of Formaldehyde: A Departure from Reasoned Decisionmaking." *Harvard Environmental Law Review* 7:297–370.

Douglas, M., and Wildavsky, A. (1982). *Risk and Culture*. Berkeley: University of California Press.

Feyerabend, P. K. (1975). *Against Method*. London: New Left Books.

Freiman, J. A., Chalmers, T. C., Smith, Jr., H., and Kuebler, R. R. (1978). "The Importance of Beta, the Type II Error and Sample Size in the Design and Interpretation of the Randomized Control Trial, Survey of 71 'Negative' Trials." *New England Journal of Medicine* 299:690–94.

Funtowicz, S. O., and Ravetz, J. R. (1985). "Three Types of Risk Assessment: A Methodological Analysis." In *Risk Analysis in the Private Sector*, ed. C. Whipple and V. T. Covello, pp. 217–31. New York: Plenum.

Hamlin, C. (1986). "Scientific Method and Expert Witnessing: Victorian Perspectives on a Modern Problem." *Social Studies of Science* 16:485–513.

Kaplan, S., and Garrick, B. J. (1981). "On the Quantitative Definition of Risk." *Risk Analysis* 1:11–27.

Jasanoff, S. (1991). "Acceptable Evidence in a Pluralistic Society." In Mayo and Hollander (1991), pp. 29–47.

Kempthorne, O., and Folks, L. (1971). *Probability, Statistics, and Data Analysis*. Ames: Iowa State University Press.

Kuhn, T. S. (1962). *The Structure of Scientific Revolutions*. Chicago: University of Chicago Press.

Lash, J., Gillman, K., and Sheridan, D. (1984). *A Season of Spoils: The Reagan Administration's Attack on the Environment*. New York: Pantheon Books.

Laudan, L. (1989). "The Sins of the Fathers . . . Positivist Origins of Post-Positivist Relativisms." In Laudan (1996).

Laudan, L. (1996). *Beyond Positivism and Relativism*. Boulder, Colo.: Westview Press.

Lynn, F. M. (1986). "The Interplay of Science and Values in Assessing and Regulating Environmental Risks." *Science, Technology, & Human Values* 11:40–50.

Mayo, D. (1985). "Increasing Public Participation in Controversies Involving Hazards: The Value of Metastatistical Rules." *Science, Technology, & Human Values* 10:55–68.

Mayo, D. (1988). "Toward a More Objective Understanding of the Evidence of Carcinogenic Risk." In *PSA 1988*, vol. 2, ed. A. Fine and J. Leplin, pp. 489–503. East Lansing, Mich.: Philosophy of Science Association.

Mayo, D. (1996). *Error and the Growth of Experimental Knowledge*. Chicago: University of Chicago Press.

Mayo, D., and Hollander, R. (1991). *Acceptable Evidence, Science and Values in Risk Management*. New York: Oxford University Press.

National Academy of Sciences–National Research Council (NAS–NRC) (1983). *Risk Assessment in the Federal Government: Managing the Process*. Washington, D.C.: National Academy Press.

Poole, C. (1987). "Beyond the Confidence Interval." *American Journal of Public Health* 77:195–99.

Roberts, M., Thomas, S., and Dowling, M. (1984). "Mapping Scientific Disputes That Affect Public Policymaking." *Science, Technology, & Human Values* 9:112–22.

Ruckelshaus, W. (1983). "Risk, Science and Public Policy." *Science* 221:1026–28.

Rushefsky, M. E. (1984). "The Misuse of Science in Governmental Decision-making." *Science, Technology, & Human Values* 9:47–59.

Russell, M., and Gruber, M. (1987). "Risk Assessment in Environmental Policy-making." *Science* 236:286–90.

Shapere, D. (1986). "External and Internal Factors in the Development of Science." *Science and Technology Studies* 4:1–9.

Silbergeld, E. (1991). "Risk Assessment and Risk Management: An Uneasy Divorce." In Mayo and Hollander (1991), pp. 99–114.

U.S. Congress. House of Representatives. Committee on Science and Technology. *Formaldehyde: Review of Scientific Basis of EPA's Carcinogenic Risk Assessment* (1982). Hearing before the Subcommittee on Investigations and Oversight. 97th Cong., 2d sess. May 20.

Weinberg, A. (1972). "Science and Trans-Science." *Minerva* 10:209–22.

Wynne, B. (1982). "Institutional Mythologies and Dual Societies in the Management of Risk." *The Risk Analysis Controversy: An Institutional Perspective,* ed. H. C. Kunreuther and E. V. Ley, pp. 127–43. Berlin: Springer-Verlag.

Study Questions

1. What is the main point of Mayo's article? Why do you think it might be correct? Why do you think it might be wrong? Explain and defend your answers.
2. What assumptions does Mayo make about technology? about ethics? Do you think these assumptions are correct? Explain and defend your answers.
3. If society accepts Mayo's main conclusions, what consequences might follow? Would these consequences be desirable? Explain and defend your answers.
4. Explain the differences between the metascientific and the sociological views of values in assessing technological risks? Why is the distinction important?
5. What is the point of Mayo's formaldehyde example?
6. Why are there value judgments involved in interpreting the significance of negative (no harmful effects) statistical results? Explain.
7. How can methodological value judgments be an "excuse" for scientists or technology assessors to introduce bias into their evaluations? Explain.

3.6

Further Reading

Beckwith, Guy V. 1989. "Science, Technology, and Society: Considerations of Method," *Science, Technology, and Human Values* 14 (Autumn): 323–39.

Bullard, R. 1994. *Dumping in Dixie*. Boulder, Colo.: Westview Press.

Camacho, Luis. 1989. *On Technology and Values in the Social Context and Values: Perspectives of the Americas*. Lanham, Md.: University Press of America.

Ezrahi, Yaron. 1992. "Technology and the Civil Epistemology of Democracy," *Inquiry* 35, 3–4 (September–December): 363–76.

Ferre, Frederick. 1993. *Hellfire and Lightning Rods: Liberating Science, Technology, and Religion*. Hertfordshire: Orbis.

Ferre, Frederick, and Mitcham, Carl. 1989. *Ethics and Technology*. Greenwich, Conn.: Jai Press.

Freyfogle, Eric T. 1993. *Justice and the Earth: Images for Our Planetary Survival*. New York: Free Press.

Glynn, Simon. 1993. "Ways of Knowing: The Creative Process and the Design of Technology," *Journal of Applied Philosophy* 10, 2: 155–63.

Goldworth, Amnon. 1987. "The Moral Limit to Private Profit in Entrepreneurial Science," *Hastings Center Report* 17 (June): 8–10.

Goodland, Robert. 1994. "Environmental Sustainability and the Power Sector, Part I: The Concept of Sustainability," *Impact Assessment* 12, 3: 275–304.

Goodpaster, K. 1978. "On Being Morally Considerable," *Journal of Philosophy* 75: 308–25.

Grim, Patrick. 1987. "Technology and Arbitrary Decisions," *Public Affairs Quarterly* 1 (July): 43–58.

Hanafi, Hassan. 1987. "Science, Technology and Spiritual Values: Possible Models and Historical Options," *Dialectics and Humanities* 14 (Summer): 5–11.

Kahneman, Daniel, Slovic, Paul, and Tversky, Amos. 1982. *Judgment*

Under Uncertainty: Heuristics and Biases. Cambridge: Cambridge University Press.

Lemons, J., and Morgan, P. 1995. "Conservation of Biodiversity and Sustainable Development." In Lemons, J., and Brown, D. (eds.), *Sustainable Development: Science, Ethics and Public Policy.* Dordrecht, Netherlands: Kluwer.

Lenk, Hans, and Maring, Matthias. 1992. "Ecology and Ethics: Notes about Technology and Economic Consequences," *Research in Philosophy and Technology* 12: 157–76.

MacLean, Douglas. 1986. *Values at Risk.* Totowa, N.J.: Rowman & Littlefield.

Mayo, Deborah G., and Hollander, Rachelle D. 1994. *Acceptable Evidence: Science and Values in Risk Management.* New York: Oxford University Press.

McGinn, Robert E. 1994. "Technology, Demography, and the Anachronism of Traditional Rights," *Journal of Applied Philosophy* 11, 1: 57–70.

Meadows, D. H., Meadows, D. L., and Randers, J. 1992. *Beyond the Limits.* Post Mills, Vt.: Chelsea Green.

National Research Council. Commission on Behavioral and Social Sciences and Education. 1991. *People and Technology in the Workplace.* Washington, D.C.: National Academy Press.

———. Committee on the Institutional Means for Assessment of Risks to Public Health. 1983. *Risk Assessment in the Federal Government: Managing the Process.* Washington, D.C.: National Academy Press.

Posner, Richard. 1973. "Strict Liability: A Comment," *Journal of Legal Studies* 2, 1 (January): 205–21.

Radnitzky, Gerard. 1984. "Science, Technology, and Political Decision: From the Creation of a Theory to the Evaluation of the Consequences of Its Application," *Revista Portuguesa de Filosofia* 40 (July–September): 307–17.

Redclift, Michael. 1993. "Sustainable Development: Needs, Values, Rights," *Environmental Values* 2, 1 (Spring): 3–20.

Rees, W. E., and Wackernagel, M. 1994. "Ecological Footprints and Appropriated Carrying Capacity: Measuring the Natural Capital Requirements of the Human Economy." In Jannson, A. M., Hammer, M., Folke, C., and Costanza, R. (eds.), *Investing in Natural Capital: The Ecological Economics Approach to Sustainability.* Washington, D.C.: Island Press.

Restivo, Sal. 1986. "Science, Secrecy, and Democracy," *Science, Technology, and Human Values* 11 (Winter): 79–84.

Sagoff, Mark. 1995. "Carrying Capacity and Ecological Economics," *BioScience* 45, 9: 610–20.

Sanmartin, Jose. 1994. "From World3 to the Social Assessment of Technology." In Mitcham, Carl (ed.), *Philosophy of Technology in Spanish Speaking Countries.* Dordrecht: Kluwer.

Shrader-Frechette, Kristin. 1994a. *The Ethics of Scientific Research.* Savage, Lanham, Md.: Rowman and Littlefield.

———. 1994b. "Science, Risk Assessment, and the Frame Problem," *BioScience* 44, 8 (September): 544–51.

———. 1992. "Calibrating Assessors of Technological and Environmental Risk," *Research in Philosophy and Technology* 12: 147–55.

———. 1991. *Risk and Rationality.* Berkeley: University of California Press.

———. 1985. "Technological Risk and Small Probabilities," *Journal of Business Ethics* 4 (December): 431–46.

———. 1984. *Science Policy, Ethics, and Economic Methodology: Some Problems with Technology Assessment and Environmental-Impact Analysis.* Boston: Kluwer.

Slovic, Paul. 1983. *How Safe Is Safe Enough? Determinants of Perceived and Acceptable Risk.* New Haven: Yale University Press.

Smith, Michael D. 1979. "The Morality of Strict Liability in Tort," *Business and Professional Ethics* 3, 1 (December): 3–5.

Strong, David. 1994. "Challenging Technology," *Research in Philosophy and Technology* 14: 69–92.

Szawarski, Zbigniew. 1989. "Dignity and Technology," *Journal of Medical Philosophy* 14 (June): 243–49.

Taitte, W. Lawson. 1987. *Traditional Moral Values in the Age of Technology.* Dallas: University of Texas Press.

Taylor, Paul. 1986. *Respect for Nature.* Princeton: Princeton University Press.

Thompson, William B. 1991. *Controlling Technology: Contemporary Issues.* Buffalo: Prometheus.

Unger, Stephen H. 1994. *Controlling Technology: Ethics and the Responsible Engineer.* New York: Wiley.

Westra, L. 1994. "Risky Business: Corporate Responsibility and Hazardous Products," *Business Ethics Quarterly* 4 (January): 97–110.

Westra, L. and Wenz, P. (eds.). 1995. *The Faces of Environmental Racism: Issues in Global Equity.* Lanham, Md.: Rowman and Littlefield.

Yanase, Michael Mutsuo. 1987. "The Challenge of Spiritual Values to Science and Technology," *Dialectics and Humanities* 14 (Summer): 13–20.

Zimmerman, Andrew D. 1995. "Toward a More Democratic Ethic of Technological Governance," *Science, Technology, and Human Values* 20, 1 (Winter): 86–107.

Part Four

Case Studies about Technology

4.1

Introduction and Overview

Kristin Shrader-Frechette and Laura Westra

As Edmund Burke noted, example is the school of humanity, and it will learn at no other. Examples or case studies, rather than abstract or general discussions, often reveal important insights about a subject, including technology and values. This fourth part of the book is devoted to a number of case studies that show how particular values are likely to play a role in developing, implementing, and assessing a given technology.

In general, the case study method aims at testing, clarifying, amending, and evaluating examples or cases.[1] In other words, in the method of case studies one often uses informal causal, inductive, retroductive, and consequentialist analyses in order to "make sense" of a particular situation; in addition, case studies require practical insight into a situation.[2] This practical reasoning aims at guiding action rather than merely developing rules or concepts for theoretical reasoning. For example, Carl Bernstein and Bob Woodward used a popular, journalistic version of the method of case studies. They posed and assessed competing explanations for how and why the Watergate coverup occurred. They also suggested, if only briefly, how their explanations might apply to other situations of political power and coverup.

Chapter 4.2, "Engineering Design & Research and Social Responsibility" by Carl Mitcham, focuses on the values that are part of engineering research and that determine engineering technology. Mitcham argues that three distinct ideas, about the relationship between engineering research and values, have developed since the beginning of the engineering profession. The first emphasizes company loyalty, the second technocratic leadership, and the third social responsibility. Many of the ethical and values problems arising in connection with

257

engineering technology occur because engineers use simplified models that overlook important elements of the technology. When engineers complained about the safety hazards in the Bay Area Rapid Transit (BART) system, for example, they argued that the piecemeal testing of the system did not adequately represent the problems that could occur in the real world. The testing overlooked important sources of risks. Mitcham's article also shows, through three examples, how important values issues can arise in designing technology for developing countries, for the military, and for mass consumer products. He closes his article with some suggestions for making the design of engineering technology more sensitive to ethical issues.

In chapter 4.3, Richard De George chronicles a famous failure of automobile technology: the case of the Ford Pinto. After discussing the facts of the Pinto case, De George asks about the ethical responsibility of the Ford engineers who were responsible for the risky automobile. The case shows not only that the engineers were at fault but also that economic values like the profit motive took precedence over the value of protecting public health and safety. Part of the problem lies with uncritical use of the method of benefit-cost analysis. To counteract problematic industry values, De George discusses the ethical values of whistleblowing and the circumstances under which employees ought to use it. He concludes that good technologies require moral structures and organizations, not just technological enterprises that are concerned with being in conformity with federal regulations.

In chapter 4.4, "Computers and Privacy," Stacey Edgar discusses the value of privacy, why it is essential to physical and mental survival, and how contemporary computer technology threatens privacy. After providing several examples of violations of privacy associated with computers, Edgar surveys some recent computer-related proposals that could threaten privacy. These include a national computerized criminal history system, the Clipper chip proposal, and computer matching programs. Next Edgar outlines some of the protections for privacy, including various U.S. laws (such as the Privacy Act of 1974) and the text of crucial legal decisions that affect privacy. Closing her article with a brief discussion of the importance of privacy, Edgar explains that choice is essential to being human and that violations of people's privacy often wrongly affect their choices and their developing individual responsibility.

Chapter 4.5, by David Lorge Parnas and Danny Cohen, comprises a case study on computer and military technology. They address the so-called "star wars," or Strategic Defense Initiative (SDI), technology for defending the United States against incoming missiles. Parnas explains why he is opposed to the SDI, and he argues that the use of nuclear weapons as a deterrent is immoral. Explaining the limits of software technology, he claims that scientists and engineers who do defense work cannot ignore the ethical and value aspects of their technologies. He says that they must be sure they are solving the "real" military problems, not just accepting high-paying jobs or satisfying their employers. He accuses developers of military technologies of being blind and wasting taxpayers' money because of their inattention to values issues raised by the technology.

Taking an opposite point of view, Danny Cohen argues that the SDI is reasonably feasible. Like Parnas, Cohen believes that scientists, engineers, and technical experts have ethical responsibilities, but he argues that their ethical opinions ought not have the status of facts. He also gives examples of situations in which pessimistic projections were wrong, and he argues that opponents of SDI are pessimistic.

In "Nuclear Technology and Radioactive Waste," chapter 4.6, Kristin Shrader-Frechette argues that commercial nuclear technology arose only because of problematic bias values that influenced its development. As a consequence, society now faces the many evaluative and scientific issues associated with developing technology to manage the dangerous nuclear waste over the next million years. Shrader-Frechette argues that this management will be extraordinarily difficult because technology for nuclear waste faces a number of technical, epistemological, and ethical problems. To resolve the social, ethical, and safety problems associated with implementing technologies for dealing with radioactive waste, Shrader-Frechette proposed two ethical solutions: attempting to secure the consent of those likely to be affected by the waste and compensating those who bear additional risks because of the waste.

David Pimentel and his associates argue in chapter 4.7, "Assessment of Environmental and Economic Impacts of Pesticide Use," that assessors have not adequately evaluated the use of chemical pesticides. These assessments ignore pesticide-related deaths and illnesses, and Pimentel et al. argue that virtually everyone everywhere is exposed to some pesticide residues in food, the atmosphere, and water. These problems continue, in part, they say, because there are few public health data about pesticide usage, even in the United States. His conclusion is that assessors need to examine the social and indirect costs associated with pesticide technology or they will continue to devalue human and environmental goals.

In chapter 4.8, "Ethical Issues in Human Genome Research," Thomas Murray argues that research on the human genome will raise at least three important classes of ethical issues: (1) what and who will control genetic information; (2) how and why should manipulation of particular genotypes and phenotypes occur; and (3) what are the genetic contributions to ethically and politically significant traits and behaviors? Murray concludes by arguing that "the sciences of inequality," like genetics, need not challenge core ethical doctrines like equal treatment under law. He argues that scientists and policy makers must work to see that ethical values are not disrupted by the use of the new technology.

Laura Westra, in chapter 4.9, argues that agricultural biotechnology raises a host of ethical and values questions, in part because these technologies are tested and controlled by those who have the most to gain from them. She also argues that although these agricultural biotechnologies are touted as "safe" and "better," often they are safe and better only from a marketing or profit-related point of view. Westra likewise argues for adequate labeling of products made through genetic technology and for the self-determination and rights to access to information that this labeling presupposes. She concludes her essay by pointing to the questionable environmental impacts that agricultural biotechnology may have on developing nations and on their ability to free themselves from environmental racism.

Notes

1. For a discussion of the method of case studies, including its benefits and problem areas, see K. S. Shrader-Frechette and E. D. McCoy, *Method in Ecology: Strategies for Conservation* (Cambridge: Cambridge University Press, 1993), ch. 5.

2. See Thomas Donaldson, "The Case Method," in T. Donaldson and A. R. Gini (eds.), *Case Studies in Business Ethics* (Englewood Cliffs, N.J.: Prentice-Hall, 1990), 13–23.

4.2

Engineering Design Research and Social Responsibility

Carl Mitcham

Although human beings have since antiquity undertaken projects that might be interpreted as engineering works, the first engineers as such did not appear until the Renaissance. If one dates the birth of modern science as an institution from the founding of the Royal Society in 1660, engineering as a profession is best dated from the formation of the Society of Civil Engineers (or Smeatonians) some hundred years later in 1771. Since then three distinct ideas have developed about engineering ethics, each of which can influence how engineering research is done. The first emphasizes company loyalty, the second technocratic leadership, the third social responsibility.

History of Ideas in Engineering Ethics

One idea is that engineers have a basic obligation to be loyal to institutional authority. An "engineer" was originally a soldier who designed military fortifications or operated engines of war such as catapults.[1] The first school to grant the engineering degree was the Ecole Polytechnique, founded in 1794 by the Directorate shortly before Napoleon became Brigadier General of the Revolutionary Army. In the United States, the Military Academy at West Point (1802) was the first school to offer engineering degrees. Within such a context, engineers' empha-

Originally published as "Engineering Design and Social Responsibility," in *Ethics of Scientific Research*. Copyright © 1994 by Rowman and Littlefield. Published by Rowman and Littlefield, 1994.

sis on duty is not surprising. During the same period as the founding of professional engineering schools, a few designers of "public works" began to call themselves "civil engineers."[2] The creation of this civilian counterpart to military engineering initially gave no cause to alter the basic sense of engineering obligation. Civil engineering was simply peacetime military engineering, and engineers remained duty-bound to obey their employer, often some branch of the government. The late eighteenth and early nineteenth centuries also witnessed the formation of the first professional engineering societies. Early in the twentieth century such organizations began to adopt formal codes of ethics. On analogy with physicians and lawyers, whose codes prescribe a fundamental obligation to patients and clients, the early ethics codes in professional engineering—such as those formulated in 1912 by the American Institute of Electrical Engineers (later to become the Institute of Electrical and Electronic Engineers or IEEE) and in 1914 by the American Society of Civil Engineers (ASCE)—defined the primary duty of the engineer to be of service as a "faithful agent or trustee" of an employing company.

There is undoubtedly some merit in engineers adopting the related principles of obedience and loyalty. Loyalty especially is a widely recognized virtue under many circumstances. But the problem with any obediential ethics is that it opens an adherent to manipulation by external powers that may well be unjust. Even in the military, for instance, it is now common to say that one is obligated to carry out only legitimate or just orders. Physicians and lawyers, too, must be loyal to their patients and clients only to the extent that patients and clients pursue health and justice, respectively. Attempts to meet this weakness in the principles of obedience and loyalty, and to articulate a substantive or regulative ideal for the engineering profession comparable to those in medicine and in law, gave rise to what has become known as "the technocracy movement."

Technocracy is at odds with both the implicit code of obedience and the explicit code of company loyalty; it is the ideology of leadership in technological progress through pursuit of the ideal of technical efficiency. In 1895, in an ASCE presidential address, George S. Morison, one of America's premier bridge-builders, spelled out this ideal in a bold vision of the engineer as the primary agent of technical change and the main force behind human progress. In Morison's words:

> We are the priests of material development, of the work which enables other men to enjoy the fruits of the great sources of power in Nature, and of the power of mind over matter. We are the priests of the new epoch, without superstitions . . .[3]

During the first third of the twentieth century such a vision of expanded engineering activity spawned the technocracy movement and the idea that engineers should have political and economic power. Economist Thorstein Veblen, for example, argued in *The Engineers and the Price System* (1921) that freeing engineers from subservience to business interests so they could exercise their own standards of good and bad, right and wrong, would result in a stronger economy and better consumer products.[4]

Again, there are evident truths in arguments for technocratic leadership and the pursuit of efficiency. Certainly the subordination of production to short-term profit-making without any concern for the good of the product is not desirable in the long run, and inefficiency or waste readily seems to be another denomination of badness. Moreover, in a highly complex technical world it is often difficult for average citizens to know what would be in their own best interests. Nevertheless, technical decisionmaking as an end can easily be separated from general human welfare. The pursuit of technical perfection for its own sake is not always the best use of limited societal resources—such as when engineers design cars to go faster and faster, despite the speed limit of 55 miles per hour. The ideal of efficiency also virtually requires the assumption of clearly defined boundary conditions that easily exclude relevant factors, including legitimate psychological and human concerns.

The World War II mobilization of science and engineering for national purpose and the North American post-war social recovery contributed to a temporary suppression of the tension between technical and economic ends, efficiency and profit, highlighted by the technocracy movement. But opposition to nuclear weapons in the 1950s and 1960s, together with the consumer and environmental movements of the 1960s and 1970s, provoked some engineers to challenge both national and business directions. In conjunction with a renewed concern for democratic values—especially as a result of the civil-rights movement—this challenge led to new ideas about engineering ethics. In the United States the seeds of this transformation were planted in 1947 when the Engineers' Council for Professional Development (ECPD, later the Accreditation Board for Engineering and Technology or ABET) drew up the first trans-disciplinary engineering ethics code. It committed the engineer "to interest himself [or herself] in public welfare." Revisions in 1963 and 1974 strengthened this commitment to the point where the first of four "fundamental principles" requires engineers to use "their knowledge and skill for the enhancement of human welfare," and the first of seven "fundamental canons" states that "Engineers shall hold paramount the safety, health and welfare of the public. . . ."

Because the third distinct emphasis in engineering ethics, social responsibility, meets many objections that can be raised against the first two, it has been widely adopted by the professional engineering community. It also allows for retention of the most desirable elements from the two earlier theories. Engineers can maintain loyalty, for instance, but within a more inclusive framework. Now loyalty is not to an individual or corporation but to the public as a whole. Leadership in technical development likewise remains, but it is now subordinated to the common welfare, especially in regard to public health and safety.

The idea of engineers' social responsibility, however, does not necessarily involve any citizen participation in decisionmaking. An engineer committed to the promotion of public safety, health, and welfare may make decisions about technical issues in a paternalistic manner at odds with democratic ideals. Indeed, recognition of the reality that technology often brings with it not only benefits but also costs and risks argues for granting all those affected by technical decisions some say in them. As a result, some have argued for the principle of "no innovation without representation."[5] Such representation does not imply veto power, but rather intelligent and relevant lay involvement. In accordance with this participation principle, the role of the engineer as technical specialist would become less that of independent decisionmaker and more that of participant in an educational dialogue and contributor to various regulatory processes within appropriate democratic governmental structures and guidelines. Acceptance of the idea of social responsibility, especially as modified by the participation principle, has nevertheless been surrounded by considerable practical and theoretical debate. Three cases from recent engineering ethics discussions highlight these practical questions.

Three Case Studies in Engineering Design Research

A fundamental practical problem for the engineering community has been how to develop an autonomy that would enable engineers to practice a professional commitment to the primacy of public safety, health, and welfare without making them independent arbiters of the public good or subverters of democratic process. The truth is that engineers often have the power to make technical decisions for the public that can easily lead to the promotion of non-public interests. One of the most influential of these "conflicts of interests" is exemplified in the Hydrolevel case. At the same time, in comparison with scientists, engineers are "more on tap than on top." This fact has been driven home by case studies of disasters associated with design flaws in the cargo bay doors on the DC-10, the Goodrich A7D airbrakes, the gas tank on the Ford Pinto, the skywalks at the Kansas City Hyatt Regency, and so on. Two of the most widely influential cases of design flaws concern the Bay Area Rapid Transit (BART) system and the space shuttle Challenger.

In the mid-1800s, in response to an increasing number of public fatalities resulting from steamboat boiler explosions, the U.S. government undertook to regulate boiler manufacture and operation. As they evolved, the resultant regulatory agencies became the enforcers of technical standards that by the early 1900s had become the responsibility of the American Society of Mechanical Engineers (ASME). In the process, research engineers, in a benevolent technocratic manner, had clearly taken on a responsibility for helping to protect public safety and welfare.[6] By the mid-1900s this responsibility and relationship had become perhaps too well established. In the late 1960s research at a small engineering firm, Hydrolevel Corporation, developed a new type of low-water fuel cutoff device for steam boilers that threatened the business of McDonnell and Miller, Inc. (M&M), the primary supplier of such devices. When an appeal was made to the ASME for an interpretation of section HG-605a (a 43-word paragraph) in its 18,000-page *Boiler and Pressure Vessel Code* in order to certify the new Hydrolevel design, ASME members who were also involved with M&M acted to secure a negative response. The result was a lawsuit that went all the way to the Supreme Court, which in 1982 ruled that ASME had violated the Sherman Anti-Trust Act. (Hydrolevel also eventually went out of business because it could not market its new product.)[7] A commitment to protect public safety and welfare through technical ideals and engineering autonomy had been used as a subterfuge for protecting the welfare of a private corporation.

Another famous case of design flaws arose in the late 1960s when metropolitan San Francisco decided to create the Bay Area Rapid Transit (BART) system. Designed to provide the most advanced rapid transit in the world, BART would eliminate both operators and conductors in favor of an automatic train control (ATC) technology. Construction began in 1964. At the end of 1971, almost three years behind schedule and considerably over budget, BART was finally nearing the first stage of completion. However, Holger Hjortsvang, a research engineer working on the ATC, became seriously concerned about BART's design and testing—especially the attempt to deal with design problems in a very complex project on a piecemeal rather than a systemic basis. Beginning in 1969 he expressed these concerns to management, and by late 1971 he found himself supported by two newly hired engineers: Max Blankenzee, a senior program analyst, and Robert Bruder, an electrical-electronics construction engineer. For months these three engineers expressed their concerns to management both orally and in writing, only to be consistently ignored. Finally, in early 1972 the engineers contacted a member of the BART District Board of Directors and gave him papers documenting their concerns. Soon afterwards, newspaper stories appeared, followed by a February meeting of the Board that yielded a split vote of confidence in BART management. Management then undertook to identify the sources for certain critical documents acquired by the Board, and fired Hjortsvang, Blankenzee, and Bruder at the beginning of March.

The engineers, however, appealed to the California Society of Professional Engineers (CSPE) for support, arguing that they were only attempting to live up to a professional code of obligation to hold "the public welfare paramount" and to "notify the proper authorities of any observed conditions which endanger public safety and health." In June CSPE submitted a report to the California State Senate that largely supported the engineers. Then in October, in dramatic confirmation of their concerns, an ATC failure caused a BART train to overrun a station, injuring four passengers and an attendant.

In 1986 the need for engineering independence again came to public attention because of the explosion of the space shuttle Challenger. Once more a major, high-tech, government-funded project was years behind schedule, considerably over budget, and thus subject to strong economic and political pressures to meet new and unrealistic deadlines. As came to light afterward, Roger Boisjoly, the Morton-Thiokol mechanical engineer in charge of research and design for the field joints on the solid rocket booster, had been questioning the safety of the Challenger's O-ring seals for almost a year. And the night prior to the January launch Boisjoly and other engineers had explicitly opposed continuing the countdown, only to have their decision overridden by senior management. Because of their testimony before the Presidential Commission during its post-disaster investigation, these engineers came under severe pressure from Morton-Thiokol. But Boisjoly, who ultimately resigned, also became an outspoken advocate for both greater autonomy in the engineering profession and the inclusion of engineering ethics in engineering curricula, thus helping to promote development of engineering ethics courses in engineering colleges throughout the United States.[8]

Uniting these three cases is not only the practical problem of promoting the right kind of engineering autonomy, but also what might be called the principle of public disclosure—a principle clearly related to that of public participation. Supporting such a principle is the argument that public good is served by a duty to disclose especially to those who might be affected by the full process of technical decisionmaking, as well as by shortcomings in areas of safety, health, and welfare.

The Perspectives of Research and Design

The problems that typically call for whistleblowing are not equally apparent in the general practice of engineering and in engineering research. Indeed, this is probably one of the things that makes research attractive to many: it avoids moral dilemmas. It is "only" research. But the term "research" can have weak and strong meanings. It can refer to (a) that aspect of the investigation of a problem that plays a continuing subsidiary role in any project, (b) the initial conceptualization and planning of a project, or (c) the specialized activity that constitutes the systematic deepening and elaboration of the engineering sciences. Of course, engineering research in the stronger senses referring to (b) and (c) is only a small part of engineering.[9]

It is worth noting that the very term "research" is distinctly modern, derived by way of the French *rechercher* (*re-*, intensifying prefix + *chercher*, to seek for) from the late Latin *recercare*. There are no classical Latin or Greek forms of the world. The "intensive searching" that implies an active or "pushy" inquiry contrasts with the more leisured and detached, observational study exhibited by premodern learning. Indeed, the term first became prominent in English during the 1600s in conjunction with the rise of modern experimental science. But not until the early 1900s did "research" become associated with engineering to designate work directed "in an industrial context" toward "the innovation, introduction, and improvement of products and processes."[10] Research directly for "making" is almost 300 hundred years younger than research for knowing.

From the perspective of engineering research in this strong sense, the codes of professional ethics appear singularly weak if not irrelevant. They focus on what most engineers do, which is simply not research. Unlike scientific research, engineering research occurs on the margins—at one end of a spectrum of activities ranging from research and development through construction or production to operation, management, and maintenance. Only a few engineers engage primarily in research, although a much larger proportion no doubt include research as some limited component of their engineering work.

Design, not research, unifies the various engineering activities. The research engineer investigates new principles and processes which a development engineer can utilize for designing the prototype of a new device. The production engineer can then modify the prototype design so that it is more easily manufactured, while operational and maintenance engineers can use practical experience to shape the design to fit given applications. Throughout all these processes, the central activity is designing, which must take into account production, marketing, maintenance, and use. Designing, in turn, is taken into account by research. Engineering research exists to contribute not to knowledge *per se* but to the design activity, that is, to "the creation of systems, devices, and processes useful to, and sought by, society."[11]

The exact character of design activity is the subject of debate. For present purposes it is not necessary to explore all aspects of these debates, but simply to note that design involves a kind of making, or making in miniature,[12] and as such has immediate impact on the world, however small that impact might be. Unlike science, then, which can plausibly claim that the outcome of its central activity, research, is knowledge—which can only have an indirect or mediated impact on the world—the outcome of *engineering design* is a physical object that perforce becomes part of the physical world. If it succeeds, it may even become a very big part of the physical world.

What is distinctive about *engineering research* as research is what has already been called its pushy character, its determination to discover and apply "new facts, techniques, and natural laws."[13] The application of techniques in engineering is further motivated by a "creative imagination" that "is always dissatisfied with present methods and equipment," and that ever seeks "newer, cheaper, better means of using natural sources of energy and materials to improve the standard of living and diminish toil."[14] The engineer's uneasiness or restlessness imparts to and picks up from research a more profoundly active character—so that engineering research is more active than scientific research, and engineering based on or utilizing research is more active than engineering otherwise. Engineering research, oriented toward providing more effective foundations and support for design, both takes on and transmits the engineer's restlessness into effective and powerful means of transforming projects. Whereas science might be said to take the world into the laboratory, engineering research takes the laboratory into the world. Indeed, it eventually makes of the world a laboratory.

The method of "testing to destruction," in which researchers intentionally load or operate materials or devices until failure, in order to discover their limits, is a particularly revealing form of engineering research. One may instructively compare testing to destruction with the systematic observation of biological field work—and even with the controlled measurement of isolated events under varied conditions in physics. The latter case, however, where scientific experiments constrain nature in order to confirm a hypothesis, as with Galileo's inclined planes, simply involves the altering of a natural motion for effective observation, without any explicit intent to destroy. Testing to destruction is, of course, closely related to traditional "cut and try" methods of construction; the difference is that, with testing to destruction, these methods are developed and pursued systematically. Moreover, with systematic development, testing to destruction readily leaves the engineering laboratory and becomes part of general engineering practice. This is the upshot of civil engineer Henry Petroski's book, *To Engineer Is Human* (1985). According to Petroski, "the concept of failure . . . is central to understanding engineering." Although

colossal disasters . . . are ultimately failures of design, . . . the lessons learned from those disasters can do more to advance engineering knowledge than all the successful machines and structures in the world. Indeed, failures appear to be inevitable in the wake of prolonged success, which encourages lower margins of safety. Failures in turn lead to greater safety margins and, hence, new periods of success. To understand what engineering is and what engineers do is to understand how failures can happen and how they can contribute more than successes to advance technology.[15]

Because public testing or "using to failure" is a normal part of the engineering-society interaction, the philosopher-engineer team of Mike Martin and Roland Schinzinger argue that engineering is properly described as "social experimentation." According to Martin and Schinzinger, although experimentation is crucial to engineering, what is involved

is not . . . an experiment conducted solely in a laboratory under controlled conditions. Rather, it is an experiment on a social scale involving human subjects.[16]

This transformation of the world into a laboratory, in which engineering design research reaches out through engineering practice, is the foundation of the unique ethical challenges numerous authors identify as the special burden of technological society. Because of engineering research, the practice of engineering now has a greater impact across space and through time than any other human action. It also extends more deeply into human nature, both psychologically (behavior modification) and physiologically (genetic engineering). Finally, the opening of such technical possibilities through engineering design research tends to draw human action into its vortex.[17] As the atomic scientist J. Robert Oppenheimer remarked about Edward Teller's design for an H-bomb, the possibility was so "technically sweet" it could not fail to be tried.[18] The wisdom of promoting such restlessness must be an ultimate issue of the ethics of engineering design research.

The external features of engineering design research are complemented by an "internal" trait of modeling. Modeling is not only miniature making but also a simplification with a paradoxical character. On the one side, because simplification is oriented toward making, it is not particularly concerned with truth. Less than fully true models, although conceptually shallow, can be technologically powerful or rich.[19] But engineering modeling, in giving up on explicit concern for truth, unavoidably takes on an ethical dimension. On the other side, precisely because of the simplifications of engineering research modeling, ethical reflection on all aspects of the design research becomes especially difficult. Free body diagrams, for instance, treat an object *as if* all forces were acting directly on its center of gravity, in order to model matter-force interactions. But this "as if" also denotes something that *is not*. The simplified model is not the complex reality. Not all forces act on the center of gravity of some object; many act on its surface which, given certain shapes, may subtly distort the result. To concentrate on gross problems it is not only permissible to ignore complex subtleties, but better to do so. The paradox is that precisely because models do not reflect the complexity of reality, they have the power to change what is. The Euclidean model of the landscape as a flat surface, for example, enables one more easily to divide it. Likewise, the modeling of terrestrial gravity—as independent of the sun, the moon, the planets, and all variations in geology or physical geography—makes possible calculation of the trajectories of military projectiles. The carefully defined boundary conditions within which the mechanical engineer determines efficiency, taking into account only mechanical energy and heat, but not social dislocation, pollution, or biological destruction, makes possible improvements in the strictly mechanical functioning of engines.

More generally, looking at the world as a whole as if it were a clock or the brain as if it were a computer brings with it tremendous power to transform the thing modeled precisely by overlooking the rich complexity of its reality, by abstracting from the life-world within which we actually live. It is thus no accident that engineering design research, given its inner demands for simplification, should sometimes fail to consider one of the most subtle dimensions of the life-world—ethics.

Ethics in Engineering Design Research

When engineers overlook many aspects of reality, an essentially inner feature of engineering design research, they not only harness power, but they court the danger of overlooking something important. Indeed, engineering failures typically reveal that something has been overlooked. Such failures are social as well as technical. Because failures, like successes, can be expected outcomes of the simplified modeling inherent in engineering design research, such research has also developed counter or compensatory principles. These principles force engineering projects to develop more complex models. Thus, engineering design research contains not only a movement toward simplification but also a countermovement toward "complexification." Systems engineering, interdisciplinary engineering research, the transformation of civil into environmental engineering, and multifactor technology assessments all illustrate this latter tendency.

In the BART case, for instance, Hjortsvang argued for a more complex testing of the ATC system. He was concerned that the piecemeal testing being done was not able to reveal the full potential for problems, and that isolated problems-solving did not adequately consider the ways various solutions might interact under real world conditions to render the system dysfunctional; to harm users; to cover up real design problems; and to produce bad engineering. Hjortsvang thus argued for establishing an interdisciplinary research team to oversee ATC design and development. The problem Hjortsvang pointed to is a specific instance of a more general difficulty related to "the paradox of information technology." This paradox is grounded in the fact that, beyond a certain point, human beings *"will never be able to model"* (and thereby check) in all relevant ways—particularly speed and complexity—data processing operations. Indeed, "the possibility of controlling information processing systems diminishes in proportion to the introduction of modeling or checking instances" because these actually further complicate a program.[20]

In the early 1900s the National Academy of Engineering held a symposium on "Engineering as a Social Enterprise" that argued for a sociotechnical systems interpretation of the profession. As symposium organizer, aeronautical engineer Walter Vincenti pointed out how engineers regularly have to deal with technical systems and are thus familiar with how such systems must be subdivided for analysis. "In the sociotechnical model, the entire society is visualized," according to Vincenti, "as a vast integrated system, with varied social and technical areas of human activity as major interacting subsystems." And although "this subdivision is made so that each subsystem can be analyzed in quasi isolation," for Vincenti "the crucial point" is that "analysis must be carried out . . . with attention at all times to the interactions between and constraints on the subsystems and to the eventual need to reassemble the system."[21] Only such a complex understanding can address the problems raised by previous failures in engineering simplification.

While implicit in systems and interdisciplinary engineering design research, the principle at issue here is seldom formulated. From the perspective of engineering design research, the fundamental technical obligation in the face of failure amid complexity can be phrased as: "Take more factors into account." The obligation to take more into account also has a moral dimension, not simply because it can on occasion avoid some specific harm. In fact, as in the BART case, an objection against taking more into account by means of more complex testing is almost always that it will cause one type of harm (i.e., greater financial costs) while only possibly avoiding another type of harm. The moral dimension of taking more into account is only realized by linking engineering design research into general reflections on the good. In this sense the duty to take more into account may be termed a duty *plus respicere* (from the Latin *plus*, more + *respicere*, to be concerned about).[22] Indeed, a central feature of deeply ethical behavior is that it is based on as wide a reflective base as possible. One reason why altruism is superior to selfishness is that it takes more into account—others as well as oneself—and considers a broader perspective. In the historicophilosophical development of ethics codes in professional engineering, one way of arguing the superiority of the principle of social responsibility is to maintain that it calls for more generous or inclusive reflection, and that through social responsibility company loyalty and technical efficiency are placed in a more comprehensive framework. Stating the issue even more pointedly: To take more into account in engineering will include taking ethics into account. Thus an imperative to seek to reduce ethical remoteness becomes part of the duty *plus respicere*.

Three Examples

Consider briefly three examples: engineering design research for developing countries, for the military, and for mass consumer products such as toys. Each of these cases shows the importance of considering more factors.

With regard to engineering research for developing countries, E. F. Schumacher's *Small Is Beautiful* (1973) argued that the mere transposition of hardware from a highly technological country to a much less technological country almost certainly does not constitute an effective technology transfer. As Schumacher says,

> To do justice to the real situation it is necessary to consider the reactions and capabilities of people, and not confine oneself to machinery or abstract concepts. [I]t is wrong to assume that the most sophisticated equipment, transplanted into an unsophisticated environment, will be regularly worked at full capacity. . . .[23]

The use of the terms "sophisticated" and "unsophisticated" here require qualification. Technically unsophisticated societies can be very sophisticated in other ways. But the basic point is that development—even conceived too simply as no more than a move from the use of technically simple (and labor intensive) tools to the technically complex (capital intensive) machines—can only proceed through the use of appropriately "intermediate" technologies. Such technologies must be designed by engineers who consider more than strictly technical factors, who make themselves aware of the broader social context of the countries within or for which they are working. For instance, the idea of research for the design of intermediate or appropriate technologies for developing countries, taking into account more than strictly technical factors, has become allied with concerns in developed countries for the design of environmentally sensitive technologies. The design of alternative energy technologies for developed countries, sometimes termed "soft technologies," also must look beyond presently available technical processes and expectations to consider broader issues such as long-term availability of resources, the proper allocation of social capital, and environmental impacts. This is certainly another form of taking more factors into account.[24]

With regard to military weapons in advanced technological societies, considerations of context also readily point beyond the strictly technical. Such considerations can include, for example, questions as diverse as those concerning the effective use of a weapon by a military force of some defined educational level and distinctions between offensive versus defensive capabilities. Of course, to some extent these questions involve military policy and politics, not engineering. But the research that goes into weapons development, even to be effective engineering, cannot wholly ignore such factors. No hard and fast line divides strictly technical issues (associated, for example, with the engineering design research for a wire-guided anti-tank missile) and military tactics and strategy. General design specifications devised by military strategists depend on end-user beliefs about technical capabilities, and attempts to meet such specifications can in turn alter these beliefs. Because of this, the military research design engineer will on occasion be called upon to engage in a dialogue of mutual education with military policy analysts and perhaps even politicians. Such dialogue stretches engineering practice beyond simple laboratory analysis. Indeed, as both a technical professional and a citizen in a democracy, the engineer involved in design research for the military may even have an obligation to help educate the general public about certain technical issues. Engineers have at least sometimes felt this to be the case, as indicated by the healthy debate that arose within the technical engineering community about President Reagan's Strategic Defense Initiative. Electrical engineer Stephen Unger has gone further and argued that engineers should allow an admitted plurality of personal moral values to influence the kinds of research work with which they become involved. If "conscientious objections create serious problems in staffing a project, then there is good reason indeed to pause and reconsider the arguments against that project."[25] (Somewhat ironically, Unger also thinks that the wide diversity of moral viewpoints among engineers probably undermines the possibility of any strong impact on the course of technical change.)

Engineers in one of the most civilian forms of research—high-tech toys—must likewise consider issues much broader than the simple technical functioning and safety of their designs. Such consideration is required because of the pervasive use of toys by children and the deep influence toys can have over many aspects of psychological development. Subliminal influences from "Barbie Dolls" (about sex roles), "Transformers" (about personal identity), video games (about isolation and the pleasures of violence)—and even from the pervasiveness of plastics (about the independent featurelessness of the materials of the world)—all deserve consideration. The explicit educational and habit-forming implications likewise deserve attention. The simple technical pleasures of designing and playing with toys should not overshadow such potential critical reflection, although reflection itself need not lead to a rejection of all unmediated delight in the technical. Engineers in their choices of research tasks and the kinds of project possibilities they suggest to corporate employers are simply called upon to help companies exercise true discernment in formulating their fundamental policies of producing and marketing products.

Practical Guidelines for Engineering Design Research

Summarizing and going beyond even these three rather expansive examples, civil engineer George Bugliarello has argued that the social responsibility of the engineer should include upholding human dignity, avoiding dangerous or uncontrolled side effects, making provisions for possible technological failures, avoiding the reinforcement of outworn social systems, and participating in discussions about the "why" of various technologies. For Bugliarello, "engineering can best carry out its social purpose when it is involved in the formation of the response to a social need, rather than just being called to provide a quick technological fix."[26] However, the duty to take more into account, the duty *plus respicere*, is an admittedly loose and somewhat shapeless obligation, perhaps even describable as a "soft" ethics. To take this soft ethics a bit further, the following questions might serve as useful guidelines for self-interrogation by research engineers.

Reviewing the basic argument concerning the limitations of simplification, a research engineer could ask:

- Are the models we use sufficiently complex to include a diversity of non-standard technical factors?
- Does reflective analysis include explicit consideration of ethical issues?

From the perspective of the three examples of development, weapons, and toys, design research engineers could further ask themselves such questions as:

- Have we made an effort to consider the broad social context of the engineering research, including impacts on the environment?
- Have we critically examined end-user assumptions?
- Have we undertaken the research in dialogue with personal moral principles and with the larger non-technical community?
- Have we given more direct consideration to peripheral implications of the research?

In summary, whereas the research scientist might ask: Is this knowledge significant? The research engineer could ask: Is this project worthwhile?

Throughout this chapter there has been an implicit assumption about the ways engineering differs from science. It is not only crucial to recognize the distinctive difference of engineering design research, but also equally important to attend to how this kind of research influences science. As a number of observers have noted, technological means have increasingly influenced contemporary science. From telescopes and microscopes to particle accelerators, computers, and space probes, contemporary science has become progressively dependent on the engineering research design of scientific instrumentation. As a result one can reasonably hypothesize that those duties most characteristic of ethics in engineering research—that is, duties related to participation, public disclosure, and *plus respicere*—may play a role in the ethics of scientific research as well.

Notes

1. See, e.g., Shakespeare, *Troilus and Cressida* (act 2, scene 3, line 8), where Achilles is termed "a rare engineer"; and *Hamlet* (act 3, scene 4, line 206), where: ". . . the engineer / Hoist with his own petard."
2. The term "civil engineering" continues in some European languages to denote all non-military engineering.
3. George S. Morison, "Address at the Annual Convention at the Hotel Pemberton, Hull, Mass., June 19th, 1895," *Transactions of the American Society of Civil Engineers* 33, no. 6 (June 1895): 483. This speech is revised and incorporated as chapter six into Morison's *The New Epoch—As Developed by the Manufacture of Power* (Boston: Houghton Mifflin, 1903).
4. The best single study of this period is Edwin T. Layton, Jr., *The Revolt of the Engineers: Social Responsibility and the American Engineering Profession* (Baltimore: Johns Hopkins University Press, 1986). First published 1971. The focus is on the years 1900–1945.
5. Steven L. Goldman, "No Innovation Without Representation: Technological Action in a Democratic Society," in *New Worlds, New Technologies, New Issues*, ed. Stephen H. Cutcliffe, Steven L. Goldman, Manuel Medina, and José Sanmartín (Bethlehem, PA: Lehigh University Press, 1992), pp. 148–60. See also K. S. Shrader-Frechette, *Risk and Rationality: Philosophical Foundations for Populist Reforms* (Berkeley: University of California Press, 1991).
6. For detail on this history see John G. Burke, "Bursting Boilers and the Federal Power," *Technology and Culture* 7, no. 1 (Winter 1966): 1–23. Reprinted in Marcel C. Lafollette and Jeffrey K. Stine, eds. *Technolgoy and Choice: Readings from Technology and Culture* (Chicago: University of Chicago Press, 1991), pp. 43–65.
7. For a good summary of the Hydrolevel case with references to the relevant legal documents and historical studies, see Paula Wells, Hardy Jones, and Michael Davis, *Conflicts of Interests in Engineering* (Dubuque, Iowa: Kendal-Hunt, 1986).
8. For documentation on this case, see Roger Boisjoly, "The Challenger Disaster: Moral Responsibility and the Working Engineer," in *Ethical Issues in Engineering*, ed. Deborah G. Johnson (Englewood Cliffs, NJ: Prentice-Hall, 1991), pp. 6–14.
9. The term "research" is, for instance, conspicuous by its absence in the article on "Engineering Design" (the main entry on engineering) in the *McGraw-Hill Encyclopedia of Science and Technology*, 7th edition (New York: Mc-

Graw-Hill, 1992), vol. 6, pp. 392–99, although "research engineering" is often mentioned in standard textbook introductions to the engineering profession.

10. *Oxford English Dictionary*, 2nd edition (1989), vol. 13, p. 692, col. 3, top.

11. "Engineering Design," *McGraw-Hill Encyclopedia of Science and Technology*, 7th edition (New York: McGraw-Hill, 1992), vol. 6, p. 392.

12. For an argument for this thesis plus relevant discussion of the debate concerning design, see Carl Mitcham, "Engineering as Productive Activity: Philosophical Remarks," in *Critical Perspectives on Nonacademic Science and Engineering*, ed. Paul T. Durbin (Bethlehem, PA: Lehigh University Press, 1991), pp. 80–117.

13. "Research," *McGraw-Hill Dictionary of Scientific and Technical Terms*, 3rd edition (New York: McGraw-Hill, 1984), p. 1362.

14. "Engineering," *McGraw-Hill Encyclopedia of Science and Technology*, 7th edition (New York: McGraw-Hill, 1992), vol. 6, p. 387.

15. Henry Petroski, *To Engineer Is Human: The Role of Failure in Successful Design* (New York: St. Martin's Press, 1985), p. xii.

16. Mike Martin and Roland Schinzinger, *Ethics in Engineering*, 2nd edition (New York: McGraw-Hill, 1989), p. 64.

17. For two arguments to this effect, see Hans Jonas, "Technology as a Subject for Ethics," *Social Research* 49, no. 4 (Winter 1982): 891–98; and Hans Lenk, "Notes on Extended Responsibility and Increased Technological Power," in *Philosophy and Technology*, Boston Studies in the Philosophy of Science, vol. 80, ed. Paul T. Durbin and Friedrich Rapp (Boston: D. Reidel, 1983), pp. 196–97.

18. *In the Matter of J. Robert Oppenheimer: Transcript of Hearing before Personnel Security Board*, Washington, D.C., April 12–May 6, 1954 (Washington, D.C.: U.S. Government Printing Office, 1954), p. 251.

19. Cf. Mario Bunge, *Scientific Research II* (New York: Springer, 1967), pp. 123 ff.

20. Walter C. Zimmerli, "Who Is To Blame for Data Pollution? On Individual Moral Responsibility with Information Technology," in *Philosophy and Technology II: Information Technology and Computers in Theory and Practice*, Boston Studies in the Philosophy of Science, vol. 90, ed. Carl Mitcham and Alois Huning (Boston: D. Reidel, 1986), p. 296.

21. Walter G. Vincenti, "Introduction," in *Engineering as a Social Enterprise*, ed. Hedy E. Sladovich (Washington, D.C.: National Academy Press, 1991), p. 2.

22. Note also that *respicere* is composed of *re-*, intensifying prefix + *specere*, to look at or behold, the latter of which is related to the Greek (from whence comes the English "skepticism"). Indeed, this might also be argued to be the engineering equivalent of René Dubos' motto: "Think globally, act locally."

23. E. F. Schumacher, *Small Is Beautiful* (New York: Harper & Row, 1973), p. 182.

24. For a good example of such alternative energy engineering design research, see the work of Amory Lovins, e.g., *Soft Energy Paths: Toward a Durable Peace* (San Francisco: Friends of the Earth; Cambridge, Mass.: Ballinger, 1977).

25. Stephen H. Unger, *Controlling Technology: Ethics and the Responsible Engineer* (New York: Holt, Rinehart and Winston, 1982), p. 38. See also pp. 4–6.

26. George Bugliarello, "The Social Function of Engineering: A Current Assessment," in *Engineering as a Social Enterprise*, ed. Hedy E. Sladovich (Washington, D.C.: National Academy Press, 1991), pp. 77, 81.

Study Questions

1. What is the main point of Mitcham's article? Why do you think it might be correct? Why do you think it might be wrong? Explain and defend your answers.
2. What assumptions does Mitcham make about technology? about ethics? Do you think these assumptions are correct? Explain and defend your answers.
3. If society accepts Mitcham's main conclusions, what consequences might follow? Would these consequences be desirable? Explain and defend your answers.
4. Explain how the notions of company loyalty, technocratic leadership, and social responsibility have influenced technology, especially engineering design technology.
5. Why does Mitcham use the BART example? Explain.
6. What values issues arise in connection with developing technology for developing nations, for the military, and for mass consumer products? Explain.
7. Ethically evaluate Mitcham's "practical guidelines."

4.3

Ethics and Automobile Technology: The Pinto Case

Richard T. De George

The myth that ethics has no place in engineering has been attacked, and at least in some corners of the engineering profession has been put to rest.[1] Another myth, however, is emerging to take its place—the myth of the engineer as moral hero. A litany of engineering saints is slowly taking form. The saints of the field are whistle blowers, especially those who have sacrificed all for their moral convictions. The zeal of some preachers, however, has gone too far, piling moral responsibility upon moral responsibility on the shoulders of the engineer. This emphasis, I believe, is misplaced. Though engineers are members of a profession that holds public safety paramount,[2] we cannot reasonably expect engineers to be willing to sacrifice their jobs each day for principle and to have a whistle ever by their sides ready to blow if their firm strays from what they perceive to be the morally right course of action. If this is too much to ask, however, what then is the actual ethical responsibility of engineers in a large organization?

I shall approach this question through a discussion of what has become known as the Pinto case, i.e., the trial that took place in Winamac, Indiana, and that was decided by a jury on March 16, 1980.

In August 1978 near Goshen, Indiana, three girls died of burns in a 1973 Pinto that was rammed in traffic by a van. The rear-end collapsed "like an accordion,"[3] and the gas tank erupted in flames. It was not the first such accident with the Pinto. The Pinto was introduced in 1971 and its gas tank housing was not changed until the 1977 model. Between 1971 and 1978 about fifty suits were brought against Ford in connection with rear-end accidents in the Pinto.

What made the Winamac case different from the fifty others was the fact that the State prosecutor charged Ford with three (originally four, but one was dropped) counts of reckless homicide, a *criminal* offense, under a 1977 Indiana law that made it possible to bring such criminal charges against a corporation. The penalty, if found guilty, was a maximum fine of $10,000 for each count, for a total of $30,000. The case was closely watched, since it was the first time in recent history that a corporation was charged with this criminal offense. Ford spent almost a million dollars in its defense.

With the advantage of hindsight I believe the case raised the right issue at the wrong time.

The prosecution had to show that Ford was reckless in placing the gas tank where and how it did. In order to show this the prosecution had to prove that Ford consciously disregarded harm it might cause and the disregard, according to the statutory definition of "reckless," had to involve "substantial deviation from acceptable standards of conduct."[4]

The prosecution produced seven witnesses who testified that the Pinto was moving at speeds judged to be between 15 and 35 mph when it was hit. Harly Copp, once a high ranking Ford engineer, claimed that the Pinto did not have a balanced design and that for cost reasons the gas tank could withstand only a 20 mph impact without leaking and exploding. The prosecutor, Michael Cosentino, tried to introduce evidence that Ford knew the defects of the gas tank, that its executives knew that a $6.65 part would have made the car considerably safer, and that they decided against the change in order to increase their profits.

Federal safety standards for gas tanks were not introduced until 1977. Once introduced, the National Highway Traffic Safety Administration (NHTSA) claimed a safety defect existed in the gas tanks of Pintos produced from 1971 to 1976. It ordered that Ford recall 1.9 million Pintos. Ford contested the order. Then, without ever admitting that the fuel tank was unsafe, it "voluntarily" ordered a recall. It claimed the recall was not for safety but for "reputational" reasons.[5] Agreeing to a recall in June, its first proposed modifications failed the safety standards tests, and it added a second protective shield to meet safety standards. It did not send out recall notices until August 22. The accident in question took place on August 10. The prosecutor claimed that Ford knew its fuel tank was dangerous as early as 1971 and that it did not make any changes until the 1977 model. It also knew in June of 1978 that its fuel tank did not meet federal safety standards; yet it did nothing to warn owners of this fact. Hence, the prosecution contended, Ford was guilty of reckless homicide.

The defense was led by James F. Neal who had achieved national prominence in the Watergate hearings. He produced testimony from two witnesses who were crucial to the case. They were hospital attendants who had spoken with the driver of the Pinto at the hospital before she died. They claimed she had stated that she had just had her car filled with gas. She had been in a hurry and had left the gas station without replacing the cap on her gas tank. It fell off the top of her car as she drove down the highway. She noticed this and stopped to turn around to pick it up. While stopped, her car was hit by the van. The testimony indicated that the car was stopped. If the car was hit by a van going 50 mph, then the rupture of the gas tank was to be expected. If the cap was off the fuel tank, leakage would be more than otherwise. No small vehicle was made to withstand such impact. Hence, Ford claimed, there was no recklessness involved. Neal went on to produce films of tests that indicated that the amount of damage the Pinto suffered meant that the impact must have been caused by the van's going at least 50 mph. He further argued that the Pinto gas tank was at least as safe as the gas tanks on the 1973 American Motors Gremlin, the Chevrolet Vega, the Dodge Colt, and the Toyota Corolla, all of which suffered comparable damage when hit from the rear at 50 mph. Since no federal safety standards were in effect in 1973, Ford was not reckless if its safety standards were comparable to those of similar cars made by competitors; that standard represented the state of the art at that time, and it would be inappropriate to apply 1977 standards to a 1973 car.[6]

The jury deliberated for four days and finally came up with a verdict of not guilty. When the verdict was announced at a meeting of the Ford Board of Directors then taking place, the members broke out in a cheer.[7]

These are the facts of the case. I do not wish to second-guess the jury. Based on my reading of the case, I think they arrived at a proper decision, given the evidence. Nor do I wish to comment adversely on the judge's ruling that prevented the prosecution from introducing about 40% of his case because the evidence referred to 1971 and 1972 models of the Pinto and not the 1973 model.[8]

The issue of Ford's being guilty of acting recklessly can, I think, be made plausible, as I shall indicate shortly. But the successful strategy argued by the defense in this case hinged on the Pinto in question being hit by a van at 50 mph. At that speed, the defense successfully argued, the gas tank of any subcompact would rupture. Hence that accident did not show that the Pinto was less safe than other subcompacts or that Ford acted recklessly. To show that would require an accident that took place at no more than 20 mph.

The contents of the Ford documents that Prosecutor Cosentino was not allowed to present in court were published in the *Chicago Tribune* on October 13, 1979. If they are accurate, they tend to show grounds for the charge of recklessness.

Ford had produced a safe gas tank mounted over the rear axle in its 1969 Capri in Europe. It tested that tank in the Capri. In its over-the-axle position, it withstood impacts of up to 30 mph. Mounted behind the axle, it was punctured by projecting bolts when hit from the rear at 20 mph. A $6.65 part would help make the tank safer. In its 1971 Pinto, Ford chose to place the gas tank behind the rear axle without the extra part. A Ford memo indicates that in this position the Pinto has more trunk space, and that production costs would be less than in the over-the-axle position. These considerations won out.[9]

The Pinto was first tested it seems in 1971, after the 1971 model was produced, for rear-end crash tolerance. It was found that the tank ruptured when hit from the rear at 20 mph. This should have been no surprise, since the Capri tank in that position had ruptured at 20 mph. A memo recommends that rather than making any changes Ford should wait until 1976 when the government was expected to introduce fuel tank standards. By delaying making any change, Ford could save $20.9 million, since the change would average about $10 per car.[10]

In the Winamac case Ford claimed correctly that there were no federal safety standards in 1973. But it defended itself against recklessness by claiming its car was comparable to other subcompacts at that time. All the defense showed, however, was that all the subcompacts were unsafe when hit at 50 mph. Since the other subcompacts were not forced to recall their cars in 1978, there is *prima facie* evidence that Ford's Pinto gas tank mounting was substandard. The Ford documents tend to show Ford knew the danger it was inflicting on Ford owners; yet it did nothing, for profit reasons. How short-sighted those reasons were is demonstrated by the fact that the Pinto thus far in litigation and recalls alone has cost Ford $50 million. Some forty suits are still to be settled. And these figures do not take into account the loss of sales due to bad publicity.

Given these facts, what are we to say about the Ford engineers? Where were they when all this was going on, and what is their responsibility for the Pinto? The answer, I suggest, is that they were where they were supposed to be, doing what they were supposed to be doing. They were performing tests, designing the Pinto, making reports. But do they have no moral responsibility for the products they design?

What after all is the moral responsibility of engineers in a large corporation? By way of reply, let me emphasize that no engineer can morally do what is immoral. If commanded to do what he should not morally do, he must resist and refuse. But in the Ford Pinto situation no engineer was told to produce a gas tank that would explode and kill people. The engineers were not instructed to make an unsafe car. They were morally responsible for knowing the state of the art, including that connected with placing and mounting gas tanks. We can assume that the Ford engineers were cognizant of the state of the art in producing the model they did. When tests were made in 1970 and 1971, and a memo was written stating that a $6.65 modification could make the gas tank safer,[11] that was an engineering assessment. Whichever engineer proposed the modification and initiated the memo acted ethically in doing so. The next step, the administrative decision not to make the modification was, with hindsight, a poor one in almost every way. It ended up costing Ford a great deal more not to put in the part than it would have cost to put it in. Ford still claims today that its gas tank was as safe as the accepted standards of the industry at that time.[12] It must say so, otherwise the suits pending against it will skyrocket. That it was not as safe seems borne out by the fact that only the Pinto of all the subcompacts failed to pass the 30 mph rear impact NHTSA test.

But the question of wrongdoing or of malicious intent or of recklessness is not so easily solved. Suppose the ordinary person were told when buying a Pinto that if he paid an extra $6.65 he could increase the safety of the vehicle so that it could withstand a 30 mph rear-end impact rather than a 20 mph impact, and that the odds of suffering a rear-end impact of between 20 and 30 mph was 1 in 250,000. Would we call him or her reckless if he or she declined to pay the extra $6.65? I am not sure how to answer that question. Was it reckless of Ford to wish to save the $6.65 per car and increase the risk for the consumer? Here I am inclined to be clearer in my own mind. If I choose to take a risk to save $6.65, it is my risk and my $6.65. But if Ford saves the $6.65 and I take the risk, then I clearly lose. Does Ford have the right to do that without informing me, if the going standard of safety of subcompacts is safety in a rear-end collision up to 30 mph? I think not. I admit, however, that the case is not clear-cut, even if we add that during 1976 and 1977 Pintos suffered 13 firey fatal rear-end collisions, more than double that of other U.S. comparable cars. The VW Rabbit and Toyota Corrola suffered none.[13]

Yet, if we are to morally fault anyone for the decision not to add the part, we would censure not the Ford engineers but the Ford executives, because it was not an engineering but an executive decision.

My reason for taking this view is that an engineer cannot be expected and cannot have the responsibility to second-guess managerial decisions. He is responsible for bringing the facts to the attention of those who need them to make decisions. But the input of engineers is only one of many factors that go to make up managerial decisions. During the trial, the defense called as a witness Francis Olsen, the assistant chief engineer in charge of design at Ford, who testified that he bought a 1973 Pinto for his eighteen-year-old daughter, kept it a year, and then traded it in for a 1974 Pinto which he kept two years.[14] His testimony and his actions were presented as an indication that the Ford engineers had confidence in the Pinto's safety. At least this one had enough confidence in it to give it to his daughter. Some engineers at Ford may have felt that the car could have been safer. But this is true of almost every automobile. Engineers in large firms have an ethical responsibility to do their jobs as best they can, to report their observations about safety and improvement of safety to management. But they do not have the obligation to insist that their perceptions or their standards be accepted. They are not paid to do that, they are not expected to do that, and they have no moral or ethical obligation to do that.

In addition to doing their jobs, engineers can plausibly be said to have an obligation of loyalty to their employers, and firms have a right to a certain amount of confidentiality concerning their internal operations. At the same time engineers are required by their professional ethical codes to hold the safety of the public paramount. Where these obligations conflict, the need for and justification of whistle blowing arises.[15] If we admit the obligations on both sides, I would suggest as a rule of thumb that engineers and other workers in a large corporation are morally *permitted* to go public with information about the safety of a product if the following conditions are met:

1) if the harm that will be done by the product to the public is serious and considerable;
2) if they make their concerns known to their superiors; and
3) if, getting no satisfaction from their immediate superiors, they exhaust the channels available within the corporation, including going to the board of directors.

If they still get no action, I believe they are morally *permitted* to make public their views; but they are not morally *obliged* to do so. Harly Copp, a former Ford executive and engineer, in fact did criticize the Pinto from the start and testified for the prosecution against Ford at the Winamac trial.[16] He left the company and voiced his criticism. The criticism was taken up by Ralph Nader and others. In the long run it led to the Winamac trial and probably helped in a number of other suits filed against Ford. Though I admire Mr. Copp for his actions, assuming they were done from moral motives, I do not think such action was morally required, nor do I think the other engineers at Ford were morally deficient in not doing likewise.

For an engineer to have a moral *obligation* to bring his case for safety to the public, I think two other conditions have to be fulfilled, in addition to the three mentioned above.[17]

4) He must have documented evidence that would convince a reasonable, impartial observer that his view of the situation is correct and the company policy wrong.

Such evidence is obviously very difficult to obtain and produce. Such evidence, however, takes an engineer's concern out of the realm of the subjective and precludes that concern from being simply one person's opinion based on a limited point of view. Unless such evidence is available, there is little likelihood that the concerned engineer's view will win the day simply by public exposure. If the testimony of Francis Olsen is accurate, then even among the engineers at Ford there was disagreement about the safety of the Pinto.

5) There must be strong evidence that making the information public will in fact prevent the threatened serious harm.

This means both that before going public the engineer should know what source (government, newspaper, columnist, TV reporter) will make use of his evidence and how it will be handled. He should also have good reason to believe that it will result in the kind of change or result that he believes is morally appropriate. None of this was the case in the Pinto situation. After much public discussion, five model years, and failure to pass national safety standards tests, Ford plausibly defends its original claim that the gas tank was acceptably safe. If there is little likelihood of his success, there is no moral obligation for the engineer to go public. For the harm he or she personally incurs is not offset by the good such action achieves.[18]

My first substantive conclusion is that Ford engineers had no moral *obligation* to do more than they did in this case.

My second claim is that though engineers in large organizations should have a say in setting safety standards and producing cost-benefit analyses, they need not have the last word. My reasons are two. First, while the degree of risk, e.g., in a car, is an engineering problem, the acceptability of risk is not. Second, an engineering cost-benefit analysis does not include all the factors appropriate in making a policy decision, either on the corporate or the social level. Safety is one factor in an engineering design. Yet clearly it is only one factor. A Mercedes-Benz 280 is presumably safer than a Ford Pinto. But the difference in price is considerable. To make a Pinto as safe as a Mercedes it would probably have to cost a comparable amount. In making cars as in making many other objects some balance has to be reached between safety and cost. The final decision on where to draw the balance is not only an engineering decision. It is also a managerial decision, and probably even more appropriately a social decision.

The difficulty of setting standards raises two pertinent issues. The first concerns federal safety standards. The second concerns cost-benefit analyses. The state of the art of engineering technology determines a floor below which no manufacturer should ethically go. Whether the Pinto fell below that floor, we have already seen, is a controverted question. If the cost of achieving greater safety is considerable—and I do not think $6.65 is considerable—there is a built-in temptation for a producer to skim more than he should and more than he might like. The best way to remove that temptation is for there to be a national set of standards. Engineers can determine what the state of the art is, what is possible, and what the cost of producing safety is. A panel of informed people, not necessarily engineers, should decide what is acceptable risk and hence what acceptable minimum standards are. Both the minimum standards and the standards attained by a given car should be a matter of record that goes with each car. A safer car may well cost more. But unless a customer knows how much safety he is buying for his money, he may not know which car he wants to buy. This information, I believe, is information a car buyer is entitled to have.

In 1978, after the publicity that Ford received with the Pinto and the controversy surrounding it, the sales of Pintos fell dramatically. This was an indication that consumers preferred a safer car for comparable money, and they went to the competition. The state of Oregon took all its Pintos out of its fleet and sold them off. To the surprise of one dealer involved in selling turned-in Pintos, they went for between $1000 and $1800.[19] The conclusion we correctly draw is that there was a market for a car with a dubious safety record even though the price was much lower than for safer cars and lower than Ford's manufacturing price.

The second issue is the way cost-benefit analyses are produced and used. I have already mentioned one cost-benefit analysis used by Ford, namely, the projection that by not adding a part and by placing the gas tank in the rear the company could save $20.9 million. The projection, I noted, was grossly mistaken for it did not consider litigation, recalls, and bad publicity which have already cost Ford over $50 million. A second type of cost-benefit analysis sometimes estimates the number and costs of suits that will have to be paid, adds to it fines, and deducts that total amount from the total saved by a particular practice. If the figure is positive, it is more profitable not to make a safety change than to make it.

A third type of cost-benefit analysis, which Ford and other auto companies produce, estimates the cost and benefits of specific changes in their automobiles. One study, for instance, deals with the cost-benefit analysis relating to fuel leakage associated with static rollover. The unit cost of the part is $11. If that is included in 12.5 million cars, the total cost is $137 million. That part will prevent 180 burn deaths, 180 serious burn injuries and 2100 burned vehicles. Assigning a cost of $200,000 per death, $67,000 per major injury, and $700 per vehicle, the benefit is $49.5 million. The cost-benefit ratio is slightly over 3–1.[20]

If this analysis is compared with a similar cost-benefit analysis for a rear-end collision, it is possible to see how much safety is achieved per dollar spent. This use is legitimate and helpful. But the procedure is open to very serious criticism if used not in a comparative but in an absolute manner.

The analysis ignores many factors, such as the human suffering of the victim and of his or her family. It equates human life to $200,000, which is based on average lost future wages. Any figure here is questionable, except for comparative purposes, in which case as long as the same figure is used it does not change the information as to relative benefit per dollar. The ratio, however, has no *absolute* meaning, and no decision can properly be based on the fact that the resulting ratio of cost to benefit in the above example is 3 to 1. Even more important, how can this figure or ratio be compared with the cost of styling? Should the $11 per unit to reduce death and injury from roll-over be weighed against a comparable $11 in rear-end collision or $11 in changed styling? Who decides how much more to put into safety and how much more to put into styling? What is the rationale for the decision?

In the past consumers have not been given an opportunity to vote on the matter. The automobile industry has decided what will sell and what will not, and has decided how much to put on safety. American car dealers have not typically put much emphasis on safety features in selling their cars. The assumption that American drivers are more interested in styling than safety is a decision that has been made for them, not by them. Engineers can and do play an important role in making cost-benefit analyses. They are better equipped than anyone else to figure risks and cost. But they are not better equipped to figure the acceptability of risk, or the amount that people should be willing to pay to eliminate such risk. Neither, however, are the managers of automobile corporations. The amount of acceptable risk is a public decision that can and should be made by representatives of the public or by the public itself.

Since cost-benefit analyses of the types I have mentioned are typical of those used in the auto industry, and since they are inadequate ways of judging the safety a car should have, given the state of the art, it is clear that the automobile companies should not have the last word or the exclusive word in how much safety to provide. There must be national standards set and enforced. The National Highway Traffic Safety Administration was established in 1966 to set standards. Thus far only two major standards have been established and implemented: the 1972 side impact standard and the 1977 gasoline tank safety standard. Rather than dictate standards, however, in which process it is subject to lobbying, it can mandate minimum standards and also require auto manufacturers to inform the public about the safety quotient of each car, just as it now requires each car to specify the miles per gallon it is capable of achieving. Such an approach would put the onus for basic safety on the manufacturers, but it would also make additional safety a feature of consumer interest and competition.

Engineers in large corporations have an important role to play. That role, however, is not usually to set policy or to decide on the acceptability of risk. Their knowledge and expertise are important both to the companies for which they work and to the public. But they are not morally responsible for policies and decisions beyond their competence and control. Does this view, however, let engineers off the moral hook too easily?

To return briefly to the Pinto story once more, Ford wanted a subcompact to fend off the competition of Japanese imports. The order came down to produce a car of 2,000 pounds or less that would cost $2000 or less in time for the 1971 model. This allowed only 25 months instead of the usual 43 months for design and production of a new car.[21] The engineers were squeezed from the start. Perhaps this is why they did not test the gas tank for rear-end collision impact until the car was produced.

Should the engineers have refused the order to produce the car in 25 months? Should they have resigned, or leaked the story to the newspapers? Should they have refused to speed up their usual routine? Should they have complained to their professional society that they were being asked to do the impossible—if it were to be done right? I am not in a position to say what they should have done. But with the advantage of hindsight, I suggest we should ask not only what they should have done. We should especially ask what changes can be made to prevent engineers from being squeezed in this way in the future.

Engineering ethics should not take as its goal the producing of moral heroes. Rather it should consider what forces operate to encourage engineers to act as they feel they should not; what structural or other features of a large corporation squeeze them until their consciences hurt? Those features should then be examined, evaluated, and changes proposed and made. Lobbying by engineering organizations would be appropriate, and legislation should be passed if necessary. In general I tend to favor voluntary means where possible. But where that is utopian, then legislation is a necessary alternative.

The need for whistle blowing in a firm indicates that a change is necessary. How can we preclude the necessity for blowing the whistle?

The Winamac Pinto case suggests some external and internal modifications. It was the first case to be tried under a 1977 Indiana law making it possible to try corporations as well as individuals for the criminal offenses of reckless homicide. In bringing the charges against Ford, Prosecutor Michael Cosentino acted courageously, even if it turned out to have been a poor case for such a precedent-setting trial. But the law concerning reckless homicide, for instance, which was the charge in question, had not been rewritten with the corporation in mind. The penalty, since corporations cannot go to jail, was the maximum fine of $10,000 per count—hardly a significant amount when contrasted with the 1977 income of Ford International which was $11.1 billion in revenues and $750 million in profits. What Mr. Cosentino did *not* do was file charges against individuals in the Ford Company who were responsible for the decisions he claimed were reckless. Had highly placed officials been charged, the message would have gotten through to management across the country that individuals cannot hide behind corporate shields in their decisions if they are indeed reckless, put too low a price on life and human suffering, and sacrifice it too cheaply for profits.

A bill was recently proposed in Congress requiring managers to disclose the existence of life-threatening defects to the appropriate federal agency.[22] Failure to do so and attempts to conceal defects could result in fines of $50,000 or imprisonment for a minimum of two years, or both. The fine in corporate terms is negligible. But imprisonment for members of management is not.

Some argue that increased litigation for product liability is the way to get results in safety. Heavy damages yield quicker changes than criminal proceedings. Ford agreed to the Pinto recall shortly after a California jury awarded damages of $127.8 million after a youth was burned over 95% of his body. Later the sum was reduced, on appeal, to $6.3 million.[23] But criminal proceedings make the litigation easier, which is why Ford spent $1,000,000 in its defense to avoid paying $30,000 in fines.[24] The possibility of going to jail for one's actions, however, should have a salutary effect. If someone, the president of a company in default of anyone else, were to be charged in criminal suit, presidents would soon know whom they can and should hold responsible below them. One of the difficulties in a large corporation is knowing who is responsible for particular decisions. If the president were held responsible, outside pressure would build to reorganize the corporation so that responsibility was assigned and assumed.

If a corporation wishes to be moral or if society or engineers wish to apply pressure for organizational changes such that the corporation acts morally and responds to the moral conscience of engineers and others within the organization, then changes must be made. Unless those at the top set a moral tone, unless they insist on moral conduct, unless they punish immoral conduct and reward moral conduct, the corporation will function without considering the morality of questions and of corporate actions. It may by accident rather than by intent avoid immoral actions, though in the long run this is unlikely.

Ford's management was interested only in meeting federal standards and having these as low as possible. Individual federal standards should be both developed and enforced. Federal fines for violations should not be token but comparable to damages paid in civil suits and should be paid to all those suffering damage from violations.[25]

Independent engineers or engineering societies—if the latter are not co-opted by auto manufacturers—can play a significant role in supplying information on the state of the art and the level of technical feasibility available. They can also develop the safety index I suggested earlier, which would represent the relative and comparative safety of an automobile. Competition has worked successfully in many areas. Why not in the area of safety? Engineers who work for auto manufacturers will then have to make and report the results of standard tests such as the ability to withstand rear-end impact. If such information is required data for a safety index to be affixed to the windshield of each new car, engineers will not be squeezed by management in the area of safety.

The means by which engineers with ethical concerns can get a fair hearing without endangering their jobs or blowing the whistle must be made part of a corporation's organizational structure. An outside board member with primary responsibility for investigating and responding to such ethical concerns might be legally required. When this is joined with the legislation pending in Congress which I mentioned, the dynamics for ethics in the organization will be significantly improved. Another way of achieving a similar end is by providing an inspector general for all corporations with an annual net income of over $1 billion. An independent committee of an engineering association might be formed to investigate charges made by engineers concerning the safety of a product on which they are working;[26] a company that did not allow an appropriate investigation of employee charges would become subject to cover-up proceedings. Those in the engineering industry can suggest and work to implement other ideas. I have elsewhere outlined a set of ten such changes for the ethical corporation.[27]

In addition to asking how an engineer should respond to moral quandaries and dilemmas, and rather than asking how to educate or train engineers to be moral heroes, those in engineering ethics should ask how large organizations can be changed so that they do not squeeze engineers in moral dilemmas, place them in the position of facing moral quandaries, and make them feel that they must blow the whistle.

The time has come to go beyond sensitizing students to moral issues and solving and resolving the old, standard cases. The next and very important questions to be asked as we discuss each case is how organizational structures can be changed so that no engineer will ever again have to face *that* case.

Many of the issues of engineering ethics within a corporate setting concern the ethics of organizational structure, questions of public policy, and so questions that frequently are amenable to solution only on a scale larger than the individual—on the scale of organization and law. The ethical responsibilities of the engineer in a large organization have as much to do with the organization as with the engineer. They can be most fruitfully approached by considering from a moral point of view not only the individual engineer but the framework within which he or she works. We not only need moral people. Even more importantly we need moral structures and organizations. Only by paying more attention to these can we adequately resolve the questions of the ethical responsibility of engineers in large organizations.

Notes

1. The body of literature on engineering ethics is now substantive and impressive. See *A Selected Annotated Bibliography of Professional Ethics and Social Responsibility in Engineering*, compiled by Robert F. Ladenson, James Choromokos, Ernest d'Anjou, Martin Pimsler, and Howard Rosen (Chicago: Center for the Study of Ethics in the Professions, Illinois Institute of Technology, 1980).

A useful two-volume collection of readings and cases is also available: Robert J. Baum and Albert Flores, *Ethical Problems in Engineering*, 2nd edition (Troy, N.Y.: Rensselaer Polytechnic Institute, Center for the Study of the Human Dimensions of Science and Technology, 1980). See also Robert J. Baum's *Ethics and Engineering Curricula* (Hastings-on-Hudson, N.Y.: Hastings Center, 1980).

2. See, for example, the first canon of the 1974 Engineers Council for Professional Development Code, the first canon of the National Council of Engineering Examiners Code, and the draft (by A. Oldenquist and E. Slowter) of a "Code of Ethics for the Engineering Profession" (all reprinted in Baum and Flores, *Ethical Problems in Engineering*).

3. Details of the incident presented in this paper are based on testimony at the trial. Accounts of the trial as well as background reports were carried by both the *New York Times* and the *Chicago Tribune*.

4. *New York Times*, February 17, 1980, IV, p. 9.

5. *New York Times*, February 21, 1980, p. A6. *Fortune*, September 11, 1978, p. 42.

6. *New York Times*, March 14, 1980, p. 1

7. *Time*, March 24, 1980, p. 24.

8. *New York Times*, January 16, 1980, p. 16; February 7, 1980, p. 16.

9. *Chicago Tribune*, October 13, 1979, p. 1, and Section 2, p. 12.

10. *Chicago Tribune*, October 13, 1979, p. 1; *New York Times*, October 14, 1979, p. 26.

11. *New York Times*, February 4, 1980, p. 12.

12. *New York Times*, June 10, 1978, p. 1; *Chicago Tribune*, October 13, 1979, p. 1, and Section 2, p. 12. The continuous claim has been that the Pinto poses "No serious hazards."

13. *New York Times*, October 26, 1978, 103.

14. *New York Times*, February 20, 1980, p. A16.

15. For a discussion of the conflict, see Sissela Bok, "Whistleblowing and Professional Responsibility," *New York University Educational Quarterly*, pp. 2–10. For detailed case studies see, Ralph Nader, Peter J. Petkas, and Kate Blackwell, *Whistle Blowing* (New York: Grossman Publishers, 1972); Charles Peters and Taylor Branch, *Blowing the Whistle: Dissent in the Public Interest* (New York: Praeger Publishers, 1972); and Robert M. Anderson, Robert Perrucci, Dan E. Schendel and Leon E. Trachtman, *Divided Loyalties: Whistle-Blowing at BART* (West Lafayette, Indiana: Purdue University, 1980).

16. *New York Times*, February 4, 1980, p. 12.

17. The position I present here is developed more full in my book *Business Ethics* (New York: Macmillan, 1981). It differs somewhat from the dominant view expressed in the existing literature in that I consider whistle blowing an extreme measure that is morally obligatory only if the stringent conditions set forth are satisfied. Cf. Kenneth D. Walters, "Your Employees' Right to Blow the Whistle," *Harvard Business Review*, July–August, 1975.

18. On the dangers incurred by whistle blowers, see Gene James, "Whistle-Blowing: Its Nature and Justification," *Philosophy in Context*, 10 (1980), pp. 99–

117, which examines the legal context of whistle blowing; Peter Raven-Hansen, "Dos and Don'ts for Whistleblowers: Planning for Trouble," *Technology Review*, May 1980, pp. 34–44, which suggests how to blow the whistle; Helen Dudar, "The Price of Blowing the Whistle," *The New York Times Magazine*, 30 October, 1977, which examines the results for whistleblowers; David W. Ewing, "Canning Directions," *Harpers*, August, 1979, pp. 17–22, which indicates "how the government rids itself of troublemakers" and how legislation protecting whistleblowers can be circumvented; and Report by the U.S. General Accounting Office, "The Office of the Special Counsel Can Improve Its Management of Whistleblower Cases," December 30, 1980 (FPCD–81–10).

19. *New York Times*, April 21, 1978, IV, p. 1, 18.

20. See Mark Dowie, "Pinto Madness," *Mother Jones*, September/October, 1977, pp. 24–28.

21. *Chicago Tribune*, October 13, 1979, Section 2, p. 12.

22. *New York Times*, March 16, 1980, IV, p. 20.

23. *New York Times*, February 8, 1978, p. 8.

24. *New York Times*, February 17, 1980, IV, p. 9; January 6, 1980, p. 24; *Time*, March 24, 1980, p. 24.

25. *The Wall Street Journal*, August 7, 1980, p. 7, reported that the Ford Motor Company "agreed to pay a total of $22,500 to the families of three Indiana teen-age girls killed in the crash of a Ford Pinto nearly two years ago. . . . A Ford spokesman said the settlement was made without any admission of liability. He speculated that the relatively small settlement may have been influenced by certain Indiana laws which severely restrict the amount of damages victims or their families can recover in civil cases alleging wrongful death."

26. A number of engineers have been arguing for a more active role by engineering societies in backing up individual engineers in their attempts to act responsibly. See Edwin Layton, *Revolt of the Engineers* (Cleveland: Case Western Reserve, 1971); Stephen H. Unger, "Engineering Societies and the Responsible Engineer," *Annals of the New York Academy of Sciences*, 196 (1973), pp. 433–37 (reprinted in Baum and Flores, *Ethical Problems in Engineering*, pp. 56–59); and Robert Perrucci and Joel Gerstl, *Profession Without Community: Engineers in American Society* (New York: Random House, 1969).

27. Richard T. De George, "Responding to the Mandate for Social Responsibility," *Guidelines for Business When Societal Demands Conflict* (Washington, D.C.: Council for Better Business Bureaus, 1978), pp. 60–80.

Study Questions

1. What is the main point of De George's article? Why do you think it might be correct? Why do you think it might be wrong? Explain and defend your answers.

2. What assumptions does De George make about technology? about ethics? Do you think these assumptions are correct? Explain and defend your answers.

3. If society accepts De George's main conclusions, what consequences might follow? Would these consequences be desirable? Explain and defend your answers.

4. Is De George right in saying that engineers do not have a responsibility to second-guess managers who could be doing something wrong? Explain and defend your response.

5. Summarize some of the problems with benefit-cost analysis, as De George describes them. How could benefit-cost analyses be improved?

6. Does the "organizational structure" of business and technology need to change in order to avoid problems like that of the Pinto? Why or why not?

4.4

Computers and Privacy

Stacey L. Edgar

> I give the fight up: let there be an end,
> A privacy, an obscure nook for me.
> I want to be forgotten even by God.
> Robert Browning, *Paracelsus*

> The grave's a fine and private place,
> But none I think do there embrace.
> Andrew Marvell, "To His Coy Mistress"

Human beings may invade your privacy, but in our modern technological society it is a variety of fancy monitoring devices and computerized record-keeping and record-processing procedures—initiated, run, and analyzed by humans—that pose the greatest threats to personal privacy. One cannot help but think of the two-way telescreens in George Orwell's *1984* or the "Boss" monitoring Charlie Chaplin's bathroom break in *Modern Times*, if the potential of much of today's technology is examined closely.[1]

To pursue our analogy with guns a bit further, defenders of the right to own guns say, "It is not guns that kill people—people kill people." Similarly, "it is not computers that invade privacy—people invade privacy." However, the storage, processing, and monitoring capabilities of today's computers make these invasions of privacy much easier and perhaps more likely.

Originally published as "Privacy," chapter 7 in *Morality and Machine: Perspectives in Computer Ethics*. Reprinted with permission of the author and publisher. Published by Jones and Bartlett Publishers, Inc. Copyright © 1997.

Why Is Privacy of Value?

John Locke, in the *Second Treatise of Government*, argued that anyone who threatens your (private) property potentially threatens your life, and so you are justified in using the same measures against such a person (including killing him or her) that are permissible in protecting your life. To Locke, the most fundamental human rights were to life, liberty, and property. The defense of private property put forward by Locke connected it with one's right to own one's own body and any natural extensions of that body through one's labor.

The right to one's own, if valid, can be interpreted to include one's thoughts and actions. Thus one would seem to have a right to protect those thoughts and actions against unwanted intrusions, in the same way that no one should violate someone else's body (by rape or murder, for example). The logical exceptions to this are if one's actions infringe on another's rights; then they are subject to control by others. We have a notion of having a personal "space" into which we can invite others but into which no one should trespass unbidden. Yet modern surveillance techniques and methods of record analysis and "matching" pose a threat to invade our personal spaces, often without our knowledge and certainly without our permission.

The Stoics found, in bad times, that even though their bodies might be in chains, their minds could still be free, and it was on this presupposition that they turned inward to engage in the freedom of their thoughts. If one's mental activity is also threatened, there is nothing left to make life worthwhile.

In order to be fully developed human beings, we must be free to choose. Kant says that the moral individual must be rational and *autonomous*. I cannot be autonomous if there are outside forces directing my decisions. My choices must be private to be truly *mine*, to be truly *choices*. One of the highest-level goods is the exercise of the will, displaying courage or fortitude. These are not group activities. One's courage, one's knowledge, and even one's pleasure are all personal, private. They could not exist without privacy. Your courage may help others, and you can share your knowledge, but they originate and develop in your own private world. Thus if privacy is not intrinsically valuable in itself, it makes possible achievements that *are* intrinsically good.

Kant also stressed the value of a *person* as an end in itself, and that persons should never be used as means. Constant surveillance diminishes your personhood. The world is no longer as you think it is, but rather a mini-world under the microscope of some larger world. You become a *means* to some other end—that of governmental power or corporate profit. Even if you do not ever know you are being observed, the value of what you hold in high regard is diminished, made into a cold statistic. If you *do* realize that you are being observed, then this affects your actions and you are no longer autonomous.

Charles Fried argues that privacy is essential to the fundamental good of friendship. Friendship entails love, trust, and mutual respect, and these can only occur in a *rational context* of privacy.[2] Trust presumes that some things are secret or confidential and will not be revealed; if there is no privacy, there can be no such confidences. A love relationship involves a certain intimacy that is just between two people; if all of their actions, conversations, and special gifts are made public, something valuable has been lost.

Privacy may simply be necessary to mental survival, just as the body needs sleep. Much of the time we are on a public stage, performing in class, in front of a classroom, at work for an employer, or acting in a play. This requires concentration and effort; we work hard to put forward a certain appearance and a high level of performance to others. It would be very difficult, if not impossible, to do this all of the time. Thus we need some privacy in which to rest, to kick off our shoes, put our feet up, and do and think what we please, *in private*. We are talking about preservation of mental health here.

A utilitarian argument can be made that invasions of privacy cause, overall, more harm than good. There is the potential for inaccurate information being spread around about a person, or even information that is accurate, but is no one else's concern, being misused. Think carefully about the sorts of instances in which you value your privacy. Your bank account? Your grades? (Of course, you may have to disclose them to a potential employer or graduate school, and the administration knows; but colleges, under federal law, must protect your records from casual prying.) Your bathroom activities? (Presumably, you would prefer not to have these broadcast over the airwaves.) Your sex life, or lack thereof? (There are legal protections in most states for what consenting adults do in the privacy of their own homes; these legal protections are grounded in fundamental ethical principles, which recognize privacy as a good, or as contributing to other goods.) Your medical records? Your private fantasies? What else would you add to the list?

Just what is it about each of the preceding cases that you value and that would be compromised if your privacy in these respects was invaded? A totalitarian government, like that of Nazi Germany or in Orwell's *1984*, might want to know everything about its citizens. Those citizens would then cease to be autonomous, freely choosing beings; they would be more like animals in a laboratory or cogs in a machine.

Alexis de Tocqueville, writing about his observations of the United States in 1832, described the sort of despotism that he feared could overcome democratic nations:

> Above this race of men stands an immense and tutelary power, which takes upon itself alone to secure their gratifications, and to watch over their fate. That power is absolute, minute, regular, provident, and mild. It would be like the authority of a parent, if, like that authority, its object was to prepare men for manhood; but it seeks, on the contrary, to keep them in perpetual childhood: it is well content that the people should rejoice, provided they think of nothing but rejoicing. For their happiness such a government willingly labors, but it chooses to be the sole agent and the only arbiter of that happiness; it provides for their security, foresees and supplies their necessities, facilitates their pleasures, manages their principal concerns, directs their industry, regulates the descent of property, and subdivides their inheritances: what remains, but to spare them all the care of thinking and all the trouble of living? . . .
>
> It covers the surface of society with a network of small complicated rules, minute and uniform, through which the most original minds and the most energetic characters cannot penetrate, to rise above the crowd. The will of man is not shattered, but softened, bent, and guided; men are seldom forced by it to act, but they are constantly restrained from acting: such a power does not destroy, but it prevents existence; it does not tyrannize, but it compresses, enervates, extinguishes, and stupefies a people, till each nation is reduced to be nothing better than a flock of timid and industrious animals, of which the government is the shepherd. (Tocqueville, 1956, pp. 303–304)

You probably remember what it was like to be a young child. You don't get any privacy. Adults are constantly hovering over you, afraid you will hurt yourself or some object and telling you "no" a lot. As one grows up, one is given more and more autonomy, control of one's own life, and *privacy*. Who would want to give this up by going back to total surveillance?

"I'll be Watching You"

Gary T. Marx uses the lyrics from the rock song "Every Breath You Take" (The Police) to point up the scope of current surveillance in our society:

every breath you take	[breath analyzer]
every move you make	[motion detection]
every bond you break	[polygraph]

every step you take	[electronic anklet]
every single day	[continuous monitoring]
every word you say	[bugs, wiretaps, mikes]
every night you stay . . .	[light amplifier]
every vow you break . . .	[voice-stress analyzer]
every smile you fake	[brain wave analysis]
every claim you stake . . .	[computer matching]
I'll be watching you.	[video surveillance][3]

This is a very dramatic way to point up what Marx calls "the new surveillance." This might also be called *dataveillance*; we are living in a world where data, or information, has become the most active commodity. One buys and sells information—on which products the affluent neighborhoods are buying, on the "best buys" in goods (*Comsumer Reports*), on who are the most likely prospects to buy into a land deal, and so on.

Information about you is sold; the magazine you subscribe to sells your mailing address to other magazines, or to those who sell products that connect with the theme of the magazine. There is an increasing demand for information that can be found in government records, and a recent *New York Times* article comments that, "for government, selling data can provide extra revenue in times of tight budgets."[4] Vance Packard had noted as early as 1964, in *The Naked Society*, that privacy was becoming harder to maintain, and surveillance was becoming more pervasive.

Some Noteworthy Violations of Privacy Involving Computers

Ed Pankau, the president of Inter-Tect, a Houston investigative agency, wrote a book called *How to Investigate by Computer*. In it, he describes how easy it is to sit at a computer terminal and gather information about someone, and how lucrative it can be for an agency like his. He says that much of his business comes from single women who want potential romantic partners investigated. "For $500, Inter-Tect investigative agency in Houston promises to verify within a week a person's age, ownership of businesses, history of bankruptcies, if there are any tax liens, appearance in newspaper articles, as well as divorces and children."[5]

Procter & Gamble, disturbed when they discovered that someone had leaked confidential company information to the *Wall Street Journal*, got a court order for the phone company in Cincinnati, Ohio, to turn over the phone numbers of anyone who had called the home or office of the *Journal* reporter on the article. This involved scanning millions of calls made by 650,000 phone units.[6] However, this sort of search is made incredibly easy by computerization. What is scary is the impact on private lives of the ease of getting such information.

In California in 1989, Rebecca Schaeffer, a television actress, was murdered by a man who got her home address through the Department of Motor Vehicles records. Since then, California has restricted access to DMV files, but they can still be accessed readily by car dealers, banks, insurance companies, and process servers. Since anyone can register as a process server for under $100, this does not afford much protection of such files.[7]

Although this is not an invasion of *personal* privacy, it makes an interesting example of the power of computer correlation. The scholars who have been working with the Dead Sea Scrolls had very slowly been releasing the documents as they finished working with them and had denied other groups any access to them. However, they did publish concordances of some of the scrolls that had not yet been published. The concordances list all the significant words and their place of occurrence in the text. A group at Union Hebrew College in Cincinnati "reverse-engineered" the concordance, by computer, to recreate one of the unreleased texts.[8] Again, this would not have been possible without the use of a computer. What the Dead Sea Scrolls scholars had hoped to keep private until they were ready to publish was made public through the use of a program that simply was able to manipulate and correlate the partial information they had released.

Software Engineering Notes reported on a middle-school computer in Burbank that calls the homes of absent students each night until it gets a live voice or an answering machine, to check on the validity of the absences.[9] This kind of surveillance could easily be extended to other areas—the workplace in particular.

Records disclosed in 1992 indicate that many employees ("dozens") of the Los Angeles Police Department had been using the police computer files to investigate babysitters, house sitters, and others (potential dates?) for personal reasons.[10] When we think of people in the police departments and in the tax return offices[11] *snooping* into our files for their own personal reasons, when they are not the ones authorized to process those records, it should send chills up and down our spines. Who is snooping into our bank records, our charge-card records, and our phone records, and for what reasons?

In *The Naked Consumer*, Erik Larson describes an interview he had with Jonathan Robbin, president of the Virginia "target-marketing company" Claritas. The company uses computerized *cluster analysis* of census data into different neighborhood types and then examines (and predicts) their consumer habits. They call this analysis *geodemographics*;

some of the categories identified are Blue Blood Estates (the richest), Gray Power (the oldest), Public Assistance (the poorest), Pools & Patios, Shotguns & Pickups, Tobacco Roads, and Bohemian Mix (Larson, 1992, pp. 46–47). Claritas has many customers, including the U.S. Army, ABC, American Express, Publishers Clearing House, and big manufacturers of consumer goods such as Coca-Cola, Colgate-Palmolive, and R.J. Reynolds Tobacco. Robbins calls his computer system "a 'proesthetic' device for the mind" (Larson, 1992, p. 47)—one that allows corporate minds to "perceive" new targets. For example, the new Saturn division of GM used the Claritas program PRIZM (Potential Rating Index for Zip Markets) "as part of its vast market research campaign to figure out what kind of car to build and how to sell it" (Larson, 1992, p. 48).

Lotus Corporation was going to market a software package called Lotus Marketplace: Households, which was a database containing information on 120,000,000 Americans, including names, addresses, incomes, consumer preferences, and other personal data. In January 1991, however, Lotus withdrew the package due to an avalanche of complaints about its invasion of privacy.

In *Privacy for Sale*, Jeffrey Rothfeder describes how easy it is to access credit rating information on anyone; he was able to get extensive information on Dan Quayle and Dan Rather. He describes how Oral Roberts was able to buy lists of debtors whose accounts were overdue sixty days or more. Roberts used these lists to send letters addressing the person by first name, saying that he is a friend, and that "it's time to get out from under a load of debt, a financial bondage." The letter continued, commiserating with the recipient's financial burden, and then gets to the point: to plant a seed to get out of the financial bondage by sending a gift of $100 to Oral Roberts, so he could intercede with God and "begin the war on your debt" (Rothfeder, 1992, pp. 23–24).

Rothfeder discusses the fact that your credit reports might not be private, but your video-rental records are. This is because of a personal incident involving a government official. When Robert Bork was a candidate for the Supreme Court in 1987, "an enterprising reporter" at *City Paper*, a Washington weekly, went to Bork's local video store and got a list of the movies that Bork had rented recently (nothing salacious; mostly John Wayne). The paper published the list of titles anyway, and "lawmakers were outraged—and quickly passed the Video Privacy Protection Act of 1988" (Rothfeder, 1992, p. 27).

Rothfeder describes a service called PhoneFile where, for a fee, you can get the address, telephone number, and "household makeup" of almost anyone in the United States, from just the name and the state of residence (1992, p. 29). A New York City woman named Karen

Hochmann received a call in Fall of 1988 from a salesman for ITT who wanted her to choose their long-distance service. When she said no, that she didn't make many out-of-town calls, the salesman said, "I'm surprised to hear you say that; I see from your phone records that you frequently call Newark, Delaware, and Stamford, Connecticut." Hochmann was "shocked" and "scared" by this blatant invasion of her personal records (Rothfeder, 1992, p. 89).

Rothfeder reports on the case of a young man whose American Express card was cancelled because the company determined that he did not have enough money in his bank balance to pay his March bill (1992, p. 17). Apparently, American Express made regular checks into people's financial accounts to determine the likelihood of their being able to pay their AmEx tabs.

There exist databases on employees in heavy industries like construction, oil, and gas which are accessed regularly by employers (for example, EIS—Employers Information Service in Louisiana). They contain information such as which employees have demanded worker's compensation; the employer consulting such a database may then decide not to hire such a potential employee (Rothfeder, 1992, pp. 154–157). There also apparently are databases maintained by landlords which can be used to determine that they do not want to rent to people who show up in the database as having complained to previous landlords (even if such complaints are legitimate about lack of heat or water or rodent control).

The coming together of the information superhighway and the proposals for national health care are going to raise serious problems in terms of the confidentiality of medical records. There is a principle that goes back to Hippocrates in the fifth century B.C. that a doctor should hold a patient's medical information to be confidential. But the movement to store patient records in electronic form, and to transfer them from one computer to another, perhaps across considerable physical distances, raises all of the difficulties that we have already seen with the security of any electronic data. A report was produced in 1991 called *The Computerized Patient Record: An Essential Technology for Health Care*; this raises serious issues about the accuracy, access, and privacy of such a record.[12]

Some people are so disturbed by the dehumanization they believe is being brought about by computers that they mount physical attacks against the machines. A "Committee for the Liquidation of Computers" bombed the government computer center in Toulouse, France (January 28, 1983) and there were bombings of the Dusseldorf, Germany, offices of IBM and Control Data Corporation in 1982; the Sperry

Rand office in West Berlin; and a Harrison, New York, office of IBM (Eaton, 1986, p. 133).

In 1974, Senators Barry Goldwater and Charles Percy proposed an amendment to a bill in Congress to stop the use of a Social Security number as a universal identifier, to "stop this drift towards reducing each person to a number":

> Once the social security number is set as a universal identifier, each person would leave a trail of personal data behind him for all of his life which could be immediately reassembled to confront him. Once we can be identified to the administration in government or in business by an exclusive number, we can be pinpointed wherever we are, we can be more easily manipulated, we can be more easily conditioned and we can be more easily coerced.[13]

The Pythagoreans thought that everything was number, especially souls (or personal identities). There is an old phrase, "I've got your number!", which implies that you know everything there is to know about that person. In the cases of concern here, the government and big business would have our numbers and so control of our personal destinies.

In Aleksandr Solzhenitsyn's *Cancer Ward*, a view from the Stalinist Soviet Union gives us a chilling, but similar picture:

> As every man goes through life he fills in a number of forms, each containing a number of questions. . . . There are thus hundreds of little threads radiating from every man, millions of threads in all. If these threads were suddenly to become visible, the whole sky would look like a spider's web, and if they materialized like rubber bands, buses and trams and even people would lose the ability to move and the wind would be unable to carry torn-up newspapers or autumn leaves along the streets of the city.

This is also reminiscent of the network of rules that Tocqueville described; is it today the network of computers containing information on all the members of our society?

A National Computerized Criminal History (CCH) System?

In 1986, Kenneth C. Laudon wrote a book called *Dossier Society: Value Choices in the Design of National Information Systems*. His purpose was primarily to examine the FBI's plan for a national computerized criminal history (CCH) system, and the impact it would have on our society and on personal privacy. He first examines the positive side of such a proposal—"the professional record-keeper vision," as he calls it. This would foresee "a more rational world in which instantly available and accurate information would be used to spare the innocent and punish the guilty" (Laudon, 1986, p. 18). He says that this vision is held primarily by police, district attorneys, and criminal courts.

The question of whether such information could be accurate and instantly available should be examined in the light of the performance of similar smaller systems; the odds of increasing errors and decreasing reliability go up alarmingly as the size of a system increases. As Laudon points out:

> The significance of a national CCH extends beyond the treatment of persons with a prior criminal record. Creating a single system is a multijurisdictional, multiorganizational effort which requires linking more than 60,000 criminal justice agencies with more than 500,000 workers, thousands of other government agencies, and private employers, from the local school district to the Bank of America, who will use the system for employment screening. (Laudon, 1986, p. 16)

Two questions immediately come to mind about such a system: (1) What kind of reliability can it possibly have (and what will be the magnitude of the consequences of system failures or reporting errors)? (2) Is it appropriate, or even legal, for a system designed for one purpose (criminal record-keeping for the justice system) to be used for such a different purpose (employment screening)?

Laudon also notes that, in addition to the fingerprint records on roughly 36 million people with criminal records, the FBI has fingerprints of over twice as many citizens who do not have any criminal record, but have had their fingerprints taken for the armed forces, work in nuclear plants, work on defense contracts, for any work requiring a security clearance for a job, and others. With the amazing capabilities of computerized fingerprint identifications, it would be quite feasible to run *all* of these prints, and not just those of people with criminal records, in any investigation. This would greatly increase the chances of "false positives" and the resulting invasion of innocent people's privacy and equanimity.

The second perspective regarding the national CCH, Laudon writes, is the "dossier society vision," held by members of the ACLU (American Civil Liberties Union), state and federal legislative research staffs, and defense lawyers. Proponents of this vision think in terms of a system with "imperfect information and incomplete knowledge," a "runaway" system that would be out of anyone's control. Linking this information (see the upcoming section on "computer matching") with that from other agencies such as the Social Security Administration, the IRS, and the Department of Defense, could create "a caste of unemployable people, largely composed of minorities and poor people, who will no longer be able to rehabilitate themselves or find gainful employment in even the most remote communities" (Laudon, 1986, pp. 20–21). He also points out that the present system already discriminates against those who live in poor and ghetto neighborhoods, where the police may regularly make a pass down the street in the summer and haul everyone in for questioning, thus creating "criminal records" for many who violated no laws (Laudon, 1986, pp. 226–227).

. . . Laudon pessimistically concludes, "In the absence of new legislation, it is clear that national information systems . . . will continue to develop in a manner which ensures the dominance of efficiency and security over freedom and liberty" (Laudon, 1986, p. 367). This is a frightening prospect. Those running these systems would claim that they are the guards of our domestic and national security; but the old question echoes, "Who will guard the guards?" Hobbes would respond that it must be the Leviathan. What other possibilities are there?

The FBI Wiretap Law

A hotly contested FBI proposal called the Wiretap Bill, or the DT (Digital Telephony) bill, finally passed the Congress *unanimously* on October 7, 1994. The bill mandates that all communications carriers must provide "wiretap-ready" equipment. The purpose of this is to facilitate the FBI's implementation of any wiretaps that are approved by the courts. The bill was strongly opposed by the Computer Professionals for Social Responsibility (CPSR), the Voters Telecomm Watch (VTM), the ACLU, and the Electronic Privacy Information Center (EPIC), among others, and much support for this opposition was marshaled in terms of letters and e-mail messages to congressional representatives.

CPSR sent out a list of "100 Reasons to Oppose the FBI Wiretap Bill"; for example, Reason 29 was that the bill contains inadequate privacy protection for private e-mail records. The estimated cost of enacting the law is (according to a CPSR report) $500 million, which will be borne by "government, industry, and consumers." This is just another instance of an action of government that will infringe on a right of the public (in this case, privacy) and require the public to pay for it.

The Clipper Chip Proposal

Another recent government proposal is the "Clipper chip," a device to establish a national data encryption standard. The device would sell for $30 (or less in large quantities) and allow the encoding of messages sent electronically, to avoid tampering. The objectors (some of whom are referred to as "cypherpunks") fear that the standard, developed by NSA, will have a "trap door" that will allow the government to eavesdrop on any transmissions it chooses.[14]

Some fear that the Clipper chip will create an "information snooperhighway," but NSA assures that Clipper is "your friend." CPSR is at the forefront of the opposition, and made available an electronic petition against Clipper; the petition goes to the president, because the proposal does not require congressional approval, and states reasons for the opposition.

The Clipper petition says that the proposal was developed in secret by federal agencies (NSA and the FBI) whose main interest is in surveillance, not in privacy. The plan will create obstacles to businesses wanting to compete in international markets and discourage the development of better technologies to enhance privacy. "Citizens who anticipate that the progress of technology will enhance personal privacy will find their expectations unfulfilled." The petition goes on to say that the signees fear that if the proposal and the standard are adopted, "even on a voluntary basis, privacy protection will be diminished, innovation will be slowed, government accountability will be lessened, and the openness necessary to ensure the successful development of the nation's communications infrastructure will be threatened" (CPSR, Clipper petition).

There is a moral dilemma arising between the desire for an open society and the need to reduce crime and disorder. This tension is never easily resolved. The ideal would be to have a society in which all people minded their own business, supported the state, and were truly just (Plato's ideal state). Since the real world is not like that, there must be some laws and restrictions to control wrongdoing. However, the laws must not suppress all freedom and undermine what openness is possible. The Wiretap Bill and the Clipper chip lean too far in the direction of suppressing fundamental goods such as freedom. If all communications can in principle be monitored, that creates a closed society, not an open one. All of the arguments against invasions of privacy in the beginning of this chapter apply.

Computerized Credit

Many people today use credit cards to buy various products and services. Credit cards can even be used in many grocery stores and liquor stores now. The information from any of these credit cards, plus data on any loans from banks or other financial institutions, goes into several huge centralized credit bureaus. An institution (or, as Rothfeder demonstrated, an individual) can request this information on someone, usually if it is considering giving the person a loan or a credit line. The Federal Trade Commission puts out informational pamphlets on credit bureaus so that individuals will know what they are, how they operate, and what the individual's rights are with respect to credit bureaus. A credit bureau is a kind of "clearinghouse" for information on a person's financial history. The Fair Credit Reporting Act of 1971 attempted to regulate the credit bureaus' operation and provide some privacy for the personal information they handle. However, in reality there is very little private about this data. The regulations *do* allow an individual access to the information and the opportunity to request correction of any errors.

One of the largest credit bureaus is TRW. It will provide you, on request, with one free copy of your credit report per year (but a free one might take three to four weeks to reach you; one you pay for [$8.68 last time I checked] comes within a week). To request a copy of your credit report, you must send a letter (with your check) indicating your full name, current mailing address (and other addresses within the past five years), your Social Security number, and your date of birth. Send this to TRW, P.O. Box 2106, Allen, Texas 75002. The report comes with a copy of your credit rights and how to go about resolving any disputes.

The report I received contained a listing of recent (it seemed to cover about the past six years) charge accounts and bank loans. It contained a description of each account (revolving credit, car loan, mortgage, etc.), the current balance, and whether all payments had been made on time. It also contained a list of inquiries made into the credit record. Most of them were inquiries from financial institutions determining whether to send me an offer of a new charge card ("to develop a list of names for a credit offer or service"). Thus TRW does not have to notify you every time an inquiry is made, but when you receive a credit report you get a listing of recent inquirers into your record.

Rothfeder says that TRW "upgrades and expands its files with information it purchases from the Census Bureau, motor vehicle agencies, magazine subscription services, telephone white pages, insurance companies, and as many as sixty other sources. 'We buy all the data we can legally buy,' says Dennis Benner, a TRW vice president" (Rothfeder, 1992, p. 38). TRW also has a Highly Affluent Consumer database, from which it sells names, phone numbers, and addresses of people categorized by income to anyone (catalog company, telemarketer, or political group) that will pay for it (Rothfeder, 1992, p. 97).

TRW will also screen out names of those who are late in payments or likely to go bankrupt!

The storage, transfer, and analysis of all of this information is made possible by computers. Computers have thus presented us with the ethical concern of whether the handling and disseminating of this information creates a breach of our moral rights—in this case, the right of privacy.

Caller ID

A new service called "Caller ID" *sells* information on who you are (your phone number) when you make a phone call. Some states (such as California) have prohibited "Caller ID," but it is in place in many other states, including New York. The original idea behind "Caller ID" was to track nuisance callers (but they can just call from pay phones), and perhaps to allow call recipients to screen their calls. However, in practice it represents an invasion of the caller's privacy, the right not to have one's number identified; many people pay to have unlisted phone numbers for just that reason. CPSR has spoken out against it: "Caller ID, for example, reduced the privacy of telephone customers and was opposed by consumers and state regulators."[15]

There also is an option called "Call Return," which allows anyone you have called to call you back, whether they answered your call or not. In New York, there is a feature called "Per-Call Restrict," where you can block display of your number on the recipient's "Caller-ID" device, if you press *67 on a touch-tone phone, or dial 1167 on a rotary phone, before you place your call. That seems like an annoyance at the very least. In something that should by rights be private, you have to go to an extra effort every time you place a call to ensure that privacy. And, given that feature, what nuisance caller (against whom the ID system was initially devised) would not use the Call-Restrict option to prevent identification?

To those who think that Caller ID is just an innocent convenience, consider some cases where it might be damaging. (1) A professor, who values her privacy and quiet time, out of decency calls a student who has left a message of panic regarding tomorrow's exam. After the professor spends half an hour on the phone answering questions, the student lets others know that the professor helped and gives out her (unlisted) phone number, obtained through Caller ID. The professor (who had more than adequate office hours all semester to help students) is harassed all evening with phone calls from other students in the class. (2) A consumer calls to make a telephone inquiry about the price of a product, but decides not to buy. The organization that took the call begins phoning her every week, trying to talk her into other products. (3) An informant, who wishes to remain anonymous, calls the police department to report a crime. The police record from Caller ID is used to track down the informant, and the phone number and address also are picked up by a nosy reporter, who puts the person's name in a newspaper article.

Computer Matching

Computer matching involves combining information from several databases to look for patterns—of fraud, criminal activity, and so on. Generally, it is used to detect people who illegally take government money from more than one source—they are termed "welfare double-dippers."

The Reagan Administration was quick to capitalize on the capabilities of computer matching. In March 1981, President Reagan issued an executive order establishing a President's Council on Integrity and Efficiency (PCIE) to "revitaliz[e] matching to fight waste, fraud and abuse" (*Computer Matching and Privacy*, 1986, p. 56). This government committee has seemed to be primarily concerned with assessing the cost-effectiveness of matching programs, rather than with looking into how they may constitute serious invasions of personal privacy. The Long-Term Computer Matching Project has published a quarterly newsletter, developed standardized formats for facilitating matching, created a "manager's guide" to computer matching, established Project Clean Data (a program that identifies incorrect or bogus Social Security numbers), and looked into problems of security with the systems in which matching takes place, in order to promote "efficient and secure matches" (*Computer Matching*, 1986, pp. 59–60).

Some of the government agencies involved in computer matching include Defense, Veterans Affairs, HEW, HUD, the Postal Service, Agriculture, Health and Human Services, Justice, Social Security, Selective Service, NASA, and Transportation. Examples of matches performed include locating wanted and missing persons files with federal and state databases, detecting underreporting of income by those receiving mortgage or rent subsidies (or food stamps or other aid), locating absent parents with child support obligations, checking wage reports against those receiving unemployment, and so on (*Computer Matching*, 1986, pp. 236–246). One can see that the possibilities are practically limitless. Some of these matches may be perfectly justifiable, as in cases where Social Security fraud is being perpetrated against the government. But it will be so easy to cross the line into blatant invasions of personal privacy, and the lines are not clearly drawn, or are not drawn at all.

The American Bar Association (ABA) has identified seven distinct areas in which computer matching programs impinge on privacy:

1. *Fourth Amendment.* Privacy advocates have argued that matching of databases containing personal information constitutes a violation of the Fourth Amendment right to security against unreasonable searches. Supporters of computer matching argue that the programs do not search in general, but rather "scan" for records of particular persons who are applying for benefits. That, of course, does not mean that such programs could not be easily modified to search in general.

2. *Privacy Act.* Some claim that computer matching programs violate the Privacy Act of 1974, which provides that personal information collected for one purpose cannot be used for a different purpose without notice to the subject and the subject's consent. However, the Privacy Act has a "routine use" clause that has been used to bypass this consideration. "Congress has in fact authorized most of the ongoing matching programs, overriding whatever protection the Privacy Act afforded" (*Computer Matching*, 1986, p. 97).

3. *Fair Information Practice Principle.* Even if the law is not violated by current computer matching practice, it does violate the fair information practice that information collected for one purpose should not be used for another purpose without due notice and consent. This is reflected in the 1979 ABA resolution: "There must be a way for an individual to prevent information obtained for one purpose from being used or made available for other purposes without his or her consent" (*Computer Matching*, 1986, p. 97).

4. *Due Process.* Violations of "due process" have occurred in computer matching cases. For example, in Massachusetts in 1982, the state matched welfare records against bank accounts and terminated benefits without notice. In several cases, the "matches" were incorrect, and in others there were legitimate reasons for the money in the bank accounts that did not violate welfare restrictions.

5. *Data Security.* Unauthorized access to the databases in which this matching information is stored constitutes a real threat to the personal privacy of those whose records are stored there, and the government has not developed security measures adequate to protect them (*Computer Matching*, 1986, p. 98).

6. *Data Merging.* The government can query any number of files on a citizen; this leads to the possibility of a merged file in a new database containing information from various sources (which could well violate areas 2 and 3). For example, recent rulings allow the IRS to deduct

debts owed the government from tax refunds, and the IRS was proposing (in 1986) to establish a "master debtor file" (*Computer Matching*, 1986, p. 98).

7. *Fundamental Privacy Rights.* What is to keep the government from linking personal data on medical records, financial records, reading and viewing preferences, and political and religious activities? This is all unsettlingly reminiscent of Orwell's *1984*. Ronald L. Plesser, who made the presentation to Congress as designee of the American Bar Association, said: "Substantive limits on data linkage have to be established. At some point, privacy and autonomy outweigh government efficiency" (*Computer Matching*, 1986, p. 99; general discussion, pp. 92–103).

When New York City and New Jersey recently compared their welfare databases, they discovered 425 people who were regularly collecting welfare in both areas, using fake IDs, to the tune of about $1 million in unauthorized cost to the welfare system. The system is currently experimenting with using fingerprints as IDs for welfare recipients. Critics complain that it will discourage poor people from getting aid, since it carries with it the stigma of a criminal context. Soon after the fingerprint experiment was implemented, over 3,000 people dropped out of the system.[16]

Some might just say "Good! Three thousand fewer leeches on society." The question comes down to the real purpose of our welfare system. Do we really want to help people in need, or not? A number of the ethical systems we examined would support this as the right thing to do, if one cannot have an ideal society in which there is enough for everyone to prosper. The utilitarian principle is clear: we should do what fosters the greatest good (pleasure and absence of pain) for the greatest number of people. Kant would emphasize that rational beings must be treated as ends in themselves, and this would include awarding them respect and humane treatment, since *they* represent what is intrinsically valuable, not dollars. Certainly it develops virtuous character to be generous. And, on the goods view, we should do what maximizes the highest goods, such as friendship, knowledge, satisfaction, health, and life.

As it is now, people who go for legitimate welfare support, or for temporary unemployment compensation when they unexpectedly lose their jobs in an unstable economy, are made to feel like dirt and are brought back repeatedly to fill out repetitive forms. If we add fingerprint identification, they will surely be made to feel like criminals, as the article suggests. If we *intend* to help those who need help, then we should do it with open hearts and efficient procedures, not grudgingly and in a demeaning way.

The Worst Scenario

We have seen how records on us are kept, analyzed, sold, and transmitted. Most of this information is financial, but there are other records on us as well. *USA Today* estimated recently that the average adult American has computerized files in at least twenty-five different databases. In 1990, the General Accounting Office reported that there were at least 2,006 federal computer databases, 1,656 of which were covered by the Privacy Act. These databases were under the various cabinet departments (Defense, Health and Human Services, Justice, Agriculture, Labor, Treasury, Interior, Transportation, Commerce, Energy, Veterans Affairs, Education, and Housing and Urban Development—in order from most systems to least), and various independent agencies such as the Environmental Protection Agency, NASA, the FCC, the Nuclear Regulatory Commission, and the Federal Reserve System, to name a few. The odds are that your name (and data) shows up in a number of those databases. When the databases start "talking to" each other, they are doing "matching" on you for various purposes.

In the Foreword to David Burnham's *The Rise of the Computer State*, Walter Cronkite wrote:

> Orwell, with his vivid imagination, was unable to foresee the actual shape of the threat that would exist in 1984. It turns out to be the ubiquitous computer and its ancillary communication networks. Without the malign intent of any government system or would-be dictator, our privacy is being invaded, and more and more of the experiences which should be solely our own are finding their way into electronic files that the curious can scrutinize at the punch of a button.
>
> The airline companies have a computer record of our travels—where we went and how long we stayed and, possibly, with whom we traveled. The car rental firms have a computer record of the days and distances we went afield. Hotel computers can fill in a myriad of detail about our stays away from home, and the credit card computers know a great deal about the meals we ate, and with how many guests.
>
> The computer files at the Internal Revenue Service, the Census Bureau, the Social Security Administration, the various security agencies such as the Federal Bureau of Investigation and our own insurance companies know everything there is to know about our economic, social and marital status, even down to our past illnesses and the state of our health.

If—or is it when?—these computers are permitted to talk to one another, when they are interlinked, they can spew out a roomful of data on each of us that will leave us naked before whoever gains access to the information. (Burnham, 1983/1980, pp. vii–viii)

Cronkite, Burnham, and many others have seen the threat to our privacy and our autonomy that these system present. Yet our complex world could not run as it does now without them. It has been suggested that the best way to cripple a country would be to take out its financial networks, and these are now totally dependent on computers. But this leaves us with the question (a very important and a very difficult one) of how we can maintain some level of autonomy and some measure of privacy for ourselves and others in the face of this ever-growing "spider's web" (as Solzhenitsyn described it) of communication and *control*. How can we remain *persons* (with the rights and responsibilities that are entailed) when we are being reduced to mere *numbers* every day?

Dennie Van Tassel wrote a very clever scenario that he devised for the collection he co-edited, *The Compleat Computer* (which is highly recommended reading). The piece is called "Daily Surveillance Sheet, 1989, From a Nationwide Data Bank" (Van Tassel and Van Tassel, 1983/1976, pp. 216–217). In it he describes the "Confidential" report on "Harry B. Slow," which tracks his every move through the use of his many credit cards. All purchases and bank transactions are listed, and from this record the computer analyzes that Harry drinks too much, is probably overweight, has a weakness for young blondes, and is probably having an affair with his secretary (a blonde).

The scenario is amusing, but it is also frightening. We are moving into an age where most of our purchases and activities can be tied to a record with our name (or number) on it, through credit cards, bank accounts, library cards, video rental card, and other identifiers. We tend not to carry a lot of cash (fear of being robbed, perhaps) and so pay for most items by a traceable credit card. My grocery store has a record of what I buy because, even though I usually pay in cash, to receive their special discounts on items each week, I have to use a special coded card that identifies me as a member of their "Club." Thus they have a record of my buying patterns to analyze and sell to various companies that manufacture the products I buy.

Van Tassel has shown us a picture that is not far from the truth today and that will most likely be very close to the truth in the future. Does this present a moral problem and, if so, is there anything we can do about it? Our political systems should have moral underpinnings, and it is through those political systems that we can try to make a difference. One of the problems is that many of the "technocrats" creating and using the computer systems that are invading privacy are not part of the government and so are not accountable to the public as the government should be.

Some of you will become part of the technological force that creates and maintains elaborate computer systems, and so you, at least, will be aware of the dangers involved and your corresponding responsibilities. Others may not be so aware, and so the answer must lie in education and in making groups accountable. Some impact can come from the public "voting with its feet" and not buying certain products, or making public complaints, such as those against the Lotus Marketplace: Households project or the Clipper chip. But government must also be sensitized to the threats to the freedoms promised by the Constitution, freedoms which are its responsibility to protect.

What Protections Are There for Privacy?

The United States Constitution, in its Bill of Rights, has two Amendments which might be thought to relate to the privacy issue—the First and the Fourth. The First Amendment guarantees freedom of religion, speech, the press, and assembly. However, none of those freedoms mentions privacy or necessarily entails privacy. The Fourth Amendment says: "The right of the people to be secure in their persons, houses, papers, and effects against unreasonable searches and seizures, shall not be violated." This relates to prohibiting anyone from looking for and/or taking anything from you or your home without a warrant issued upon demonstration of *probable cause*. However, it does not say anything regarding information about you that is not on your person or in your home, but which should be kept private, or at least not made public (such as your tax return, medical records, driving record, or the like). Notice also that your garbage is not considered private; it is no longer in your home, and someone can search it without violating any laws.

The point is that your tax return is no one's business but yours and the IRS's, your medical files are no one's business but yours (or yours and your doctor's), and your driving record is no one's business but yours and the Motor Vehicle Bureau's. No one outside of the offices mentioned has any *need to know*, or any *right to know*, such information about you. Yet there are cases of people working at the IRS "snooping" into people's files (not the ones they are assigned to work on, but those of neighbors, or famous people), of medical information being sold to drug or insurance companies, and of drivers' records being available for a small fee to anyone in thirty-four states.[17] Medical records of political candidates have been compromised to leak stories about psychiatric counseling, suicide attempts, or other damaging information. The press will claim the public's "right to know" in cases of public figures, but the fact remains that a fundamental guarantee of the privacy of information between patient and doctor has been violated.

In the *Roe v. Wade* (410 U.S. 113, 1973, p. 152) abortion decision, Justice Harry A. Blackman stated the opinion:

> The Constitution does not explicitly mention any right of privacy. In a line of decisions, however, . . . the Court has recognized that a right of personal privacy, or a guarantee of certain zones or areas of privacy, does exist under the Constitution. In varying contexts, the Court or individual Justices have, indeed, found the roots of that right in the First Amendment, . . . in the Fourth and Fifth Amendments, . . . in the Ninth Amendment, . . . or in the concept of liberty guaranteed by the first section of the Fourteenth Amendment . . . These decisions make it clear that only personal rights can be deemed "fundamental" or "implicit in the concept of ordered liberty" . . . are included in this guarantee of personal privacy. (Cited in Freedman, 1987, p. 73)

If there is any suggestion in the Fifth Amendment, it would have to lie in the right of a person not to be compelled to be a witness against himself (or herself). The Ninth Amendment states: "The enumeration in the Constitution, of certain rights, shall not be construed to deny or disparage others retained by the people." This is too vague or too broad to be helpful. If it is interpreted to support privacy, it could also be construed to support an individual's right to a free car. The Fourteenth Amendment states that all citizens have the rights of citizens, but does not spell these out beyond the rights to life, liberty, and property. It does not seem that a right to privacy can be inferred as strongly from these Amendments (Five, Nine, and Fourteen) as from Amendments One and Four.

Warren and Brandeis Make a Case for Privacy

The beginning of the Court's recognition of the right to privacy lies in the article, entitled "The Right to Privacy," written by Samuel D. Warren and Louis D. Brandeis in 1890.[18] They suggest that the rights to life, liberty, and property as previously interpreted were seen to relate only to the *physical* side of human beings, and do not take into account their *spiritual* side. Gradually this came to be recognized, and "now the right to life has come to mean the right to enjoy life—the right to be let alone;

the right to liberty secures the exercise of extensive civil privileges; and the term 'property' has grown to comprise every form of possession—intangible as well as tangible" (Warren and Brandeis, cited in Schoeman, 1984, p. 75).

Recognition of the "legal value of sensations" gave rise to laws against assault, nuisance, libel, and alienation of affections. The rights to property were legally extended to processes of the mind, works of literature and art, good will, trade secrets, and trademarks. They suggest that this development of the law was "inevitable"; the heightened intellect brought about by civilization "made it clear . . . that only a part of the pain, pleasure, and profit of life lay in physical things" (p. 76). Their concern is for the "right to be let alone"—to be free of the intrusions of unauthorized photographs and yellow journalism.

Warren and Brandeis point out that the law of libel only covers broadcasting of information that could injure the person in his or her relations with other people; but there are other personal items of information that no one else should be able to communicate, and libel does not cover this. The very occasion of the article was untoward prying into the lives of Mr. and Mrs. Warren, and the wedding of their daughter, by the Boston newspapers. William L. Prosser, in commenting on privacy in the context of the Warren and Brandeis article, suggests that four torts (wrongful actions for which civil suit can be brought) arose out of their position and subsequent court decisions:

1. Intrusion upon the plaintiff's seclusion or solitude, or into his private affairs.
2. Public disclosure of embarrassing private facts about the plaintiff.
3. Publicity which places the plaintiff in a false light in the public eye.
4. Appropriation, for the defendant's advantage, of the plaintiff's name or likeness. (Prosser, in Schoeman, 1984, p. 107)

Warren and Brandeis close by saying that "the protection of society must come through a recognition of the rights of the individual." Each individual is *responsible* for his or her own actions only; the individual should resist what he or she considers wrong, with a weapon for defense.

Has he then such a weapon? It is believed that the common law provides him with one, forged in the slow fire of the centuries, and today fitly tempered to his hand. The common law has always recognized a man's house as his castle, impregnable, often, even to its own officers engaged in the execution of its commands. Shall the courts thus close the front entrance to constituted authority, and open wide the back door to idle or prurient curiosity? (Warren and Brandeis, cited in Schoeman, 1984, p. 90)

The subtitle to "The Right to Privacy" is "The Implicit Made Explicit." It would seem that what Warren and Brandeis tried to do, and whose lead most courts subsequently followed, was to point out the strong connection between holding an individual *responsible* for actions and allowing him or her the *privacy* of thought and control of action to exercise such responsibility.

The Warren and Brandeis analysis deals only with the harm that can be done if someone uses information about you, and suggests that merely possessing that information is not a harm. However, one can think of many instances when someone merely possessing private information about you would present, at the very least, an embarrassment. It is analogous to the cases discussed at the beginning of the chapter where, even if you are not aware that you are being monitored, harm is done because you are diminished as a person and a free agent.

Existing Privacy Legislation

The *Fair Credit Reporting Act* (15 U.S.C. 1681) All credit agencies must make their records available to the person whose report it is; they must have established means for correcting faulty information; and they are to make the information available only to *authorized* inquirers.

The *Crime Control Act of 1973* Any state information agency developed with federal funds must provide adequate security for the privacy of the information it stores.

The *Family Educational Right and Privacy Act of 1974* (20 U.S.C. 1232g) Educational institutions must grant students and/or their parents access to student records, provide means for correcting any errors in the records, and make such information available only to authorized third parties.

The *Privacy Act of 1974* (5 U.S.C. 552a) This act restricts the collection, use, and disclosure of personal information, and gives individuals the right to see and correct such information.

Sam Ervin had felt that the legislation required the establishment of a Privacy Commission. The report of the Senate Committee on Government Operations (September 1974) stated:

It is not enough to tell agencies to gather and keep only data for whatever they deem is in their intended use, and then to pit the individual against government, armed only with the power to inspect his file, and the right to challenge it in court if he has the resources and the will to do so.

To leave the situation there is to shirk the duty of Congress to protect freedom from the incursions by the arbitrary exercise of the power of government and to provide for the fair and responsible use of such power. (U.S. Senate 1976: 9–12; cited in Burnham, 1983/1980, p. 364)

The *Tax Reform Act of 1976* (26 U.S.C. 6103)   Tax information must remain confidential and not be released for nontax purposes. (However, there has been an ever-increasing list of exceptions to the release restriction since 1976.)

The *Right to Financial Privacy Act of 1978* (12 U.S.C. 3401)   The act establishes some privacy with regard to personal bank records, but provides means for federal agencies to access such records.

The *Protection of Pupil Rights Act of 1978* (20 U.S.C. 1232h)   The act gives parents the right to examine educational materials being used in the schools, and prohibits intrusive psychological testing of students.

The *Privacy Protection Act of 1980* (42 U.S.C. 2000aa)   Government agents may not conduct unannounced searches of news offices or files if no crime is suspected of anyone.

The *Electronic Funds Transfer Act of 1980*   Customers must be notified of any third-party access to their accounts.

The *Debt Collection Act of 1982* (Public Law 97-365)   Federal agencies must observe *due process* (of notification, etc.) before releasing bad-debt data to credit bureaus.

The *Congressional Reports Elimination Act of 1983*   This act eliminated the requirement under the Privacy Act for all agencies to republish all of their systems notices every year.

The *Cable Communications Policy Act of 1984* (Public Law 98-549).   A cable subscriber must be informed of any personal data collected (and when), and the use and availability there will be of such information.[19]

The Importance of Privacy

Not only do all of the databases discussed have information on private citizens that can be "matched" and from which patterns of behavior can be inferred, but there are the problems of *security* of the information (from parties who should not have access to it), whether anyone has a right to that data, and the *accuracy* of the information. . . .

Rothfeder (1989) reports that a GAO survey made in 1990 of the TECS II (Treasury Enforcement Communications Systems, Version II) revealed errors in 59 percent of the records it examined. There is good reason to believe that the other computerized records (in the justice system, NCIC, voter records, credit bureaus, the IRS, the welfare system, and many more) are also filled with errors. . . .

Thus the problem becomes more than just one of an invasion of privacy; it is also one of "them" having lots of information on individual citizens, which "they" will use for various purposes to manipulate and tax them, but that much of that data is wrong (in small or in large ways).

One objection raised to the demand for privacy says that anyone who needs privacy must be planning to do something that he or she should not do and so needs the protection of privacy to carry it out. We might call this the "ring of Gyges" objection. Glaucon, in the *Republic*, claimed that if the "just" individual and the "unjust" individual both had a ring of Gyges that would make them invisible at will, they would both behave exactly the same way (taking advantage of others for their own gain). Thus the objectors to demands for privacy say that the only reason anyone would want the "cloak of invisibility" that privacy offers is that she must be up to no good.[20]

It is true that privacy helps to protect the actions of the thief, or the murderer, or the terrorist. But it would be an unfortunate conclusion to draw that therefore no one should be allowed any privacy. Such a world has already been described by George Orwell and others, and it is not a world that anyone would want to live in.

Some can be immune to punishment without needing privacy, if they are sufficiently powerful. Not everyone who wants privacy is planning to do evil; you might, for example, want to plan a surprise birthday party for someone. And not everyone who is planning to do evil needs privacy; Hitler carried out much of his evil in full public view. Many criminals want the notoriety they feel they have earned. Think of examples of hackers like Kevin Mittnick, who seem to glory in the challenge and the recognition.

If we act on the assumption of a world where everyone is selfishly out for personal gain, with no scruples, then we would seem to have no alternative to the society described in Hobbes' *Leviathan*. Then we could let a supercomputer like the one in the movie *Colossus: The Forbin Project* monitor all of our actions and punish misbehavior. But in such a world there would be no creativity, no flowering of the human spirit;

we would all become cogs in a big machine, obediently doing what we are told by Colossus. Hopefully, this is not our world, and we do have the human right to exercise our choices.

In the Canadian report on *The Nature of Privacy* by the Privacy and Computer Task Force, the following insightful statement is made:

> If we go about observing a man's conduct against his will the consequence of such observation is that either the man's conduct is altered or his perception of himself as a moral agent is altered. The notion of altering conduct or self-perception against the will of the moral agent is offensive to our sense of human dignity. If, through a monitoring device, we are able to regulate or indeed to follow the conduct of a person it is obvious that we are in effect compromising his responsibility as a chooser of projects in the world, thereby delimiting or influencing the kinds of choices that we make available to the doer. (Weisstaub and Gotlieb, 1973, p. 46)

It is clear that, in order to be a moral agent, in order to be *responsible* for what we do, we must be in control of our own choices, autonomous. If you are being constantly monitored in everything you do, that monitoring will affect your behavior. It is sort of like the Heisenberg Uncertainty Principle in physics; when you try to measure something, you disturb what it is you are trying to measure, and so you do not get a true reading.

The real choices you make are internal, and you need to know that they are truly your own choices (not determined by some external pressure or other). You cannot have that assurance if everything you do is public. If it is, you will always wonder, did I just do that because Dad was watching me? Because I was afraid of being punished if I didn't do it? Then it is not really *my* choice.

Augustine pointed out that it is better to have choice than not to have it, even though that means you may make some bad choices. Similarly, it is better to have privacy than not to have it, even if sometimes others (and even you) misuse it. If we do not have it, just as if we do not truly have choice, we are not fully human, and we cannot fulfill our human potential (which Aristotle would say is to *choose* the life of understanding, which is a very private thing, though we may choose to share it).

Notes

1. Ithiel de Sola Pool wrote, in the first chapter of *Technologies of Freedom*, entitled "A Shadow Darkens": "Electronic modes of communication . . . are moving to center stage. The new communication technologies have not inher-

ited all the legal immunities that were won for the old. When wires, radio waves, satellites, and computers became major vehicles of discourse, regulation seemed to be a technical necessity. And so, as speech increasingly flows over those electronic media, the five-century growth of an unabridged right of citizens to speak without controls may be endangered" (Pool, 1983, p. 1). We should be deeply concerned about anything that threatens our right to freedom of speech or our privacy. These are not the same, but we may require privacy to protect our freedom of thought and speech.

2. Charles F. Fried, *An Anatomy of Values* (Cambridge: Harvard University Press, 1970), p. 9.

3. G. Marx, "I'll Be Watching You: The New Surveillance," *Dissent* (Winter 1985), 26.

4. Sellers of Government Data Thrive," *New York Times*, 26 December 1991, Sec. D, p. 2, col. 2.

5. *Software Engineering Notes* 15, no. 1 (Jan. 1990), 4.

6. *Software Engineering Notes* 16, no. 4 (Oct. 1991), 8.

7. Ibid., p. 12.

8. Ibid., pp. 11–12.

9. *Software Engineering Notes* 17, no. 1 (Jan. 1992), 6–7.

10. "Police Computers Used for Improper Tasks," *New York Times*, 1 November 1992, Sec. 1, p. 38, col. 1.

11. Stephen Barr, "Probe Finds IRS Workers Were 'Browsing' in Files," *The Washington Post*, 3 August 1993, Sec. A, p. 1, col. 1.

12. See the article on CPR by Scott Wallace (1994) and the article by Sheri Alpert (1993) on "Smart Cards, Smarter Policy."

13. 94th Congress, 2nd Session, *Legislative History of the Privacy Act of 1974*, 759. In Eaton, 1986, p. 152.

14. The interested reader is referred to the articles by John Markoff (1991), "Wrestling Over the Key to the Codes," and Steven Levy (1994), "Cypherpunks vs. Uncle Sam: Battle of the Clipper Chip."

15. *The CPSR Newsletter* 11, no. 4, and 12, no. 1 (Winter 1994), 23.

16. Mitch Betts, "Computer Matching Nabs Double-Dippers." *Computerworld* 28, no. 16 (April 28, 1994), 90.

17. Mitch Betts, "Driver Privacy on the Way?", *Computerworld*, 28 February 1994, 29.

18. Brandeis was to serve as an Associate Justice of the Supreme Court from 1916 to 1939.

19. The main source of this list of legislation affecting privacy was Kenneth C. Laudon, *Dossier Society* (1986).

20. There is another side to the story of the ring of Gyges, as you may recall. Socrates says that the truly just individual will behave the same way with or without the ring. One is truly just from within, not because of external sanctions. Kant would tie it to *intentions*. Two individuals may do the same thing, that we view as a just act (such as helping a little old lady across the street), but one does it from ulterior motives (hoping to be remembered in her will) while the other does it from Kant's pure duty—just because it is the right thing

to do. Only the second person's act has moral worth. Its worth arose because the person acted autonomously (governing her own actions), *and* acted from a good will. However, she may well value her privacy and not want her act spread over the newspaper headlines.

References and Recommended Readings

Alpert, Sheri. "Smart Cards, Smarter Policy: Medical Records, Privacy, and Health Care Reform." *Hastings Center Report* (Hastings-on-Hudson) (November–December 1993), 13–23.

Branscomb, Lewis M., and Anne M. Branscomb. "To Tap or Not to Tap." *Communications of the ACM* 36, no. 3 (March 1993), 36–73. See also Rotenberg (1992), Rivest (1993), and Marx (1993).

Brown, Geoffrey, *The Information Game: Ethical Issues in a Microchip World*. Atlantic Highlands, N.J.: Humanities Press, 1990, Chapters 3, 5, and 6.

Burnham, David. *The Rise of the Computer State: The Threat to Our Freedoms, Our Ethics and Our Democratic Process*. Foreword by Walter Cronkite. New York: Random House, 1983/1980.

Center for Philosophy and Public Policy. "Privacy in the Computer Age." *Philosophy & Public Policy* (College Park, Md.: University of Maryland) 4, no. 3 (Fall 1984), 1–5.

Computer Matching and Privacy Protection Act of 1986: Hearing Before the Subcommittee on Oversight of Government Management of the Committee on Governmental Affairs, United States Senate (99th Congress). September 16, 1986. Washington, D.C.: U.S. Government Printing Office, 1986.

Cotton, Paul. "Confidentiality: A Sacred Trust Under Siege." *Medical World News* (March 27, 1989), 55–60.

Dejoie, Roy, George Fowler, and David Paradice, eds. *Ethical Issues in Information Systems*. Boston: Boyd & Fraser, 1991.

Dunlop, Charles, and Rob Kling, eds. *Computerization and Controversy: Value Conflicts and Social Choices*. Boston: Academic Press, 1991, Part V.

Eaton, Joseph W. *Card-Carrying Americans: Privacy, Security, and the National ID Card Debate*. Totowa, N.J.: Rowman & Littlefield, 1986.

Ermann, M. David, Mary B. Williams, and Claudio Gutierrez, eds. *Computers, Ethics, & Society*. New York: Oxford, 1990, Chapters 4, 5, 10, 11, and 12.

Flaherty, David. *Protecting Privacy in Surveillance Societies*. Chapel Hill: University of North Carolina Press, 1989.

Forester, Tom, and Perry Morrison. *Computer Ethics: Cautionary Tales and Ethical Dilemmas in Computing*. 2d ed. Cambridge, Mass.: The MIT Press, 1994/1990, Chapter 6.

Freedman, Warren. *The Right of Privacy in the Computer Age*. New York: Quorum Books, 1987.

GAO. "Government Computers and Privacy." GAO/IMTEC-90-70BR. Washington, D.C.: U.S. General Accounting Office, 1990.

Gellman, H. S. *Electronic Banking Systems and Their Effects on Privacy: A Study by the Privacy and Computer Task Force.* Department of Communications/Department of Justice, Canada.

Gould, Carol C., ed. *The Information Web: Ethical and Social Implications of Computer Networking.* Boulder, Col.: Westview Press, 1989, Chapters 1–5.

Hunter, Larry. "Public Image." *Whole Earth Review* (Dec. 1984–Jan. 1985), 32–40.

Johnson, Deborah G. *Computer Ethics.* 2d ed. Englewood Cliffs, N.J.: Prentice-Hall, 1994/1985, Chapter 5.

Johnson, Deborah G., and John W. Snapper, eds. *Ethical Issues in the Use of Computers.* Belmont, Cal.: Wadsworth, 1985, Part 3.

Jordan, F. J. E. *Privacy, Computer Data Banks, Communications, and the Constitution: A Study by the Privacy and Computer Task Force.* Department of Communications/Department of Justice, Canada.

Kant, Immanuel. *Foundations of the Metaphysics of Morals.* 1785.

Kimball, Peter. *The File.* San Diego, Cal.: Harcourt Brace Jovanovich, 1983.

Lacayo, Richard. "Nowhere to Hide." *Time,* 11 November 1991, 34–40.

Larson, Erik. *The Naked Consumer: How Our Private Lives Become Public Commodities.* New York: Henry Holt, 1992.

Laudon, Kenneth. *Dossier Society: Value Choices in the Design of National Information Systems.* New York: Columbia University Press, 1986.

Lewis, Peter H. "Censors Become a Force on Cyberspace Frontier." *New York Times,* 29 June 1994, A1, D5.

Levy, Steven. "The Cypherpunks vs. Uncle Sam: The Battle of the Clipper Chip." *New York Times Magazine,* 12 June 1994, 44–51, 60, 70.

Markoff, John. "Europe's Plans to Protect Privacy Worry Business." *New York Times,* 11 April 1991, pp. A1, D6.

Markoff, John. "Wrestling Over the Key to the Codes." *New York Times,* 9 May 1993, Sec. 3, p. 9.

Marx, Gary T. "I'll Be Watching You: The New Surveillance." *Dissent* (Winter 1985), 26–34.

Marx, Gary T. "To Tap or Not to Tap." *Communications of the ACM* 36, no. 3 (March 1993), 41. See also Branscomb & Branscomb (1993), Rotenberg (1993), and Rivest (1993).

Miller, Arthur. "Computers and Privacy." In Dejoie et al. (1991), pp. 118–133.

Miller, Michael. "Lawmakers Begin to Heed Calls to Protect Privacy and Civil Liberties as Computer Usage Explodes." *Wall Street Journal,* 11 April 1991, A16.

Mitchum, Carl, and Alois Huning, eds. *Philosophy and Technology II: Information Technology and Computers in Theory and Practice.* Dordrecht: D. Reidel, 1986.

Moshowitz, Abbe. *Conquest of Will: Information Processing in Human Affairs.* Reading, Mass.: Addison-Wesley, 1976.

Perrolle, Judith A. *Computers and Social Change: Information, Property, and Power.* Belmont, Cal.: Wadsworth, 1987.

Pool, Ithiel de Sola. *Technologies of Freedom: On Free Speech in an Electronic Age.* Cambridge, Mass.: The Belknap Press of Harvard University Press, 1983.

Privacy Commissioner of Canada. *The Privacy Act.* Ottawa, Ontario: Minister of Supply and Services Canada, 1987.

Rachels, James. "Why Privacy Is Important." *Philosophy & Public Affairs* 4, no. 4 (Summer 1975), 323–333. Also in Johnson & Snapper (1985), in Dejoie et al. (1991), and in Schoeman (1984).

Rivest, Ronald L. "To Tap or Not to Tap." *Communications of the ACM* 36, no. 3 (March 1993), 39–40. See also Branscomb & Branscomb (1993), Rotenberg (1993), and Marx (1993).

Ross, W. D. *The Right and the Good.* Oxford: Clarendon Press, 1930.

Roszak, Theodore. *The Cult of Information.* New York: Pantheon, 1986, Chapters 8 and 9.

Rotenberg, Marc. "Protecting Privacy" ("Inside RISKS" column). *Communications of the ACM* 35, no. 4 (April 1992), 164.

Rotenberg, Marc. "To Tap or Not to Tap." *Communications of the ACM* 36, no. 3 (March 1993), 36–39. See also Branscomb & Branscomb (1993), Rivest (1993), and Marx (1993).

Rothfeder, Jeffrey. "Is Nothing Private? Computers Hold Lots of Data on You—And There Are Few Limits On Its Use." *Business Week*, 4 September 1989, 74–82.

Rothfeder, Jeffrey. *Privacy for Sale: How Computerization Has Made Everyone's Private Life an Open Secret.* New York: Simon & Schuster, 1992.

Rule, James B. "Where Does It End?" The Public Invasion of Privacy." *Commonweal*, 14 February 1992, 14–16.

Savage, John E., Susan Magidson, and Alex M. Stein. *The Mystical Machine: Issues and Ideas in Computing.* Reading, Mass.: Addison-Wesley, 1986, Chapter 14.

Schoeman, Ferdinand David, ed. *Philosophical Dimensions of Privacy: An Anthology.* Cambridge: Cambridge University Press, 1984.

Shattuck, John, and Richard Kusserow. "Computer Matching: Should It Be Banned?" *Communications of the ACM* 27, no. 6 (June 1984), 537–545.

Shattuck, John, and Muriel Morisey Spence. "The Dangers of Information Control." *Technology Review* (April 1988), 62–73.

Simons, G. L. *Privacy in the Computer Age.* Manchester, England: The National Computing Centre Limited, 1982.

Software Engineering Notes.

Thompson, Judith Jarvis. "The Right to Privacy." In Schoeman (1984), pp. 272–289.

Tocqueville, Alexis de. *Democracy in America.* Ed. Richard D. Heffner. New York: Mentor/NAL, 1956.

Tuerkheimer, Frank M. "The Underpinnings of Privacy Protection." *Communications of the ACM* 36, no. 8 (August 1993), 69–73.

Van Tassel, Dennie L., and Cynthia L. Van Tassel, eds. *The Compleat Computer.* 2d ed. Chicago: SRA, 1983/1976.

Wallace, Scott. "The Computerized Patient Record." *Byte* (May 1994), 67–75.

Warren, Samuel D., and Louis D. Brandeis. "The Right to Privacy." *Harvard Law Review* 4, no. 5 (December 15, 1890), 193–220. Also in Schoeman (1984), pp. 75–103, and in Johnson and Snapper (1985), pp. 172–183.

Ware, Willis, "Information Systems Security and Privacy." *Communications of the ACM* 27, no. 4 (April 1984), 315–321.

Weisstaub, D. N., and C. C. Gottlieb. *The Nature of Privacy: A Study by the Privacy and Computer Task Force.* Department of Communications/Department of Justice, Canada, 1973.

Winner, Langdon. "Just Me and My Machine: The New Solipsism." *Whole Earth Review* (December 1984–January 1985), 29.

Study Questions

1. What is the main point of Edgar's article? Why do you think it might be correct? Why do you think it might be wrong? Explain and defend your answers.

2. What assumptions does Edgar make about technology? about ethics? about privacy? Do you think these assumptions are correct? Explain and defend your answers.

3. If society accepts Edgar's main conclusions, what consequences might follow? Would these consequences be desirable? Explain and defend your answers.

4. Employers now use computers to monitor the rate of work of employees (such as the speed of keyboard operators or the number of rest stops of truckers). Do you think that this computer monitoring is ethically defensible?

5. Who has computerized information about you? How accessible is it by other people? Could release of any of this information hurt you?

6. Would an egalitarian or a utilitarian ethical theorist be more likely to favor extensive rights to privacy? Explain.

4.5

Ethics and Military Technology: Star Wars

David Lorge Parnas and Danny Cohen

"CON"—David Lorge Parnas

Introduction

In May of 1985 I was asked by the Strategic Defense Initiative Organization, the group within the Office of the U.S. Secretary of Defense that is responsible for the "Star Wars" program, to serve on a $1,000(US)/day advisory panel, the SDIO Panel on Computing in Support of Battle Management. The panel was to make recommendations about a research and development program to solve the computational problems inherent in space-based defense systems.

Like President Reagan, I consider the use of nuclear weapons as a deterrent to be dangerous and immoral. If there is a way to make nuclear weapons impotent and obsolete and end the fear of nuclear weapons, there is nothing I would rather work on. However, two months later I had resigned from the panel. I have since become an active opponent of the SDI. The purpose of this article is to explain why I am opposed to the program. I begin by stating my personal views on defense work and professional responsibility.

© 1987 by The Regents of the University of California. Published by the University of California Institute on Global Conflict and Cooperation, 1987. Originally published as "SDI: One View of Professional Responsibility," in *SDI Two Views of Professional Responsibility*.

My View of Professional Responsibility

My decision to resign from the panel was consistent with long-held views about the individual responsibility of a professional. I believe that professionals have responsibilities that go beyond an obligation to satisfy the short-term demands of their immediate employer.

As a professional:

1. I am responsible for my own actions and cannot rely on any external authority to make my decisions for me.
2. I cannot ignore ethical and moral issues. I must devote some of my energy to deciding whether the task that I have been given is of benefit to society.
3. I must make sure that I am solving the real problem, not simply providing short-term satisfaction to my supervisor.

Some have held that a professional is a "team player" and would not "blow the whistle" on his colleagues and employer. I disagree. As the Challenger incident demonstrates, such action is sometimes necessary. One's obligations as a professional precede other obligations. One must not enter into contracts that conflict with one's professional obligations.

My Views on Defense Work

Many opponents of SDI oppose all military development. I am not one of them. I have been a consultant to the Department of Defense and other components of the defense industry since 1971. I am considered an expert on the organization of large software systems and I lead the U.S. Navy's Software Cost Reduction project at the Naval Research Laboratory. Although I have friends who argue that "people of conscience" should not work on weapons, I maintain that it is vital that people with a strong sense of social responsibility continue to work within the military industrial complex. I do not want to see that power completely in the hands of people who are not conscious of social responsibility.

My own views on military work are close to those of Albert Einstein. Einstein, who called himself a militant pacifist, at one time held the view that scientists should refuse to contribute to arms development. Later in his life he concluded that to hold to a "no arms" policy would be to place the world at the mercy of its worst enemies. Each country has a right to be protected from those who use force, or the threat of force, to impose their will on others. Force can morally be used only against those persons who are themselves using force. Weapons development should be limited to weapons that are suitable for that use. Neither the present arms spiral nor nuclear weapons are consistent with Einstein's principles. One of our greatest scientists, he knew that international security requires progress in political education, not weapons technology.[1]

What Is SDI?

SDI, popularly known as "Star Wars," was initiated by a 1983 presidential speech calling on scientists to free us from the fear of nuclear weapons. President Reagan directed the Pentagon to search for a way to make nuclear strategic missiles impotent and obsolete. In response, SDIO has embarked upon a project to develop a network of satellites carrying sensors, weapons, and computers to detect ICBMs and intercept them before they can do much damage. In addition to sponsoring work on the basic technologies of sensors and weapons, SDI has funded a number of Phase I "architecture studies," each of which proposes a basic design for the system. The best of these have been selected and the contractors are now proceeding to "Phase II," a more detailed design.

My Early Doubts

As a scientist, I wondered whether technology offered us a way to meet these goals. My own research has centered on computer software and I have used military software in some of my research. My experience with computer-controlled weapon systems led me to wonder whether any such system could meet the requirements set forth by President Reagan.

I also had doubts about conflict of interest. I have a project within the U.S. Navy that could benefit from SDI funding. I suggested to the panel organizer that this conflict might disqualify me. He assured me that if I did not have such a conflict, they would not want me on the panel. He pointed out that the other panelists, employees of defense contractors and university professors dependent on DoD funds for their research, had similar conflicts. Readers should think about such conflicts the next time they hear of a panel of "distinguished experts."

My Work for the Panel

The first meeting increased my doubts. In spite of the high rate of pay, the meeting was poorly prepared; presentations were at a disturbingly unprofessional level. Technical terms were used without definition; numbers were used without supporting evidence. The participants appeared predisposed to discuss many of the interesting, but tractable, technical problems in space-based missile defense while ignoring the basic problems and "big picture." Everyone seemed to have a pet project of their own that they thought should be funded.

At the end of the meeting we were asked to prepare position papers describing research problems that must be solved in order to build an effective and trustworthy shield against nuclear missiles. I spent the weeks after the meeting writing up those problems and trying to convince myself that SDIO-supported research could solve those problems. I failed! I could not convince myself that it would be possible to build a system that we could trust or that it would be useful to build a system that we did not trust.

Why Trustworthiness Is Essential to President Reagan's Goals

If the U.S. does not trust SDI it will not abandon deterrence and nuclear missiles. Even if the U.S. did not trust its shield, the USSR could not assume that SDI would be completely ineffective. Seeing both a "shield" and missiles, it would feel impelled to improve its offensive forces in an effort to compensate for SDI. The U.S., not trusting its defense, would feel a need to build still more nuclear missiles to compensate for the increased Soviet strength. The arms race would speed up. Further, because NATO would be wasting an immense amount of effort on a system it couldn't trust, we would see a weakening of our relative strength. Instead of the safer world that President Reagan envisions, we would have a far more dangerous situation. Thus, the issue of our trust in the system is critical. Unless the shield is trustworthy, it will not benefit any country.

The Role of Computers in "Star Wars"

SDI discussions often ignore computers, focusing on new developments in sensors and weapons. However, the sensors will produce vast amounts of raw data that computers must process and analyze. Computers must detect missile firings, determine the source of the attack, and compute the attacking trajectories. Computers must discriminate between threatening warheads and decoys designed to confuse our defensive system. Computers will aim and fire the weapons. All the weapons and sensors will be useless if the computers do not function properly. Software is the glue that holds such systems together. If the software is not trustworthy, the system is not trustworthy.

The Limits of Software Technology

Computer specialists know that software is always the most troublesome component in systems that depend on computer control. Traditional engineering products can be verified by a combination of mathematical analysis, case analysis, and prolonged testing of the complete product under realistic operating conditions. Without such validation, we cannot trust the product. None of these validation methods works well for software. Mathematical proofs verify only abstractions of small programs in restricted languages. Testing and case analysis sufficient to ensure trustworthiness take too much time. As E.W. Dijkstra has said, "Testing can show the presence of bugs, never their absence."

The lack of validation methods explains why we cannot expect a real program to work properly the first time it is really used. This is confirmed by practical experience. We can build adequately reliable software systems, but they become reliable only after extensive use in the field. Although responsible developers perform many tests, including simulations, before releasing their software, serious problems always remain when the first customers use the product. The test designers overlook the same problems as the software designers overlook. No experienced person trusts a software system before it has seen extensive use under actual operating conditions.

Why Software for SDI Is Especially Difficult

SDI is far more difficult than any software system we have ever attempted. Some of the reasons are listed below. A more complete discussion can be found in "Software Aspects of Strategic Defense Systems."[2]

SDI software must be based on assumptions about target and decoy characteristics; those characteristics are controlled by the attacker. We cannot rely upon our information about them. The dependence of any program on those assumptions is a rich source of effective countermeasures. Espionage could render the whole multi-billion-dollar system worthless without our knowledge. It could show an attacker how to exploit the inevitable differences between the computer model on which the program is based and the real world.

The techniques used to provide high reliability in other systems are hard to apply for SDI. In space, the redundancy required for high reliability is unusually expensive. The dependence of SDI on communicating computers in satellites makes it unusually vulnerable. High reliability can be achieved only if failures of individual components are statistically independent; for a system subject to coordinated attacks, that is not the case.

Overloading the system will always be a potent countermeasure because any computer system will have a limited capacity and even crude decoys would consume computer capacity. An overloaded system must either ignore some of the objects it should track, or fail completely. For SDI, either is catastrophic.

Satellites will be in fixed orbits that will not allow the same one to both track a missile from its launch and destroy it. Responsibility for tracking a missile will transfer from one satellite to another. Because of noise caused by the battle and enemy interference, a satellite will require data from other satellites to assist in tracking and discrimination. The result is a distributed real-time data base. For the shield to be effective, the data will have to be kept up-to-date and consistent in real-time. To do that, satellite clocks will have to be accurately synchronized. None of this can be done when the network's components and communication links are unreliable; unreliability must be expected during a real battle in which an enemy would attack the network. Damaged stations are likely to inject inaccurate or false data into the data base.

Realistic testing of the integrated hardware and software is impossible. Thorough testing would require "practice" nuclear wars including attacks that partially damage the satellites. Our experience tells us that many potential problems would not be revealed by lesser measures such as component testing, simulations, or small-scale field tests.

Unlike other weapon systems, there will be no opportunity to modify the software during or after its first battle. It must work the first time.

These properties are inherent in the problem, not a particular system design. As we will see below, they cannot be evaded by proposing a new system structure.

My Decision to Act

After reaching the conclusions described above, I solicited comments from other scientists and found none that disagreed with my technical conclusions. Instead, they told me that the program should be continued, not because it would free us from the fear of nuclear weapons, but because the research money would advance the state of computer science. I disagree with that statement, but I also consider it irrelevant. Taking money allocated for developing a shield against nuclear missiles, while knowing that such a shield was impossible seemed, to me, to constitute fraud. I did not want to participate and submitted my resignation. I felt it would be unprofessional to resign without explanation and submitted my position papers to support my letter. I sent copies to a number of government officials and friends but did not send them to the press until they had been sent to reporters by others. They have since been widely published.[2]

SDIO's Reaction

The SDIO's reaction to my resignation transformed my stand on SDI from a passive refusal to participate, to an active opposition. Neither SDIO nor the other panelists reacted with a serious and scientific discussion of the technical problems that I raised.

The first reaction came from one of the panel organizers. He asked me to reconsider, but not because he disagreed with my technical conclusions. He accepted my view that an effective shield was unlikely, but argued that the money was going to be spent and I should help to see it well spent. There was no further reaction from SDIO until a *New York Times* reporter called. Then, the only reaction that I received was a telephone call demanding to know who had sent the material to the *Times*.

After the story broke, the statements made to the press seemed, to me, to be designed to mislead, rather than inform, the public. Examples are given below. When I observed that SDIO was engaged in "damage control," rather than a serious consideration of my arguments, I felt that I should inform the public and its representatives of my own view. I want the public to understand that no trustworthy shield will result from the SDIO-sponsored work. I want them to understand that technology offers no magic that will eliminate the fear of nuclear weapons. I feel that to be part of my personal professional responsibility as a scientist and an educator.

Democracy can only work if the public is accurately informed. Most of the statements made by SDIO supporters seem designed to mislead the public. For example, one SDIO scientist told the press that there could be 100,000 errors in the software and it would still work properly. Strictly speaking this statement is true. If one picks one's errors very carefully, they won't matter much. However, a single error caused the complete failure of a Venus probe many years ago. I find it hard to believe that the SDIO spokesman was not aware of that.

Another panelist repeatedly told the press that there was no fundamental law of computer science that said the problem could not be solved. Again, strictly speaking, the statement is true but it does not counter my arguments. I did not say that a correct program was impossible; I said that it was impossible that we would trust the program. It is not impossible that such a program would work the first time it was used; it is also not impossible that 10,000 monkeys would reproduce the works of Shakespeare if allowed to type for 5 years. Both are highly unlikely. However, we could tell when the monkeys have succeeded; there is no way that we could verify that the SDI software was adequate.

Another form of disinformation was the statement that I, and other SDI critics, were demanding perfection. Nowhere have I demanded perfection. To trust the software we merely need to know that the software is free of catastrophic flaws, flaws that could cause massive failure or that could be exploited by a sophisticated enemy. That is certainly easier to achieve than perfection, but there is no way to know when we have achieved it.

A common characteristic of all these statements is that they argue with statements other than the ones that I published in my papers. In fact, in some cases SDIO officials dispute statements made by earlier panels or by other SDIO officials rather than debate the points that I made.

The "90%" Distraction

One of the most prevalent arguments in support of SDI suggests that if there are 3 layers, each 90% effective, the overall "leakage" would be less than 1% because the effectiveness multiplies. This argument is accepted by many people who do not have scientific training. However,

1. there is no basis for the 90% figure; an SDI official told me it was picked for purpose of illustration,

2. the argument assumes that the performance of each layer is independent of the others when it is clear that there are many links,
3. it is not valid to rate the effectiveness of such systems by a single "percentage." Such statistics are only useful for describing a random process. Any space battle would be a battle between two skilled opponents. A simple "percentage" figure is no more valid for such systems than it is as a way of rating chess players. The performance of defensive systems depends on the opponent's tactics. Many defensive systems have been completely defeated by a sophisticated opponent who found an effective countermeasure.

The "Loose Coordination" Distraction

The most sophisticated response was made by the remaining members of SDIO's Panel on Computing in Support of Battle Management, which named itself the Eastport group, in December. This group of SDI proponents wrote that the system structures proposed by the best Phase I contractors, those being elaborated in Phase II, would not work because the software could not be built or tested. They said that these "architectures" called for excessively tight coordination between the "battle stations," i.e., excessive communication, and they proposed that new Phase I studies be started. However, they disputed my conclusions, arguing that the software difficulties could be overcome using "loose coordination."[3]

The Eastport Report neither defines its terms nor describes the structure that it had in mind. Parts of the report imply that "loose coordination" can be achieved by reducing the communication between the stations. Later sections of the report discuss the need for extensive communication in the battle station network, contradicting some statements in the earlier section. However, the essence of their argument is that SDI could be trustworthy if each battle station functioned autonomously, i.e., without depending on help from others.

The Eastport group's argument is based on four unstated assumptions:

1. Battle stations do not need data from other satellites to perform their basic functions.
2. An individual battle station is a small software project that will not run into the software difficulties described above.
3. The only interaction between the stations is by explicit communication. This assumption is needed to conclude that test results about a single station allow one to infer the behavior of the complete system.

4. A collection of communicating systems differs in fundamental ways from a single system.

All of these assumptions are false!

1. The data from other satellites is essential for accurate tracking and for discriminating between warheads and decoys in the presence of noise.
2. Each battle station has to perform all the functions of the whole system. The original arguments apply to it. Each one is unlikely to work, impossible to test in actual operating conditions, and, consequently, impossible to trust. Far easier projects have failed.
3. Battle stations interact through weapons and sensors as well as through their shared targets. The weapons might affect the data produced by the sensors. For example, destruction of a single warhead or decoy might produce noise that makes tracking of other objects impossible. If we got a single station working perfectly in isolation, it might fail completely when operating near others. The failure of one station might cause others to fail because of overload. Only a real battle would give us confidence that such interactions would not occur.
4. A collection of communicating programs is mathematically equivalent to a single program. In practice, distribution makes the problem harder, not easier.

Restricting the communication between the satellites does not solve the problem. There is still no way to know the effectiveness of the system and it would not be trusted. Further, the restrictions on communication are likely to reduce the effectiveness of the system. I assume that this is why none of the Phase I contractors chose such an approach.

The first claim is appealing and reminiscent of arguments made in the 60s and 70s about modular programming.[4] Unfortunately, experience has shown that modular programming is an effective technique for making errors easier to correct, not for eliminating errors. Modular programming does not solve the problems described earlier in this paper. None of those arguments were based on an assumption of tight coupling; some of the arguments do assume that there will be data passed from one satellite to another. The Eastport Report, like earlier reports, supports that assumption.

The Eastport group is correct when it says that designs calling for extensive data communication between the battle stations are unlikely to work. However, the Phase I contractors were also right when they assumed that without such communication the system could not be effective.

The Ultimate Response: Redefining the Problem

The issue of SDI software was debated in March 1986 at an IEEE Computer Conference. While two of us argued that SDI could not be trusted, the two SDI supporters argued that that did not matter. Rather than argue the computer science issues, they tried to use strategic arguments to say that a shield need not be considered trustworthy. One of them argued, most eloquently, that the president's "impotent and obsolete" terminology was technical nonsense. He suggested that we ignore what "the president's speechwriters" had to say and look at what was actually feasible. Others argued that increased uncertainty is a good thing—quite a contrast to President Reagan's promise of increased security.

In fact, the ultimate response of the computer scientists working on SDI is to redefine the problem in such a way that there is a trivial solution and improvement is always possible. Such a problem is the ideal project for government sponsorship. The contractor can always show both progress and the need for further work. Contracts will be renewed indefinitely!

Those working on the project often disparage statements made by the president and his most vocal supporters, stating that SDIO scientists and officials are not responsible for such statements. However, the general public remains unaware of their position and believe that the president's goals are the goals of those who are doing the scientific work.

Is SDIO-Sponsored Work of Good Quality?

Although the Eastport panel was unequivocally supportive of continuing SDI, its criticisms of the Phase I studies were quite harsh. They assert that those studies, costing US$1,000,000 each, overlooked elementary problems that were discussed in earlier studies. If the Eastport group is correct, the SDIO contractors and the SDIO evaluators must be considered incompetent. If the Eastport group's criticisms were unjustified, or if their alternative is unworkable, their competence must be questioned.

Although I do not have access to much of the SDIO-sponsored work in my field, I have had a chance to study some of it. What I have seen makes big promises, but is of low quality. Because it has bypassed the usual scientific review processes, it overstates its accomplishments and makes no real scientific contribution.

Do Those Who Take SDIO Funds Really Disagree with Me?

I have discussed my views with many who work on SDIO-funded projects. Few of them disagree with my technical conclusions. In fact, since the story became public, two SDIO contractors and two DoD agencies have sought my advice. My position on this subject has not made them doubt my competence. Those who accept SDIO money give a variety of excuses. "The money is going to be spent anyway, shouldn't we use it well?" "We can use the money to solve other problems." "The money will be good for computer science." I have also discussed the problems with scientists at the Los Alamos and Sandia National Laboratories. Here, too, I found no substantive disagreement with my analysis. Instead, I was told that the project offered lots of challenging problems for physicists.

In November I read an interview with a leading German supporter of Star Wars. He made it clear that he thought of SDI as a way of injecting funds into high technology and not as a military project. He even said that he would probably be opposed to participation in any deployment should it come to pass.[5]

The Blind Led by Those with Their Eyes Shut

My years as a consultant in the defense field have shown me that unprofessional behavior is common. When consulting, I often find people doing something foolish. Knowing that the person involved is quite competent, I may say something like "You know that's not the right way to do that." "Of course" is the response, "but this is what the customer asked for." "Is your customer a computer scientist? Does he know what he is asking?" ask I. "No" is the simple reply. "Why don't you tell him?" elicits the response, "At XYZ Corporation, we don't tell our customers that what they want is wrong. We get contracts."

That may be a businesslike attitude but it is not a professional one. It misleads the government into wasting taxpayers' money.

The Role of Academic Institutions

Traditionally, universities provide tenure and academic freedom so that faculty members can speak out on issues such as these. Many have done just that. Unfortunately, at U.S. universities there are institutional pressures in favor of accepting research funds from any source. A researcher's ability to attract funds is taken as a measure of his ability.

The president of a major university in the U.S. recently explained his acceptance of a DoD institute on campus by saying, "As a practical matter, it is important to realize that the Department of Defense is a major administrator of research funds. In fact, the department has more research funds at its disposal than any other organization in the country . . . increases in research funding in significant amounts can be received only on the basis of defense-related appropriations."

Should We Pursue SDI for Reasons Other Than the President's?

I consider such rationalizations to be both unprofessional and dangerous. SDI endangers the safety of the world. By working on SDI, these scientists allow themselves to be counted among those who believe that the program can succeed. If they are truly professionals, they must make it very clear that an effective shield is unlikely and a trustworthy one impossible. The issue of more money for high technology should be debated without the smokescreen of SDI. I can think of no research that is so important that it justifies pretending that an ABM system can bring security to populations. Good research stands on its own merits; poor research must masquerade as something else.

I believe in research; I believe that technology can improve our world in many ways. I also agree with Professor Makowski of the Technion who wrote, "Overfunded research is like heroin, it leads to addiction, weakens the mind, and leads to prostitution." Many research fields in the U.S. are now clearly overfunded, largely because of DoD agencies. I believe we are witnessing the proof of Professor Makowski's statement.

My Advice to Others about Participation in Defense Projects

I believe it quite appropriate for a professional to devote his energies to making the people of his land more secure. In contrast, it is not professional to accept employment doing things that do not advance the legitimate defense interests of that country. If the project would not be effective, or if, in his opinion, it goes beyond the legitimate defense needs of the country, a professional should not participate. Too many do not ask such questions. They ask only how they can get another contract.

It is a truism that if each of us lives as if what we do does matter, the world will be a far better place than it now is. The cause of many serious problems in our world is that many of us act as if our actions do not matter. Our streets are littered, our environment polluted, and children are neglected because we underestimate our individual responsibility.

The arguments given to me for continuation of the SDI program are examples of such thinking. "The government has decided, we cannot change it." "The money will be spent, all you can do is make good use of it." "The system will be built, you cannot change that." "Your resignation will not stop the program."

It is true my decision not to toss trash on the ground will not eliminate litter. However, if we are to eliminate litter, I must decide not to toss trash on the ground. We all make a difference.

Similarly, my decision not to participate in SDI will not stop this misguided program. However, if everyone who knows that the program will not lead to a trustworthy shield against nuclear weapons refuses to participate, there will be no program. Every individual's decision is important.

It is not necessary for computer scientists to take a political position, they need only be true to their professional responsibilities. If the public were aware of the technical facts, if they knew how unlikely it is that such a shield would be effective, public support would evaporate. We do not need to tell the public not to build SDI. We only need to help them to understand why it won't be an effective and trustworthy shield.

References

1. Albert Einstein, Sigmund Freud, "Warum Krieg." Diogenes Verlag, Zuerich, 1972.

2. Parnas, D. L. "Software Aspects of Strategic Defense Systems," *American Scientist*, September–October 1985, pp. 432–440. Also published in German in *Kursbuch 83, Krieg and Frieden—Streit um SDI*, by Kursbuch/Rotbuch Verlag, March 1986, and in Dutch in *Informatie*, Nr. 3, March 1986, pp. 175–186.

3. Eastport Group "Summer Study 1985," A Report to the Director— Strategic Defense Initiative Organization, December 1985.

4. Parnas, D. L. "On the Criteria to be Used in Decomposing Systems into Modules," *Comm. ACM* 15, 12, December 1972, pp. 1053–1058.

5. "Wer Kuscht, hat Keine Chance," *Der Spiegel*, Nr. 47, 18 November 1985.

"PRO"—Danny Cohen

Introduction

In March of 1983 President Reagan announced his goal to free us from the danger of nuclear holocaust. Within minutes the country was already polarized between supporters (who found the idea desirable) and opponents (who found the idea to be undesirable and contrived reasons why it would be technically impossible to attain).

Many professional organizations, such as physicists, physicians, and "responsible" computer scientists aligned themselves against the SDI.

Professionals have the responsibility to provide professional judgment and guidance. Is this responsibility restricted only to their domain of expertise, or does it apply to all areas of life?

It is my opinion that when professionals step off their turf, they cease to be "professionals" and become mortals, like the rest of us.

What Is SDI?

The SDI (also known as "Star Wars") is a research program directed eventually to freeing us from the fear of nuclear disaster. The SDI is not building a "shield" as some depict it. It was not "sold" to the public as an umbrella that will be in use momentarily.

In January 1985, the Administration described the program as follows:

> [SDI's] purpose is to identify ways to exploit recent advances in ballistic missile defense technologies that have potential for strengthening deterrence—and therefore increasing our security and that of our Allies. The program is designed to answer a number of fundamental scientific and engineering questions that must be addressed before the promise of these new technologies can be fully assessed. The SDI research program will provide to a future President and a future Congress the technical knowledge necessary to support a decision in the early 1990s on whether to develop and deploy such advanced defensive systems.

Originally published as "SDI: Another View on Professional Responsibility," in *SDI: Two Views of Professional Responsibility*. Copyright © 1987 by the Regents of the University of California. Published by the University of California Institute on Global Conflict and Cooperation, 1987.

The SDI is working now on *research and development* of the technology, not on *deploying* it. The answers to all questions about the performance of the SDI system depend on what the mission of the system would be, on what the future threat would be, and how (and if) we will deploy the system.

One may wonder about spending funds on developing the technology if these basic questions cannot be answered first. However, the only way to answer these questions intelligently is by starting from the knowledge about what the technology can do—knowledge that we do not yet have, about technology still under development.

The same situation is not unique to SDI. When scientists develop new wings for supersonic fighters it is too early to ask about their effect on the performance of the entire air defense system, before these wings are integrated into complete aircraft (such as the F-16) and before decisions are made about the deployment of these aircraft.

The SDIO realizes that it does not have all the answers yet, and even not all the questions. Therefore, it conducts its research in a variety of directions, and exposes every step to reviews and critique.

The Eastport Panel

In 1985, SDIO convened the Eastport panel to "devise an appropriate computational/communications response to the [strategic defense battle management] problem and make recommendations for a research and technology development program to implement the response."

The Eastport panel was not asked to be a "team player" and to rubber stamp everything that was presented to it—and it did not behave as such. The report of the panel criticized some aspects of the work performed for SDI in no uncertain terms (or "quite harsh" as Prof. Parnas said).

The panel met several times during 1985, for presentations and discussions, and its members performed related work between meetings. As the work progressed the panel chartered its way (rather than asking SDIO to direct its path), and chose the next steps of action.

Prof. Parnas was the only member that reached his conclusions before the second meeting. This is not the place to argue them. Having arrived at these conclusions, Prof. Parnas took the only possible action, and resigned from the panel.

In his description of the first meeting, Prof. Parnas recalls, "Everyone seemed to have a pet project of their own that they thought should be funded." This is true only about Prof. Parnas himself! In this first meeting he already *knew* the "solution," the right direction for SDIO to pursue. When he presented that pet of his to the rest of the panel, his general-purpose approach was immediately criticized as being equally applicable to an intensive-care unit in a hospital, for example, and not addressing the special properties and problems of the SDI system.

We, the rest of the panel, could understand his anger with our response, but were surprised by his convenient recollection of the event as one in which "Everyone seemed to have a pet project of their own."

Prof. Parnas is correct in reporting that the Eastport panel did not react with "serious and scientific discussion of the technical problems that I [Prof. Parnas] raised." The Eastport panel, like many others, found the issues that Prof. Parnas raised in the minipapers that followed his resignation letter to be *irrelevant*. For example, one of them is, "Can automatic programming solve the SDI software problem?" In it Prof. Parnas observes that "claims that have been made for automatic programming systems are greatly exaggerated." This would be relevant if SDIO (or even just the Eastport panel) had advocated the inverse, that automatic programming will solve the SDI software problem.

The Eastport panel was, in fact, astonished at the irrelevance of nearly all the topics of these minipapers, and concluded that he had used the occasion to voice all his pet opinions.

Since the subject of this publication is *professional responsibility*, it is not the right forum for arguing against the points that Prof. Parnas raises.

The Dichotomy

A few months ago a reporter called me about SDI. When I identified myself as a supporter of the program he said, "This means that you believe that it is possible."

This typical comment is due to the existing dangerous dichotomy in the country about SDI. There are those who find it desirable, and therefore worth the effort of finding out how to pursue it, and there are those who find it undesirable and therefore "know" that it is impossible, and that no research has the potential to further our knowledge on the subject.

Here is where professionals enter the picture. The general public and the media feel helpless confronting the technical issues, and are looking up to the professionals to provide guidance.

Professionals have the right and the obligation to guide the public. However, they also have the responsibility to alert the public when they step out of their area of expertise. For example, when a dentist (or a physicist, or a "responsible computer professional") discusses what the Soviet Union would do in response to the SDI, the public has the right to know that this is speculation rather than professional opinion.

The American public is exposed time and again to "experts" claiming that the SDI system cannot be accomplished because of technical difficulties. This is not the first time that experts predicted that certain things are impossible.

For example, in 1937, Prof. Hans Bethe published a short article in which he mathematically proved that "The Maximum Energy Obtained from the Cyclotron" (*Physical Review*, Vol. 52) is about 10 MV for protons, a limit that has already been exceeded by many orders of magnitude.

Incidentally, his theoretically proven limit had been exceeded even before the article was published. It was not that his mathematics was wrong—he just did not realize all the possible ways of accomplishing it.

Bethe's limited vision erred in two ways: (1) The cyclotron itself was developed, by additional invention and development, to the "synchrocyclotron" with energies in excess of 200 MV for protons; and (2) a wholly new device, the "synchrotron," was invented *to accomplish the same purpose*, that is capable of at least 2,000,000 times the "limit" Bethe "proved" in accelerating protons.

When professionals make claims about impossibilities, they have the responsibility to clarify whether their claims apply only to the methods that they examine or to all possible methods, including those that they do not envision.

The public, guided by the professionals, has the right to know the difference. It is unfortunate that the critics of SDI do not fulfill this part of their obligation as well.

Those who claimed that "you cannot hit a bullet with a bullet" (i.e., hit a reentry vehicle with a missile) never revisited the issue (at least not in public) after repeated experiments proved the possibility of doing just that.

Prof. Parnas stated that "To do that, satellite clocks will have to be accurately synchronized. None of this can be done when the network's components and communication links are unreliable." This is a typical classic mistake of predicting future technical capabilities based on what one knows today, and approaching problems with "how would I do it," a mistake that professionals are expected to avoid. Even laymen, having access only to public unclassified information, can find out that satellites already keep time within a few nanoseconds (with drift of about 3μsec per year) using today's technology. Laboratory systems have already improved this by an order of magnitude. I dare not predict what would be possible beyond the mid-1990s. This simple example shows that being a professional in one field is not an expert-license for all domains.

About Another Defense System

Gregory Fossedal has pointed out that the current criticism of SDI resembles British arguments against air-defense measures before World War II:

> In the 1930s, opponents of a British defense against German attack argued that a few firebombs could result in a "total holocaust" of London, as one member of Parliament put it. "The bomber will always get through," said Prime Minister Stanley Baldwin. Hence, they argued any defense was useless.
>
> Winston Churchill, then a discredited backbencher, saw the folly of this reasoning and decided, given that no defense is ever perfect, to see what kinds of defenses could be built. "Science is always able to provide something" he wrote in a memo in 1935.
>
> The key he said was not scientific at all, but political. "General tactical considerations, and what is technically feasible act and react upon one another. Thus, the scientists should be told what facilities the air force would like to have, and airplane design be made to fit into and implement a definite scheme of warfare."
>
> Thanks to Mr. Churchill's efforts, private scientists were able to design and build defenses—despite furious objection from Air Ministry bureaucracy. By seeing the strategic error of demanding perfection of defense, Mr. Churchill knew that the opinion of scientists who opposed him wasn't so much wrong as it was irrelevant. In fact British air defense concentrated on rapidly deploying what was available, knowing that more exotic technologies, such as radar, would come along later. (Fossedal, "A Common Thread Linking Star Wars," *Washington Times*, 23 December 1986.)

Those who found the idea of air defense impossible had the professional responsibility and the right not to work on the project, just as those who preferred to pursue it had the professional responsibility and the right to work toward that goal.

Each group should respect the rights of the other. Unfortunately, this courtesy is not always followed.

The Reliability Issue

Many of the critics repeat the argument that the expected SDI system will not be reliable due mainly to the impossibility of complete full-scale realistic testing.

This is a very important point. It is interesting that the same media that criticize SDIO on this point also criticize SDIO for its attempts to conduct simulations, experiments, and validation efforts through its future National Test Bed.

It would be more responsible not to raise the issue in a vacuum, but instead in the context of defense systems in general.

None of our major defense systems has ever been fully tested in realistic conditions—and, thank God, neither was any of our adversaries'. Many remind us that we test our ICBMs only on east to west flights (from California to Kwajalein) and have no *proof* of their ability to navigate over the pole. Luckily the same holds true for the Soviets, too (except that they fly from west to east).

We have never conducted a realistic full-scale test of our strategic forces, of NORAD, or of our strategic Attack Warning system, and I hope that we never will. Thinking about these issues does not make one feel good about defense systems in general. Does it? Does this imply that we should not develop *any* major defense system because by definition it can never be fully tested in realistic conditions? Obviously not.

We have never tested a nuclear warhead on an ICBM, but for the sake of knocking SDI the critics trust these untested ICBMs as the cornerstone of our security.

Pointing to these issues as faults unique to the SDI system is misleading and not necessarily of high professional responsibility. It implies to the general public that the government is about to replace defense systems that were fully tested under the most realistic conditions (and hence are of the utmost reliability and stability) by an unreliable system.

Those who raise the important issue of reliability should have the professional responsibility to raise it in the proper context.

The "technical" debate about the doability/trustworthiness/etc., is reminiscent of the debates in the 1950s and the 1960s about the doability/trustworthiness/etc., of the ICBMs. Many of the "responsible" professionals of those days argued about the undoability of inertial navigation. Then, as now, opinions led many who wrapped themselves in their professional credentials and claimed to "know" that it was technically impossible.

Some Key Questions

I believe that Prof. Parnas managed to find incompetent people who work for the Defense Department and for SDIO (like the person who argued in favor of 100,000 errors). I also believe that there are dishonest people employed by some contractors. Unfortunately, this is true for every profession and every field of human activity. Obviously, this should be fought against. Prof. Parnas's error is the consequent conclusion that because such people are involved in SDI, the program cannot succeed ("the blind led by those with their eyes shut"). In fact, most good activities accomplished by human beings (from medicine and welfare to space flights) are accomplished within and despite this truth; this should not be used to conclude that we cannot do anything useful until the earth is populated only by good and smart people.

Somehow it happened that words like "responsible," "concerned," and "conscious" are monopolized by the opponents of the SDI program. One may get the feeling that Soviet nuclear warheads, like the California condor, should be preserved and never be shot at.

Are we more safe with our security in the heads of our adversaries or in our own hands? Those who preach about the dangers of miscalculation, misunderstanding, overreaction, and the danger of accidental nuclear war—neglect all these arguments when it comes to our ability to *defend* ourselves rather than our "ability" to *deter* our enemies.

The question we *should* be asking is, "If we have *any* reason to believe that we might have it in our power to prevent 'The Day After' from happening, what kind of people are we if we don't have the patience, the commitment—the *humanity* to do so?"

The president has pointed out that the purpose of the SDI program is not to build a fully capable anti-ballistic system now, but rather to conduct research to support an informed decision by *another* president and *another* Congress sometime in the future—perhaps not before the turn of the century.

Our approach to solving this family of problems must be to attack it with a family of solutions. Since it seems unlikely that there exists any single solution to this thicket of problems, our objective must be to find that combination of answers that offers the greatest synergism and the highest leverage in the sum of its individual approaches.

Summary

Professionals have the right and the obligation to guide the public in their areas of expertise. They also have the obligation to alert the public to the fact that their speculations are not more profound than those by laymen. They should present their knowledge and learned results, and not neglect to describe their limitations.

Professionals, more than laymen, have the responsibility of not letting their *opinions* (e.g., about desirability) drive their *judgment* (e.g., about feasibility), and *not* follow the queen's practice in *Alice in Wonderland*, "Sentence First—Verdict Afterwards," as many have demonstrated.

RESPONSE—David L. Parnas

It is unfortunate that Dr. Cohen has chosen to attack the critics of SDI rather than their arguments. It is more unfortunate that he has chosen to base his arguments on false assumptions about their positions. The following six statements are intended to set the record straight.

1. Responsible opponents to SDI do not oppose the goal attributed to SDIO by Dr. Cohen, "free us from the danger of nuclear holocaust," or that stated by President Reagan, "make nuclear weapons impotent and obsolete." Many SDI opponents have campaigned for similar goals for decades. The arguments made against SDI are based on the inability of technology to accomplish those goals, not their undesirability.
2. No responsible professional would claim that professionals are not mortals—even on "their turf." We claim no infallibility but give the public information and opinions to the best of our ability. In contrast, several supporters of SDI have stated that their information and expertise is available only to the government, not the public.
3. No critic of SDI has ever suggested that a BMD [Ballistic Missile Defense] system is now being deployed. However, SDIO's official charter calls upon SDIO to undertake "a comprehensive program to develop key technologies . . . move the United States towards its ultimate goal of a thoroughly reliable defense." The SDIO charter, and its progress reports, makes it clear that the program is a development program, not a research program as Dr. Cohen suggests. Dr. Cohen knows that SDI is officially classified as advanced development.
4. At no point before or during the initial meeting of the SDIO Panel on Computing in Support of Battle Management, which later renamed itself the "Eastport Group," did I believe I have a solution, present a solution, or have a proposal rejected. In public debates,

Dr. Cohen has been asked to describe the solution he claims I presented, but was unable to do so.

I was asked to inform the panel about the work I have done on weapon delivery software; I presume that it was this work that led to my invitation to join the panel. Unaware that my information would be perceived as a proposal, and unaware that the panel had rejected it, I did not feel the anger attributed to me by Dr. Cohen. Even today, I do not consider comments about the broad applicability of my work to be pejorative.

5. While Dr. Cohen has claimed that the topics of my papers were considered by his panel to be irrelevant to SDI, Lt. Gen. James Abrahamson, director of the SDIO, has referred to those problems as "the long pole of the tent." The record will show that even the most "far out" area discussed in those papers, Artificial Intelligence, has been proposed as a "solution" for SDI and that SDIO has funded, and continues to fund, work in Artificial Intelligence and the other areas that I discussed. The relevance of software development methods has never been questioned by SDIO officials.

6. Most SDI critics do not "trust untested ICBMs as the cornerstone of our security." In fact, many SDI critics argue that these devices bring insecurity not security. To argue, as SDIO officials and supporters do, that opponents of SDI are in favor of a policy of nuclear deterrence, is like arguing that opponents of execution by lethal injection favor a return to the noose.

Dr. Cohen argues the feasibility of SDI by reminding us of situations where pessimistic predictions were wrong. It is useful to remember the most relevant past debate, that about the SAFEGUARD missile defense system. That system was eventually dismantled because it was found to be ineffective. The opponents of that development were clearly correct. While that certainly does not prove that SDI will fail, it does show the irrelevance of such analogies.

Dr. Cohen is quite correct in saying that professionals should clearly distinguish between statements based on professional experience and other opinions. It is our role to inform the public of things that they may not know. For example, they may not know that the well publicized failure of shuttle software some years ago was caused by exactly the type of synchronization problem that Dr. Cohen so lightly dismisses. The flaw was a design error and would not have been removed by subsequent improvements in clock accuracy.

The public might also be interested in knowing that in the last few years I have repeatedly challenged software people to name software products that functioned adequately when first given to users for actual (not test) use. The only program in question is 18,000 instructions long—a far cry from the millions of lines estimated for SDI. More shocking, an investigation reveals that all but 200 instructions of that program had actually been in use for several years before the purported first use. The 200 lines that were used for the first time dealt with a simple physical phenomenon that could be, and was, accurately simulated by an analog device for test purposes. The software in question is considered to be of exceptional quality. It is, in fact, the proverbial exception that proves the rule. The rule is that software products rarely, if ever, work the first time they are used in a new situation, and no software professional expects them to do so.

The public should not expect SDIO to produce the "thoroughly reliable and effective shield" called for by the administration.

RESPONSE—Danny Cohen

In reference to the 6 statements in Prof. Parnas's rebuttal:

1. I am in total agreement with Prof. Parnas that responsible opponents of SDI do not oppose the desirability of SDI, only its feasibility. I join Prof. Parnas in labelling those who oppose the desirability as irresponsible.
2. Prediction of the ways the Soviets will technically and politically encounter SDI is often used in arguments against SDI by many professionals, without specific disclaimers about their speculative, rather than professional, nature. Again, I join Prof. Parnas in labelling them as irresponsible.
3. The objective of the SDI program is to develop the key technologies needed for achieving its goals. Future knowledge, understanding, and evaluations of these technologies will be used by a future Congress and future president to decide about deployment.
4. Since Prof. Parnas made some personal statements here, I must set the record straight. It is not true that he was asked "to inform the panel about the work that I [Parnas] have done on weapon delivery software," as he claims. Neither is it true that "In public debates, Dr. Cohen has been asked to describe the solution he

claims I [Parnas] presented, but was unable to do so." Since all my participations in public debates have been recorded it is possible to verify this.

5. Lt. Gen. James Abrahamson's reference to the "longest pole in the tent" was made about the entire BM/C³ [Ballistic Missile/ Command, Control, Communication] problem, not about the specific points that Prof. Parnas raised, points that the panel found to be irrelevant to our discussions.

6. Many of the "professionals" criticize SDI for the infeasibility of full tests under realistic conditions, as if this is an SDI-unique problem, rather than a problem that is typical to all major defense systems.

In his rebuttal, Prof. Parnas also mentions the SAFEGUARD project. There are as many opinions about the reasons for discontinuing SAFE-GUARD as there are people who were involved. For Prof. Parnas to claim that it was due to the ineffectiveness of the system is unsubstantiated, and therefore irresponsible. In fact a very large portion of the software developed for SAFEGUARD is still in continuous use in several other defense systems.

Prof. Parnas tells us that "the public might also be interested in knowing that in the last few years I [Parnas] have repeatedly challenged software people to name software products that functioned adequately when first given to users for actual (not test) use." Apollo, Voyager, and the flight control of the shuttle are just a few of the answers to his challenge. One wonders why it is that Prof. Parnas does not consider these examples.

IGCC Titles

Research Paper Series

No. 1. Lawrence Badash, Elizabeth Hodes, and Adolph Tiddens
Nuclear Fission: Reaction to the Discovery in 1939
53 pp, 1985

No. 2. Michael D. Intriligator and Dagobert L. Brito
Arms Control: Problems and Prospects
12 pp, 1987

No. 3. G. Allen Greb
Science Advice to Presidents From Test Bans to the Strategic Defense Initiative
21 pp. 1987

Policy Paper Series

No. 1.   George A. Keyworth II
        Security and Stability: The Role for Strategic Defense
        12 pp, 1985

No. 2.   Gerald R. Ford
        The Vladivostok Negotiations and Other Events
        13 pp, 1986

No. 4.   Johan Galtung
        United States Foreign Policy: As Manifest Theology
        20 pp, 1987

No. 5.   David Lorge Parnas
        Danny Cohen
        SDI: Two Views of Professional Responsibility
        28 pp. 1987

Other Titles

Neil Joeck and Herbert F. York
Countdown on the Comprehensive Test Ban
(Joint publication with the Ploughshares Fund, Inc.)
23 pp, 1986

James M. Skelly, ed.
Sociological Perspectives on Global Conflict and Cooperation: A Research Agenda
36 pp, 1986

Gregg Herken, ed.
Historical Perspectives on Global Conflict and Cooperation
34 pp. 1987

Seymour Feshbach and Robert D. Singer, eds.
Psychological Research on International Conflict and Nuclear Arms Issues: Possible Directions
50 pp, 1987

Walter Kohn, Frank Newman, and Roger Revelle, eds.
Perspectives on the Crisis of UNESCO
71 pp. 1987

Study Questions

1. What is the main point of Parnas's and Cohen's articles? Why do you think each might be correct? Why do you think each might be wrong? Explain and defend your answers.

2. What assumptions do Parnas and Cohen make about technology? about ethics? Do you think these assumptions are correct? Explain and defend your answers.
3. If society accepts Parnas's and Cohen's main conclusions, what consequences might follow? Would these consequences be desirable? Explain and defend your answers.
4. Which view do you think is more ethically defensible, that of Parnas or that of Cohen?
5. Do scientists, engineers, and technologists have a right or a duty to ask questions about whether technologies like SDI waste taxpayers' money? Is this a political question? What values issues ought they to be concerned about discussing? Why?
6. Should a technological or scientific researcher accept funds from any source, provided the funds do not require illegal behavior? Why or why not?

Nuclear Technology and Radioactive Waste

Kristin Shrader-Frechette

The accidents at Three Mile Island and Chernobyl have slowed the development of commercial nuclear fission in most industrialized countries, although nuclear proponents are trying to develop smaller, allegedly "fail-safe" reactors. Regardless of whether or not they succeed, we will face the problem of radioactive wastes for the next million years. After a brief, "revisionist" history of the radwaste problem, I survey some of the major epistemological and ethical difficulties with storing nuclear wastes and outline four ethical dilemmas common to many technological and environmental controversies. I suggest two solutions to these ethical dilemmas and show why they are also economical and realistic proposals.

Introduction

Egyptians have been unable to protect the tombs of the Pharoahs for less than *four* thousand years, and some of them were looted within centuries. Italians have been unable to protect Renaissance art treasures for less than *one* thousand years.[1] Yet we in this generation have to be able to protect nuclear wastes, for *hundreds* of thousands to millions of years, for a period longer than all recorded history, because of their long-lived radionuclides (like plutonium 239, carbon 14, and iodine 129).[2]

Originally published as "Ethical Dilemmas and Radioactive Waste: A Survey of the Issues," in *Environmental Ethics* 13:4 (1991), pp. 327–43.

There are no plausible inductive arguments based on data for rad-waste storage for *one* century, much less data for minimum storage period of *thousands* of centuries. What data we do have is not comforting. Regulatory problems and safety questions have repeatedly forced the Energy Department to delay planning and opening facilities for nuclear waste. In the past, hundreds of people have died in radwaste accidents. The worst occurred when twenty-two square miles in Soviet Kasli were rendered uninhabitable by high-level radwaste that went critical three decades ago.[3] At the premier U.S. high-level facility, in Hanford, Washington, over 500,000 gallons of high-level waste have leaked into the soil, the Columbia River, and the Pacific Ocean.[4]

At the nuclear waste facility containing more plutonium than any other commercial site in the world, Maxey Flats, experts were wrong by six orders of magnitude when they predicted how fast the stored plutonium would migrate. They said that it would take 24,000 years for the plutonium to travel one-half inch off-site. It went two miles off-site in ten years. Current plans for future U.S. storage of high-level radioactive waste require the steel canisters to resist corrosion for as little as 300 years. Nevertheless, the U.S. Department of Energy admits that the waste will remain dangerous for longer than 10,000 years. Government experts agree that "there is no doubt that the repository will leak over the course of the next 10,000 years."[5]

The U.S. government has extrapolated, on the basis of past leaks at its nuclear waste facilities, and has said that future leaks should occur at a rate of two to three per year. Using U.S. government-estimated exposure levels (580 person rem) at each radwaste site, each existing facility could cause approximately 12 cancers and 116 genetic deaths per century,[6] and ultimately, tens of thousands of cancers per storage site.

Perhaps because they are worried about human error and about scientists' claims to store waste safely in perpetuity, virtually no one wants it in his or her backyard. Over the past two years, Congress has been besieged with more than thirty bills proposing to delay, abandon, or change the repository program established under the 1982 U.S. Nuclear Waste Policy Act.[7]

A Revisionist History of Commercial Nuclear Fission

Standard accounts of nuclear history are often misleading regarding the source and the magnitude of radwaste. Some persons have claimed that most nuclear waste arises (1) from military activities and important hospital uses of nuclear medicine and (2) because a number of utilities were eager to provide inexpensive electricity. Both these myths are untrue.

High-level radwaste is, for the most part, spent fuel rods from fission reactors and residues from fuel reprocessing.[8] Less than one percent of high-level radwaste is from medical activities.[9] Moreover, at least in the U.S., approximately half of the high-level radwaste now needing storage is from commercial nuclear fission, not military activities.[10] The commercial half of the waste, primarily spent fuel, is expected to rise dramatically by 1995, to eleven times the metric tons that now need to be stored, while the military waste will increase very slowly and remain close to current levels. By 1995, most high-level and low-level radwaste will be from commercial reactors, not military activities, and certainly not medical processes.[11]

Nor do we have the problem of nuclear waste because industry was eager to generate electricity, and fission was an economical means of doing so. At the beginning of the atomic era, industry was reluctant, both on economic and on safety grounds, to use fission to generate electricity. Worried about safety, every major U.S. corporation with nuclear interests refused to generate nuclear electricity unless some indemnity legislation was passed to protect them in the event of a major accident.[12] Commercial nuclear fission began, and was pursued, only because government hoped to justify continuing military expenditures in nuclear-related areas and to obtain weapons-grade plutonium.[13] Moreover, at least in the U.S., nuclear fission began only because government provided more than $100 billion in subsidies (for research, development, waste storage, and insurance) to the nuclear industry. It also gave the utilities a liability limit (the Price-Anderson Act) that protects licensees from ninety-nine percent of all claims in the event of a catastrophic nuclear accident.[14]

Twenty years after commercial fission reactors began operating, in 1976, the *Wall Street Journal* proclaimed them an economic disaster. Nuclear electricity has proved so costly that year-2000 projections for commercial fission reactors are now approximately one-eighth of what they were in the mid-seventies. No new reactors have been ordered in the U.S. since 1974.[15]

The few U.S. nuclear manufacturers that are still in business have survived by selling reactors to other nations, often developing countries. Yet, as I argue below, many of the commercial reactors going to these nations may not be in their best interests. Indeed, commercial nuclear fission may be the current version of infant formula. In the last two decades, U.S. and multinational corporations made great profits by exporting infant formula to developing nations, but they were able to do so only by coercive sales tactics and by misleading foreign consumers.

Some diplomats also have charged that developing nations are seeking fission-generated electricity as a subterfuge for obtaining weapons capability,[16] through the plutonium byproduct. India exploded its first nuclear bomb, for example, by using plutonium from a reactor exported by Canada. Whether or not this is the reason for the survival of nuclear fission, it is not obviously a safe, inexpensive way to boil water and run a turbine.

Radwaste Storage: Technical, Epistemological, and Ethical Problems

With this revisionist nuclear history behind us, let's look at some of the main technical problems posed by radwaste, especially high-level radwaste. Each year, each 1000-megawatt reactor discharges about 25.4 metric tons of high-level waste, spent fuel.[17] For 300 commercial reactors, worldwide, the annual high-level radwaste would be 7,620 metric tons per year. Compare this to the fact that ten micrograms of plutonium is almost certain to induce cancer, and that several grams of plutonium, dispersed in a ventilation system, are enough to cause thousands of deaths.[18] Moreover, each of the 7,620 metric tons of high-level waste produced annually has the potential to cause hundreds of millions of cancers for at least the first 300 years of storage, and then tens of millions of cancers for the next million years.[19]

These cancers could be prevented, of course, with perfect isolation of the wastes for a million years. But the U.S Environmental Protection Agency (EPA) has warned that we cannot count on institutional safeguards for the waste beyond one hundred years.[20] Moreover, there are geological and hydrological problems with all forms of storage.[21] The famous U.S. Interagency Review Group on nuclear-waste management reported that the scientific feasibility of dry storage in geologic repositories, deep in salt beds, or hard rock, "remains to be established."[22] As a result, granite storage sites in Sweden have been vetoed as unsafe, Kansas salt beds have been rejected because they are riddled with holes, and the first model U.S. repository for high-level radwaste will not be ready before 2008 or later.[23]

In the absence of proof that we can successfully store radwaste, doing so requires a great gamble—that our descendants will not breach the repositories through war, terrorism, or drilling for minerals; that water and heat will not combine to create nuclear reactors in underground waste, as already happened in the USSR; and a gamble that ice sheets and geological folding will not uncover the waste. Nuclear proponents who have ignored the waste problem have been like contractors who built houses without toilets, and then alleged that constructing the toilets would be easy.[24] Since no country has a permanent high-level disposal facility,[25] perhaps the task is not so easy as has been alleged.

Most of the epistemological difficulties with radwaste arise from the fact that secure storage cannot be guaranteed. Hence, regardless of the technology used, anyone who favors a particular method of radwaste management must use some form of the fallacy of the appeal to ignorance. Namely, "I know of no way in which containment could be breached; therefore, containment will probably not be breached." In the past, we were wrong to use an appeal to ignorance when we dumped radwaste into the sea, when we treated mastitis with radiation, when we used X-rays to determine shoe fit, and when we subjected U.S. soldiers to nuclear-test fallout during peacetime. We may likewise be wrong to use an argument from ignorance to justify storage of radwastes.

Another epistemological problem is how we can completely isolate the radwaste from the biosphere and yet monitor it to ensure that containment has not been breached. Complete isolation appears to preclude adequate monitoring, and adequate monitoring appears to preclude isolation. Also, how can we guarantee the so-called "neutrality criterion," that the levels of risk to which future generations will be subjected, because of the waste, will be no greater than those of present persons?[26] Because of the absence of good inductive evidence, any suggestion that this criterion can be met amounts to an argument from ignorance.[27]

Most of the ethical problems associated with radwaste focus on the issue of equity. Kasperson places them in three groups: locus problems, legacy problems, and labor-laity problems. The locus issues have to do with where and how to site radwaste facilities. The labor-laity problems focus on whether to maximize the safety of the public or that of radwaste workers, since both cannot be accomplished at once. The legacy problems concern exporting radwaste risks to future generations.[28]

The key question raised by legacy concerns is whether one can justify intergenerational inequity by mortgaging the future, by imposing our debts of radwaste on subsequent generations. If we saddle our descendants with our medical and financial debts of waste, then taxpayers in later centuries could be forced to pay an annual tab for radwaste storage between $3.8 and $1.9 million per reactor per year.[29] This expenditure is obviously questionable because future generations ought not be saddled with other persons' debts. Also, there is little public funding by taxpayers for decentralized energy technologies, like solar, which are less likely to burden the future.[30]

A second legacy question is what sort of criteria might justify environmentally irreversible damage to the environment, like that caused by deep-well storage of high-level radwaste. Radwaste management schemes which are irreversible theoretically impose fewer management burdens on later generations, but they also preempt future choices about how to deal with the waste. Schemes which are reversible allow for greater choices for future generations, but they also impose greater management burdens.

Perhaps the most important legacy question concerns the contribution of radwaste production and storage to the "plutonium economy" which is necessary for building nuclear weapons.[31] Still other legacy questions have to do with the use of social discount rates. Any alleged economies or safety claims associated with high-level radwaste storage are in large part questionable because of their dependence on a particular discount rate. Using a discount rate amounts to discounting future costs (of radwaste storage, like radiation-related deaths or injures) at some rate of x percent per year. Thus, at a discount rate of ten percent, effects on people's welfare twenty years from now count only for one-tenth of what effects on people's welfare count for now. With a discount rate of five percent, effects next year count for 1000 times more than effects 200 years from now. Or, more graphically, with a discount rate of five percent, a billion deaths in 400 years counts the same as one death next year. Yet, it is not obvious that the moral importance of future events, like the death of a person, declines at some x percent per year.[32] Without discounting, however, it would be impossible to justify the dangers and costs of storing radwaste for centuries.

Imposing nuclear waste on future generations might also be questionable from a practical point of view. Uranium will not be available much beyond the year 2000, even to supply existing fission reactors. Hence, after having generated tons of highly toxic waste, nuclear energy (without the breeder reactor and without fusion) will not have provided a long-term technological fix for our energy problems. From a practical point of view, it is not clear that the *temporary* benefits of nuclear fission are worth the *permanent* costs of radwaste.[33]

In addition to the legacy issues, there are locus or siting problems associated with managing radioactive waste. One of the key difficulties here is vesting, allowing a company to obtain a return on its initial capital investment in a radwaste site. Are there ethical grounds for limiting property rights, even when such limitations fly in the face of current vesting doctrine?[34]

Still other, and even more far-reaching, locus questions arise because of the emergent field of land ethics.[35] How does one justify siting any land for radwaste storage? This is a use which amounts to exercising the most extreme form of property rights, since it preempts both present and future choices about land use at that site. Another important siting or locus issue is geographical equity, especially since it is a forgone conclusion that radwaste depositories will be located in rural areas, away from major population centers.[36] Is it fair to impose a risk on a person just because she lives in a rural community rather than a large city? Likewise, is it ethical for one geographical subset of persons to receive the benefits of nuclear-generated electricity, while a much larger set of persons bears the costs? Despite the fact that utilities contribute ($1 for every 1000-kilowatt hours of electricity that they generate) toward a radwaste management fund,[37] these contributions cover only a small fraction of storage costs. The bulk of storage expenditures comes from the $100 billion in nuclear subsidies already spent by U.S. taxpayers, part of which is for nuclear waste storage.[38] In the past decade, for example, government subsidies to the nuclear industry, in the form of write-off for capital invested in plants not completed, was $4 billion in the U.S. alone.[39]

Once taxpayer subsidies for costs like waste storage and decommissioning are included in the calculations, nuclear power can be shown to be more expensive than every other energy alternative.[40] The only way to make it viable is to remove it entirely from the discipline of the market and to increase taxpayer subsidies.[41] Already this removal from the market has resulted in taxpayers and members of future generations picking up the tab for billions of dollars of waste storage that should be borne by the nuclear industry and its beneficiaries.

Philosophical Dilemmas

Rather than discuss each of these ethical problems, a task beyond the scope of a short paper, I now examine four classical philosophical dilemmas posed by radwaste storage. I call them the consent dilemma, the federalism dilemma, the threshold dilemma, and the contributor's dilemma. The consent dilemma is that siting radwaste facilities and employing waste management workers requires the consent of those put at risk; yet those most able to give free, informed consent are usually unwilling to do so, and those least able to validly consent are often willing to give alleged consent.

To see how the consent dilemma arises, consider a typical case. When West Valley, New York was proposed as a storage site for low-level radioactive waste, townspeople were eager to obtain the economic benefits they believed the facility would bring. City leaders predicted a "boomtown," and this prediction was the basis of community acceptance of the site. Although the boomtown never occurred, the facility paid twenty percent of the town and county taxes. From a superficial perspective, it might look as if West Valley citizens gave free, informed consent to the waste site. But consider what sorts of communities would be most likely to consent to a radwaste facility nearby. Probably not those whose residents had high incomes, job security, and high levels of education. Communities full of people with these characteristics would know enough to be wary of the risk, and they probably would already have excellent jobs. Hence, they would not need to take any risks in order to better themselves economically.[42]

Or consider the case of workers at another radwaste facility. A British study of 35,000 living workers from the Hanford high-level radwaste facility showed increased chromosomal damage in workers exposed to less than one-half of the annual allowable exposure of five rems of radiation.[43] Workers at the facility were justified in having these risks imposed on them, according to classical economic theory, because of an alleged compensating wage differential. According to the theory behind the alleged differential, the riskier the occupation, the higher the wage required to compensate the worker for bearing the risk, all things being equal.[44]

According to classical ethical theory, imposition of these higher workplace risks is legitimate apparently only after the worker consents, with knowledge of the risks involved, to perform the work for the agreed-upon wage. The dilemma arises once one considers who is most likely to give legitimate informed consent. It is a person who is well educated and adequately informed about the risk, especially its long-term and probabilistic effects. It is a person who is not forced under dire financial constraints to take a job which he knows is likely to harm him. Yet, sociological data reveal that, as education and income rise, persons are less willing to take risky jobs, and those who do so are primarily those who are poorly educated or financially strapped. Sociological data also show that the alleged compensating wage differential does not operate for poor, unskilled, minority, or non-unionized workers. Yet these are precisely the persons most likely to work at risky jobs like storing radwaste.[45] As a result, the very set of persons *least* able to give free, informed consent to workplace radwaste risks are precisely those who most *often* work in risky jobs and are alleged to have given consent.

If this observation about worker radwaste risk and community acceptance of a radwaste site is accurate, then medical experimentation may have something to teach us about risk assessment. We know that the promise of early release for a prisoner who consents to risky medical experimentation provides a highly coercive context which could jeopardize his legitimate consent. Likewise, high wages for a desperate worker who consents to take a risky job provides a highly coercive context which could jeopardize his legitimate consent. We must, therefore, either admit that our classical theory of free, informed consent is wrong or we must question whether those closely affected by radwaste risk genuinely consent to it.

Consider now a second classic dilemma. Liberty and grass-roots self-determination require local control of whether a radwaste facility is sited in a particular area. Yet, equality of consideration for people in all locales and the minimization of overall risk require federal control. Do we say that the local community can veto a radwaste site, even though that site may be the best in the country and may provide for the most equal protection of all people? Or do we say that the national government can impose a radwaste site on a local community, even though the imposition is at odds with their free and self-determined choice?[46] How do we resolve this dilemma? Noted law professor R. B. Stewart claims that it is impossible to maximize both liberty and environmental integrity, or both liberty and equality.[47] The dilemma cannot be resolved in the sense of maximizing all three values.[48]

A third problem, the threshold dilemma, is based on the fact that society must declare some threshold below which risk is declared to be negligible or minimal, so far as its acceptability is concerned. Typically this threshold level is set at what would cause less than a 10^{-6} increase in one's average annual probability of fatality.[49]

The reasoning behind setting such a level is that a zero-risk society is impossible, and some standard needs to be set, especially in order to determine pollution-control expenditures. Choosing the 10^{-6} standard also appears reasonable, both because society must attempt to reduce larger risks first, and because 10^{-6} is the natural hazards death rate.[50]

The dilemma arises because no threshold standard is able to provide *equal* protection from harms like radwaste to all citizens. Any such standard guarantees merely that an *average* annual probability of fatality is associated with some hazard. Because this 10^{-6} threshold seems acceptable on the average, however, does not mean that it is acceptable to each individual. Blacks, for example, face higher risks from air pollution than do whites, even though they share the same "average" exposure.[51] Likewise, those around radwaste facilities are exposed to radiation levels for average persons, not for them as individuals with unique needs.

Most civil rights, however, are not accorded on the basis of the *aver-*

age needs of persons, but on the basis of *individual* characteristics. For example, we do not accord constitutionally guaranteed civil rights to public education on the basis of average characteristics of students. If we did, then retarded children or gifted children would have rights only to education for children at the average level. Instead, we say that according "equal" civil rights to education means according "comparable education," given one's aptitudes and needs. That is why the state can provide special schools for both the retarded and the gifted.

This example from the field of education raises an interesting question for radwaste risks and indeed for all risks requiring a threshold to be set: if civil rights to education are accorded on the basis of individual, not average, characteristics, then why are civil rights to equal protection from risks not accorded on the basis of individual, rather than average, characteristics? Why is a 10^{-6} average threshold accepted for everyone, without compensation, when adopting it poses risks higher than 10^{-6} for the elderly, for children, for persons with previous exposures to radiation, for those with allergies, for persons who must lead sedentary lives, and for the poor? Which should we choose: average protection or equal protection, efficiency in regulation or equality of protection?

A fourth difficulty is what I call the contributor's dilemma. It arises out of the fact that citizens are subject to numerous small risks, e.g., to certain carcinogens, each of which is allegedly acceptable; yet, together such exposures are clearly unacceptable. In the case of radioactive waste, selected groups of citizens, such as those living near a storage facility or a radwaste transport route, are exposed to radiation. Each of these small exposures is alleged to be acceptable because it is below the threshold at which some statistically significant increase in harm occurs; yet together these exposures can cause serious damage.

Statistically speaking, twenty-five to thirty-three percent of us are going to die from cancers. The U.S. Office of Technology Assessment says that ninety percent of these cancers are environmentally induced and hence theoretically preventable.[52] Many of the cancers are obviously caused by the aggregation of numerous exposures to carcinogens, like radiation, no one exposure of which is alone alleged to be harmful. The contributor's dilemma is especially problematic in cases involving the devising of ethical regulations for small risks, like radiation. Risk assessors who condone sub-threshold risks, but who condemn the deaths caused by the aggregate of these sub-threshold risks, are something like the bandits who eat the tribesmen's lunches in the famous story of Jonathan Glover:

> Suppose a village contains 100 unarmed tribesmen eating their lunch. One hundred armed bandits descend on the village and each bandit at

gun-point takes one tribesman's lunch and eats it. The bandits then go off, each having done a discriminable amount of harm to a single tribesman. Next week, the bandits are tempted to do the same thing again, but are troubled by new-found doubts about the morality of such a raid. Their doubts are put to rest by one of their number [a government assessor]. . . . They then raid the village, tie up the tribesmen, and look at their lunches. As expected, each bowl of food contains one hundred baked beans. . . . Instead of each bandit eating a single plateful as last week, each [of the one hundred bandits] takes one bean from each plate. They leave after eating all of the beans, pleased to have done no harm, as each has done no more than subthreshold harm to each person.[53]

The obvious question raised by this example is how a risk assessor can say that sub-threshold radwaste exposures are harmless, as the data allegedly indicate, and yet that the additivity, or contribution, of these doses causes great harm. It appears that risk regulators need to amend their theory regarding synergistic or additive risks like cancer.

Two Solutions by Ethics: Consent and Compensation

Although there is inadequate space here to defend the sorts of solutions that would help resolve these four dilemmas, it is possible to outline briefly the arguments which, when developed, would provide such a defense. One solution to the consent dilemma and to the federalism dilemma is to apply the same strict standards of informed consent, already well-known in doctor-patient or experimenter-patient relationships, to the radwaste siting issue.

We all probably believe that it would be unethical for a doctor to impose a risky treatment on a patient without her free, informed consent. Yet most of us need to develop our moral sensibilities so as to see radwaste workers or radwaste siting in the same light. Until or unless a risk imposition receives the consent of those who are its potential victims, it cannot be justified. This means that those who wish to impose societal risks on others need to do whatever is necessary to compensate them to a degree adequate to obtain their free, informed consent. If no compensation is adequate to obtain free, informed consent, then it is questionable whether the risk imposition is justified.

Whenever the federalism and consent dilemmas have arisen in radwaste siting issues, these have often been resolved by means of increasing the compensation to the risk takers likely to be affected by the waste site. In some cases in which it was not possible to meet the compensatory demands of a community proposed as a radwaste site, the facility was not located. In the cases in which compensation was able to be negotiated, however, often this was sufficient to ensure community consent. Typically industry-supplied compensation in radwaste cases includes tax breaks or funding community projects or schools. According to social scientists (studying waste siting), however, one form of compensation dominates all negotiation. The one factor almost always essential to achieving local consent is giving citizens/workers funding to control health and safety monitoring at the facility. By forcing the waste managers to pay for outside monitors, citizens and workers are freed from relying on company monitoring. Once they have greater control of their safety, they appear ready to give informed consent to the risk.[54]

Compensation also provides one solution to the threshold dilemma and to the contributor's dilemma, both of which raise equity issues. By providing full compensating benefits for all unavoidable radwaste risks, industry can offset the inequities generated by some of the locus and legacy problems.[55] Admittedly, compensation of future generations who bear the brunt of the legacy problems might be difficult. Obtaining free, informed consent and guaranteeing compensation in such cases requires that the consent of future persons be obtained by means of representatives acting in their best interests. It also requires that those who store radwaste actually set up a fund for compensation of future persons possibly harmed by the waste. To the degree that we cannot guarantee that persons a million years from now will have equal opportunity to protect themselves from our radwaste hazards, then to that same degree we cannot justify generating radwaste or imposing it upon others.[56] The rationale for requiring full compensation and consent, even regarding future persons, is that any technology ought to "pay its own way." If nuclear power cannot "pay its own way," especially in regard to waste storage, it will reinforce all the old income distributions and inequities in which the poor and the uneducated bear the social costs of contemporary technology.

Also, if someone can impose a bodily risk of harm on another without his consent, then there is no right to life. If someone can profit by imposing a threat of physical harm on another without compensating him, then there is no right to due process. These considerations suggest that, if we refrain from requiring genuine informed consent and from guaranteeing complete compensation for radwaste risks, then radwaste will put more at risk than our health and that of future generations. It will put at risk our most basic rights to justice and equity.

These Ethical Solutions Make Economic Sense

Providing equity, informed consent, and full compensation regarding the risks associated with managing radwaste requires that we forgo use of nuclear power and the generation of additional radwastes. Avoiding additional nuclear plants, however, is also an economically sound decision, for a variety of reasons. Virtually all commercial nuclear construction in developed nations has come to a halt, largely because fission generation of electricity is uneconomical and likely unsafe. France is the only developed country with an ongoing nuclear program, allegedly the most successful in the world.

A closer look at the French situation, however, reveals that it is questionable. For one thing, the French use a breeder reactor, not the fission technology employed in other energy programs around the world. Most nations have decided against the breeder because it generates inordinate amounts of radioactive waste—materials for nuclear weapons—and hence creates a "plutonium economy." Moreover, the apparent reasons for the success of the French nuclear program—the fact that it is centralized and government owned—are artificial and not transferable to other nations. The French utility is protected from market forces, protected from environmentalists' criticisms, and protected from public participation in decision making. Even with these benefits, the French utility now has a debt of $30 billion, due to its commitment to nuclear power. In 1982 this debt had accumulated to $152 billion, although part of it was forgiven by the government. In light of the French deficit, developing nations, many of which are on the edge of insolvency, will be hard pressed to pay the enormous capital costs for nuclear reactors.[57] Developing nations, in particular, will probably be hard pressed to pay the costs of decommissioning and waste storage associated with their approximately thirty nuclear reactors. Some experts claim that waste storage costs will add an estimated five to ten percent to the total cost of nuclear power.[58] U.S. industry experts maintain that the decommissioning of a reactor could amount to twenty-four percent of the original construction costs. Such an estimate translates to more than $500 million for recently built facilities. If French nuclear-industry experts are correct, then decommissioning currently would run at least forty percent of initial construction costs of the reactor.[59]

Apart from paying for decommissioning the reactor and managing the waste, most nations (and especially developing ones) will likely have a difficult time paying for nuclear power at all. Since the mid-seventies, nuclear-plant construction costs have doubled every four years. More than twenty-five percent of these costs are for financing.[60] Part of the reason for these increases is that, because of additional safety measures needed in the wake of accidents, the amount of concrete, piping, and cable used in the average plant has more than doubled, while labor requirements have more than tripled.[61] No one in any country is going to build a nuclear plant without knowing its full costs, including decommissioning and waste storage.

A particular obstacle to nuclear plants in developing nations is the small size of electricity grids in most of them. If a single power plant provides more than fifteen percent of the grid's capacity, the whole system will "crash" if that plant is shut down. Yet only four or five developing countries have grids large enough for a conventional 1000 megawatt reactor. This problem could be addressed by plants smaller than the 1000 megawatt ones, but the per-kilowatt construction cost for a nuclear plant of 200 megawatts is more than twice that for a 1000 megawatt one, and even the larger ones are not cost-effective compared to other energy alternatives. Because of the capital intensity of nuclear power, it seems unattractive for debt-strapped developing nations.[62]

Right now small hydropower and cogeneration plants are all much cheaper than nuclear fission.[63] Hydropower is particularly attractive because, although North America and Europe have developed sixty percent and thirty-six percent of their hydropower potential, respectively, Asia has used just nine percent, Latin America eight percent, and Africa only five percent. In China, for example, 76,000 small hydropower plants supply almost 10,000 megawatts of power in rural areas. By the year 2000, cogeneration can account for ten times that amount in some countries.[64]

Biomass is also an inexpensive alternative for rural electrification. The greatest use of biomass residues is found in the relatively treeless plains of Northern India, Bangladesh, and China, where crop residues and dung provide as much as ninety percent of household energy in many villages and a considerable proportion in urban areas.[65] Likewise, in inner Mongolia, 2,000 small wind turbines are used for lighting, running televisions, electrifying corral fences, and projecting movies.[66]

Compared to such alternatives, nuclear power is also a questionable energy source because it creates fewer jobs and requires more dependence upon foreign companies and governments than almost any other investment a developing nation can make. Also, nuclear energy is a target for military and terrorist abuse in a politically unstable region. Besides, it is likely to serve only the minority that uses electricity. It bypasses the majority who rely on fuelwood and charcoal. All these points suggest that investment in rural electrification, using small-scale renewables, is a better way for developing nations to go.[67]

Conclusion

Apart from the ethical and economical reasons for avoiding the generation of radioactive waste, there might be a number of general philosophical considerations that suggest that the commercial, and not only the military, nuclear path is wrong. On this view, how we deal with the radwaste problem might be a barometer for the collective sanity and morality of our society. A thousand years ago, the world's finest architectural and engineering talents were mobilized to build cathedrals. It is ironic that comparable talents and even more skills are today dedicated to devising foolproof nuclear garbage dumps.[68]

Notes

1. P. Z. Grossman and E. S. Cassedy, "Cost-Benefit Analysis of Nuclear Waste Disposal," *Science, Technology, and Human Values* 10, no. 4 (Fall 1985): 49; hereafter cited as "Cost-Benefit Analysis."

2. David Hawkins, *Considerations of Environmental Protection Criteria for Radioactive Waste* (Washington, D.C.: U.S. Environmental Protection Agency, 1978), p. 1.

3. See, for example, J. Raloff, "Nuclear Waste Still Homeless," *Science News* 136, no. 3 (15 July 1989): 47; R. Monastersky, "More Questions Plague Nuclear Waste Dump," *Science News* 135, no. 25 (24 June 1989): 389; R. Monastersky, "Opening Delayed for Nuclear Waste Site," *Science News* 134, no. 13 (24 September 1988): 199; Zhores Medvedev, *Nuclear Disaster in the Urals*, trans. George Sanders (New York: Norton, 1979).

4. U.S. Energy Research and Development Administration, *Final Environmental Statement: Waste Management Operations, Hanford Reservation, Richland, Washington*, vol. 1 (ERDA-1538) (Springfield, Virginia: National Technical Information Service, 1975), pp. x–28; hereafter cited as "ERDA-1538."

5. See James Neel, "Low-Level Radioactive Waste Disposal," statement in U.S. Congress before a subcommittee of the Committee on Government Operations, House of Representatives, 94th Congress, Second Session, 23 February, 12 March, and 6 April 1976 (Washington, D.C.: U.S. Government Printing Office, 1976), p. 258. See also U.S. Geological Survey, vertical file, "Maxey Flats: Publicity" (Louisville, Kentucky: Water Resources Division, U.S. Division of the Interior, n.d.); hereafter cited as U.S.G.S.-P. (The Louisville office of the U.S.G.S. is responsible for monitoring the Maxey Flats radioactive facility.) Finally see A. Weiss and P. Columbo, *Evaluation of Isotope Migration—Land Burial*, NUREG/CR-1289 BNL-NUREG-51143 (Washington D.C.: U.S. Nuclear Regulatory Commission, 1980), p. 5. For future U.S. radwaste storage and requirements, see R. Monastersky, "The 10,000-Year Test," *Science News* 133, no. 9 (27 February 1988): 139–41. See also G. Hart, "Address to the Forum," U.S. EPA, *Proceedings of a Public Form on Environmental Protection Criteria for Radioactive Wastes* (ORP/CSD-78-2) (Washington, D.C.: U.S. Government Printing Office, May 1978), p. 6.

6. U.S. ERDA, "ERDA-1538," vol. 1, pp. x–74. See chap. 4, notes 13 and 16; vol. 1, pp. ii, 1–57. U.S. Atomic Energy Commission, *Comparative Risk-Cost-Benefit Study of Alternative Sources of Electrical Energy* (WASH-1224) (Washing-

ton, D.C.: U.S. Government Printing Office, December 1974), pp. 3–83. See also I. Amato, "Dangerous Dirt: An Eye on DOE," *Science News* 130, no. 14 (4 October 1986): 221; hereafter cited as "DOE."

7. J. Raloff and I. Peterson, "Trouble With EPA's Radwaste Rules," *Science News* 132, no. 5 (1 August 1987): 73.

8. J. P. Murray, J. J. Harrington, and R. Wilson, "Chemical and Nuclear Waste Disposal," *The Cato Journal* 2, no. 2 (Fall 1982): 569; hereafter cited as "Waste Disposal."

9. Sierra Club, *Low-Level Nuclear Waste: Options for Storage* (Buffalo: Sierra Club Radioactive Waste Campaign, 1984), p. 2.

10. J. M. Deutch and the Interagency Review Group on Nuclear Waste Management, *Report to the President* (TID-2817) (Springfield, Virginia: National Technical Information Service, October 1978); hereafter cited as *Report*.

11. D. MacLean, "Introduction," to D. Bodde and T. Cochran, "Conflicting Views on a Neutrality Criterion for Radioactive Waste Management," University of Maryland, College Park, Center for Philosophy and Public Policy, 23 February 1981, p. 3. See also IRG, *Report*, pp. D-11; D-12; D-14, and D-19.

12. W. S. Caldwell et al., 'The "Extraordinary Nuclear Occurrence" Threshold and Uncompensated Injury Under the Price-Anderson Act,' *Rutgers-Camden Law Journal* 6, no. 2 (Fall 1974): 379. See also K. Shrader-Frechette, *Nuclear Power and Public Policy* (Boston: D. Reidel, 1983), pp. 10–11; hereafter cited as *NPPP.*

13. Cited by Sheldon Novick in a taped interview with Carl Walske in *The Electric War* (San Francisco: Sierra Club Books, 1976), pp. 32–33. See also Shrader-Frechette, *NPPP,* pp. 8–9.

14. See Shrader-Frechette, *NPPP,* pp. 75–81.

15. Christopher Flavin, *Nuclear Power: The Market Test* (Washington, D.C.: Worldwatch Institute, 1983), p. 33.

16. A. Lovins and L. Lovins, *Energy/War* (San Francisco: Friends of the Earth, 1980), chaps. 2–3.

17. E. Winchester, "Nuclear Wastes," *Sierra,* July–August 1979.

18. Grossman and Cassedy, "Cost Benefit Analysis," p. 48.

19. Murray, Harrington, and Wilson, "Waste Disposal," p. 586.

20. Hawkins, *Considerations,* pp. 27–29.

21. Winchester, "Nuclear Wastes." See also R. Schneider and J. Trask, *U.S. Geological Survey Research in Radioactive Waste Disposal—Fiscal Year 1982,* Water Resources Investigations Report 84-4205 (Reston, Va.: U.S. Geological Survey, 1984), p. 38.

22. Winchester, "Nuclear Wastes."

23. Mark Crawford, "DOE, States Reheat Nuclear Waste Debate," *Science* 230, no. 4722 (11 October 1985): 151.

24. Flavin, *Nuclear Power: The Market Test,* p. 31.

25. Cynthia Polluck, *Decommissioning: Nuclear Power's Missing Link* (Washington, D.C.: Worldwatch Institute, 1986), p. 13.

26. See note 11.

27. Polluck, *Decommissioning,* p. 15.

28. R. E. Kasperson, *Equity Issues in Radioactive Waste Management* (Cambridge, Mass.: Oelgelschlager, Gunn, and Hain, 1983).

29. Shrader-Frechette, *NPPP,* pp. 57–58.

30. J. Berger, *Nuclear Power* (Palo Alto, Calif.: Ramparts, 1976), p. 150.

31. See K. Shrader-Frechette, "Nuclear Arms and Nuclear Power: Philosophical Connections," in M. Fox and L. Groarke, eds., *Nuclear War: Philosophical Perspectives* (New York: Peter Lang Publishers, 1985); S. Cohen, *Arms and Judgment* (Boulder: Westview, 1989); and K. Kipnis and D. Meyers, eds., *Political Realism and International Morality* (Boulder: Westview, 1987).

32. Derek Parfit, "Energy Policy and the Further Future," University of Maryland, College Park, Center for Philosophy and Public Policy, 23 February 1981, pp. 1–19, especially p. 1; H. S. Burness, "Risk: Accounting for the Future," *Natural Resources Journal* 21, no. 4 (October 1981): 723–34; Grossman and Cassedy, "Cost Benefit Analysis," p. 51.

33. See P. L. Joskow, "Commercial Impossibility, the Uranium Market, and the Westinghouse Case," *Journal of Legal Studies* 6, no. 1 (January 1977): 119–76.

34. See Mark Sagoff, "Property Rights and Environmental Law," *Philosophy and Public Policy* 8, no. 2 (Spring 1988): 9–12.

35. See previous note. See also L. C. Becker, *Property Rights* (Boston: Routledge and Kegan Paul, 1977).

36. P. J. Leahy, U.S. Senator, Vermont, "The Socioeconomic Effects of a Nuclear Waste Storage Site on Rural Areas and Small Communities," statement before the Subcommittee on Rural Development, Committee on Agriculture, Nutrition, and Forestry, U.S. Senate, Washington, D.C., 26 August 1980 (Washington, D.C.: U.S. Government Printing Office, 1980), p. 2.

37. Flavin, *Nuclear Power: The Market Test,* p. 31.

38. Berger, *Nuclear Power,* pp. 94–97, 106–112, 144–47. See also J. Gofman and A. Tamplin, *Poisoned Power* (Emmaus, Pa.: Rodale Press, 1971), pp. 177, 199; J. Primack and F. Von Hippel, "Nuclear Reactor Safety," *Bulletin of the Atomic Scientists* 30, no. 8 (October 1974): 7; E. Muchnicki, "The Proper Role of the Public in Nuclear Power Plant Licensing Decisions," *Atomic Energy Law Journal* 15, no. 2 (Spring 1973): 45.

39. Flavin, *Nuclear Power: The Market Test,* p. 41.

40. Ibid., esp. pp. 1–33. See also Saunders Miller, *The Economics of Nuclear and Coal Power* (New York: Praeger, 1976), p. 105. Finally, see Shrader-Frechette, *NPPP,* p. 123.

41. Flavin, *Nuclear Power: The Market Test,* p. 42.

42. See R. E. Kasperson, "The Socioeconomic Effect of a Nuclear Waste Storage Site on Rural Areas and Small Communities," Hearing before the Subcommittee on Rural Development of the Committee on Agriculture, Nutrition, and Forestry, U.S. Senate, 96th Congress, Second Session, 26 August 1980 (Washington, D.C.: U.S. Government Printing Office, 1980), pp. 61–62. For substantiation of the claim about those that consent to locating a risky facility nearby, see note 45.

43. See Amato, "DOE"; see also Dr. Alice Stewart's study cited in Winchester, "Nuclear Wastes."

44. Flavin, *Nuclear Power: The Market Test*, esp. pp. 40–42.

45. K. Shrader-Frechette, *Risk Analysis and Scientific Method* (Boston: D. Reidel, 1985), pp. 107–12; hereafter cited as *RASM*. K. Shrader-Frechette, *Science Policy, Ethics, and Economic Methodology* (Boston: D. Reidel, 1985), p. 137; hereafter cited as *SP*.

46. Shrader-Frechette, *SP*, p. 214.

47. Ibid., p. 215.

48. Ibid., p. 214. See. R. B. Stewart, "Pyramids of Sacrifice? Problems of Federalism in Mandating State Implementation of National Environmental Policy," in F. A. Strom, ed., *Land Use and Environment Law Review—1978* (New York: Clark Boardman, 1978), pp. 162–63. See also R. B. Stewart, "Paradoxes of Liberty, Integrity, and Fraternity: The Collective Nature of Environmental Quality and Judicial Review of Administrative Action," *Environmental Law 7*, no. 3 (Spring 1977): 472. Finally, see M. Markovich, "The Relationship between Equality and Local Autonomy" in W. Feinberg, ed., *Equality and Social Policy* (Urbana: University of Illinois Press, 1978), p. 96.

49. See Shrader-Frechette, *RASM*, sec. 3.3.2 of chap. 2. See C. Zracket, "Opening Remarks," in Mitre Corporation, *Symposium/Workshop . . . Risk Assessment and Governmental Decision Making* (McLean, Va.: Mitre Corporation, 1979), p. 3 (The Mitre publication will hereafter be cited as *Symposium/Workshop*). See C. Starr, "Benefit-Cost Studies in Socio-technical Systems," in Committee on Public Engineering Policy, ed., *Perspectives on Benefit-Risk Decision Making* (Washington, D.C.: National Academy of Engineering, 1972); D. Okrent and C. Whipple, *Approach to Societal Risk Acceptance Criteria and Risk Management*, PB-271 264 (Washington D.C.: U.S. Department of Commerce, 1977); A. P. Hull, "Discussion," in Mitre, *Symposium/Workshop*, pp. 171–72; and N. Rescher, *Risk: A Philosophical Introduction* (Washington D.C.: University Press of America, 1983), pp. 35–40.

50. C. Starr and C. Whipple, "Risks of Risk Decisions," *Science* 208 (6 June 1980): 1114–19. See also C. Starr, *Current Issues in Energy* (New York: Pergamon, 1979), p. 15.

51. Shrader-Frechette, *SP*, p. 134. See also Kasperson, "Socioeconomic Effect," pp. 62–63.

52. J. C. Lashof, et al., Health and Life Sciences Division of the U.S. Office of Technology Assessment (OTA), *Assessment of Technologies for Determining Cancer Risks from the Environment* (Washington D.C.: U.S. Office of Technology Assessment, 1981), pp. 3, 6.

53. Quoted by Derek Parfit, *Reasons and Persons* (Oxford: Clarendon Press, 1984), p. 511.

54. E. Peele, "Innovative Process and Inventive Solutions," unpublished manuscript, 1986, p. 6 (available from Peele at the Energy Division, Oak Ridge National Laboratory, Oak Ridge, Tennessee 37831); hereafter cited as Peele, "Innovative Process." See also E. Peele, "Mitigating Community Impacts of Energy Development," *Nuclear Technology* 44 (1979): 132–40, esp. p. 133–36.

55. W. A. O'Connor, "Incentives for the Construction of Low-Level Nuclear Waste Facilities," in *Low Level Waste*, Final Report of the National Governors

Association Task Force on Low-Level Radioactive Waste Disposal, Washington D.C., 1980 E. Peele, "Siting Strategies for an Age of Distrust," unpublished essay (available from Peele at the address listed in note 54); finally, see Peele, "Innovative Process."

56. Bodde and Cochran, "Radioactive Waste Management," p. 7.

57. Polluck, *Decommissioning*, p. 27. Flavin, *Nuclear Power: The Market Test*, pp. 45–47.

58. Flavin, *Nuclear Power: The Market Test*, p. 31.

59. Polluck, *Decommissioning*, p. 27.

60. Flavin, *Nuclear Power: The Market Test*, p. 15.

61. Ibid., p. 26.

62. Ibid., p. 52.

63. Ibid., pp. 54–59.

64. But much of this includes reliance on natural gas and some dependence on coal. See Christopher Flavin and Alan Durning, *Building on Success: The Age of Energy Efficiency* (Washington, D.C.: Worldwatch Institute, March 1988), p. 35.

65. Jessica Tuchman Mathews, *World Resources 1986* (New York: Basic Books, 1986), p. 111.

66. Kosta Tsipis, "Nuclear Power and Energy Needs of the Third World Economies," *Church and Society*, Report and Background Papers, Meeting of the Working Group, World Council of Churches, Glion, Switzerland, September 1987, pp. 227, 229.

67. Flavin, *Nuclear Power: The Market Test*, pp. 54–55.

68. Winchester, "Nuclear Wastes."

Study Questions

1. What is the main point of Shrader-Frechette's article? Why do you think it might be correct? Why do you think it might be wrong? Explain and defend your answers.

2. What assumptions does Shrader-Frechette make about technology? about ethics? Do you think these assumptions are correct? Explain and defend your answers.

3. If society accepts Shrader-Frechette's main conclusions, what consequences might follow? Would these consequences be desirable? Explain and defend your answers.

4. Did any problematic value decisions encourage the development of commercial nuclear technology? Explain.

5. List some of the technical, ethical, and epistemological problems with storing radioactive waste? Have value judgments about these problems caused assessors to ignore or minimize these problems?

6. Explain the three philosophical or evaluative dilemmas posed by radioactive waste and technologies for handling it.
7. Critically evaluate Shrader-Frechette's two proposed solutions to some of the value issues confronting technologies for storing radioactive waste.

4.7

Assessment of Environmental and Economic Impacts of Pesticide Use

David Pimentel, H. Acquay, M. Biltonen, P. Rice, M. Silva, J. Nelson, V. Lipner, S. Giordano, A. Horowitz, and M. D'Amore

Introduction

Worldwide, about 2.5 million tons of pesticides are applied each year with a purchase price of $20 billion (PN, 1990). In the United States approximately 500,000 tons of 600 different types of pesticides are used annually at a cost of $4.1 billion (including application costs) (Pimentel et al., 1991).

Despite the widespread use of pesticides in the United States, pests (principally insects, plant pathogens, and weeds) destroy 37% of all potential food and fiber crops (Pimentel, 1990). Estimates are that losses to pests would increase 10% if no pesticides were used at all; specific crop losses would range from zero to nearly 100% (Pimentel et al., 1978). Thus, pesticides make a significant contribution to maintaining world food production. In general, each dollar invested in pesticidal control returns about $4 in crops saved (Carrasco-Tauber, 1989; Pimentel et al., 1991).

Although pesticides are generally profitable, their use does not always decrease crop losses. For example, even with the 10-fold increase in insecticide use in the United States from 1945 to 1989, total crop losses from insect damage have nearly doubled from 7% to 13% (Pimentel et al., 1991). This rise in crop losses to insects is, in part, caused by changes in agricultural practices. For instance, the replacement of

---

corn-crop rotations with the continuous production of corn on about half of the original hectarage has resulted in nearly a fourfold increase in corn losses to insects despite approximately a 1,000-fold increase in insecticide use in corn production (Pimentel et al., 1991).

Most benefits of pesticides are based only on direct crop returns. Such assessments do not include the indirect environmental and economic costs associated with pesticides. To facilitate the development and implementation of a balanced, sound policy of pesticide use, these costs must be examined. Over a decade ago the U.S. Environmental Protection Agency pointed out the need for such a risk investigation (EPA, 1977). Thus far, only a few scientific papers on this complex and difficult subject have been published.

The obvious need for an updated and comprehensive study prompted this investigation of the complex of environmental and economic costs resulting from the nation's dependence on pesticides. Included in the assessment are analyses of pesticide impacts on human health; livestock and livestock product losses; increased control expenses resulting from pesticide-related destruction of natural enemies and from the development of pesticide resistance; crop pollination problems and honeybee losses; crop and crop product losses; fish, wildlife, and microorganism losses; and governmental expenditures to reduce the environmental and social costs of pesticide use.

Public Health Effects

Human pesticide poisonings and illnesses are clearly the highest price paid for pesticide use. A recent World Health Organization and United Nations Environmental Programme report (WHO/UNEP, 1989) estimated there are 1 million human pesticide poisonings each year in the world with about 20,000 deaths. In the United States, pesticide poisonings reported by the American Association of Poison Control Centers total about 67,000 each year (Litovitz et al., 1990). J. Blondell (EPA, PC [personal communication], 1990) has indicated that because of demographic gaps, this figure represents only 73% of the total. The number of accidental fatalities is about 27 per year (J. Blondell, EPA, PC, 1990).

While the developed countries annually use approximately 80% of all the pesticides produced in the world (Pimentel, 1990), less than half of the pesticide-induced deaths occur in these countries (Committee, House of Commons Agriculture, 1987). Clearly, a higher proportion of pesticide poisonings and deaths occurs in developing countries where there are inadequate occupational and other safety standards, insufficient enforcement, poor labeling of pesticides, illiteracy, inadequate protective clothing and washing facilities, and users' lack of knowledge of pesticide hazards (Bull, 1982).

Both the acute and chronic health effects of pesticides warrant concern. Unfortunately, while the acute toxicity of most pesticides is well documented (Ecobichon et al., 1990), information on chronic human illnesses resulting from pesticide exposure is not as sound (Wilkinson, 1990). Regarding cancer, the International Agency for Research on Cancer found "sufficient" evidence of carcinogenicity for 18 pesticides, and "limited" evidence of carcinogenicity for an additional 16 pesticides based on animal studies (Lijinsky, 1989; WHO/UNEP, 1989). With humans the evidence concerning cancer is mixed. For example, a recent study in Saskatchewan indicated no significant difference in non-Hodgkin's lymphoma mortality between farmers and nonfarmers (Wigle et al., 1990), whereas others have reported some human cancer difference (WHO/UNEP, 1989). A realistic estimate of the number of U.S. cases of cancer in humans due to pesticides is given by D. Schottenfeld (University of Michigan, PC, 1991), who estimated that less than 1% of the nation's cancer cases are caused by exposure to pesticides. Considering that there are approximately 1 million cancer cases/year (USBC, 1990), Schottenfeld's assessment suggests less than 10,000 cases of cancer due to pesticides per year.

Many other acute and chronic maladies are beginning to be associated with pesticide use. For example, the recently banned pesticide dibromochloropropane (DBCP) caused testicular dysfunction in animal studies (Foote et al., 1986; Sharp et al., 1986; Shaked et al., 1988) and was linked with infertility among human workers exposed to DBCP (Whorton et al., 1977; Potashnik and Yanai-Inbar, 1987). Also, a large body of evidence suggesting pesticides can produce immune dysfunction has been accumulated over recent years from animal studies (Devens et al., 1985; Olson et al., 1987; Luster et al., 1987; Thomas and House, 1989). In a study of women who had chronically ingested groundwater contaminated with low levels of aldicarb (mean = 16.6 ppb), Fiore et al. (1986) reported evidence of significantly reduced immune response, although these women did not exhibit any other overt health problems.

Of particular concern are the chronic health problems associated with effects of organophosphorous pesticides which have largely replaced the banned organochlorines (Ecobichon et al., 1990). The malady of OPIDP (organophosphate-induced delayed polyneuropathy) is well documented and includes irreversible neurological defects (Lotti, 1984). Other defects in memory, mood, and abstraction have been documented by Savage et al. (1988). Well-documented cases indicate that persistent neurotoxic effects may be present even after the termination of an acute poisoning incident (Ecobichon et al., 1990).

Such chronic health problems are a public health issue, because everyone, everywhere, is exposed to some pesticide residues in food, water, and the atmosphere. About 35% of the foods purchased by U.S. consumers have detectable levels of pesticide residues (FDA, 1990). Of this from 1% to 3% of the foods have pesticide residue levels above the legal tolerance level (Hundley et al., 1988; FDA, 1990). Residue levels may well be higher than has been recorded because the U.S. analytical methods now employed detect only about one-third of the more than 600 pesticides in use (OTA, 1988). Certainly the contamination rate is higher for fruits and vegetables because these foods receive the highest dosages of pesticides. Therefore, there are many good reasons why 97% of the public is genuinely concerned about pesticide residues in their food (FDA, 1989).

Food residue levels in developing nations often average higher than those found in developed nations, either because there are no laws governing pesticide use or because the numbers of skilled technicians available to enforce laws concerning pesticide tolerance levels in foods are inadequate or because other resources are lacking. For instance, most milk samples assayed in a study in Egypt had high residue levels (60% to 80%) of 15 pesticides included in the investigation (Dogheim et al., 1990).

In all countries, the highest levels of pesticide exposures occur in pesticide applicators, farm workers, and people who live adjacent to heavily treated agricultural land (L. W. Davis, Com. of Agr. and Hort., Agr. Chem. and Environ. Services Div., Arizona, PC, 1990). Because farmers and farm workers directly handle 70% to 80% of all pesticides used, the health of these population groups is at the greatest risk of being seriously affected by pesticides. The epidemiological evidence suggests significantly higher cancer incidence among farmers and farm workers in the United States and Europe than among non–farm workers in some areas (e.g., Sharp et al., 1986; Blair et al., 1985; Brown et al., 1990). A consistent association has been documented between lung cancer and exposure to organochlorine insecticides (Blair et al., 1990). Evidence is strong for an association between lymphomas and soft-tissue sarcomas and certain herbicides (Hoar et al., 1986; Blair and Zahm, 1990; Zahm et al., 1990).

Medical specialists are concerned about the lack of public health data about pesticide usage in the United States (GAO, 1986). Based on an investigation of 92 pesticides used on food, the GAO (1986) estimates that 62% of the data on health problems associated with registered pesticides contain little or no information on tumors and even less on birth defects.

Although no one can place a precise monetary value on a human life, the "costs" of human pesticide poisonings have been estimated. Studies done for the insurance industry have computed monetary ranges for the value of a "statistical life" at between $1.6 and $8.5 million (Fisher et al., 1989). For our assessment, we use the conservative estimate of $2 million per human life. Based on the available data, estimates indicate that human pesticide poisonings and related illnesses in the United States total about $787 million each year (Table 1).

Domestic Animal Poisonings and Contaminated Products

In addition to pesticide problems that affect humans, several thousand domestic animals are poisoned by pesticides each year, with dogs and cats representing the largest number (Table 2). For example, of 25,000 calls made to the Illinois Animal Poison Control Center in 1987, nearly 40% of all calls concerned pesticide poisonings in dogs and cats (Beasley and Trammel, 1989). Similarly, Kansas State University reported that 67% of all animal pesticide poisonings involved dogs and cats (Barton and Oehme, 1981). This is not surprising because dogs and cats usually wander freely about the home and farm and therefore have

Table 1. Estimated economic costs of human pesticide poisonings and other pesticide-related illnesses in the United States each year.

Human health effects from pesticides	Total costs ($)
Cost of hospitalized poisonings	
2380[a] × 2.84 days @ $1,000/day	6,759,000
Cost of outpatient treated poisonings	
27,000[c] × $630[b]	17,010,000
Lost work due to poisonings	
4680[a] workers × 4.7 days × $80/day	1,760,000
Pesticide cancers	
<10,000[d] cases × $70,700[c]/case	707,000,000
Cost of fatalities	
27 accidental fatalities[c] × $2 million	54,000,000
TOTAL	786,529,000

[a]Keefe et al. (1990).
[b]Includes hospitalization, forgone earnings, and transportation.
[c]J. Blondell, EPA, PC (1991).
[d]See text for details.

Table 2. Estimated domestic-animal pesticide poisonings in the United States.

Livestock	Number × 1000	$ per head	Number III[e]	$ cost per poisoning[f]	$ cost of poisonings[g]	Number Deaths[d]	$ cost of Deaths × 1,000[g]	Total $ × 1,000
Cattle	99,484[a]	607[a]	100	121.40	12,140	8	4,856	16,996
Dairy cattle	10,298[a]	900[a]	10	180.00	1,800	1	900	2,700
Dogs	52,000[c]	125[h]	50	25.00	1,250	4	500	1,750
Horses	10,600[b]	1,000[c]	11	200.00	2,200	1	1,000	3,200
Cats	55,000[c]	20[h]	50	4.00	200	4	80	280
Swine	52,485[a]	66.30[a]	53	13.26	703	4	265	968
Chickens	5,700,000[a]	2.04[a]	5,700	0.40	2,280	456	912	3,192
Turkeys	260,000[a]	10[c]	260	2.00	520	21	210	730
Sheep	10,800[a]	82.40[a]	11	16.48	181	1	82	263
TOTAL	6,250,667				21,274		8,805	30,079

[a]USDA (1989a).
[b]FAO (1986).
[c]USBC (1990).
[d]Based on a 0.008% mortality rate. (See text.)
[e]Based on a 0.1% illness rate. (See text.)
[f]Based on each animal illness costing 20% of total production value of that animal.
[g]The death of the animal equals the total value for that animal.
[h]Estimated.

greater opportunity to come into contact with pesticides than other domesticated animals.

The best estimates indicate that about 20% of the total monetary value of animal production, or about $4.2 billion, is lost to all animal illnesses, including pesticide poisonings (Gaafar et al., 1985). Colvin (1987) reported that 0.5% of animal illnesses and 0.04% of all animal deaths reported to a veterinary diagnostic laboratory were due to pesticide toxicosis. Thus, $21.3 and $8.8 million, respectively, are lost to pesticide poisonings (Table 2).

This estimate is considered low because it is based only on poisonings reported to veterinarians. Many animal deaths that occur in the home and on farms go undiagnosed and are attributed to factors other than pesticides. In addition, when a farm animal poisoning occurs and little can be done for the animal, the farmer seldom calls a veterinarian but, rather, either waits for the animal to recover or destroys it (G. Maylin, Cornell University, PC, 1977). Such cases are usually unreported.

Additional economic losses occur when meat, milk, and eggs are contaminated with pesticides. In the United States, all animals slaughtered for human comsumption, if shipped interstate, and all imported meat and poultry, must be inspected by the U.S. Department of Agriculture. This is to insure that the meat and products are wholesome, properly labeled, and do not present a health hazard. One part of this inspection, which involves monitoring meat for pesticide and other chemical residues, is the responsibility of the National Residue Program (NRP). The samples taken are intended to insure that if a chemical is present in 1% of the animals slaughtered it will be detected (NAS, 1985).

However, of more than 600 pesticides now in use, NRP tests are made for only 41 different pesticides (D. Beermann, Cornell University, PC, 1991), which have been determined by FDA, EPA, and FSIS to be of public health concern (NAS, 1985). While the monitoring program records the number and type of violations, there is no significant cost to the animal industry because the meat is generally *sold and consumed* before the test results are available. About 3% of the chickens with illegal pesticide residues are sold in the market (NAS, 1987).

Compliance sampling is designed to prevent meat and milk contamination with pesticides. When a producer is suspected of marketing contaminated livestock, the carcasses are detained until the residue analyses are reported. If there are illegal residues present, the carcasses or products are condemned and the producer is prohibited from marketing other animals until it is confirmed that all the livestock are safe (NAS, 1985). If carcasses are not suspected of being contaminated, then by the time the results of the residue tests are reported the carcasses have been sold to consumers. This points to a major deficiency in the surveillance program.

In addition to animal carcasses, pesticide-contaminated milk cannot be sold and must be disposed of. In certain incidents these losses are substantial. For example, in Oahu, Hawaii, in 1982, 80% of the milk supply, worth more than $8.5 million, was condemned by public health officials because it had been contaminated with the insecticide heptachlor (van Ravenswaay and Smith, 1986). This incident had immediate and far-reaching effects on the entire milk industry on the island. Initially, reduced milk sales due to the contaminated milk alone were estimated to cost each diary farmer $39,000. Subsequently, the structure of the island milk industry has changed. Because island milk was considered by consumers to be unsafe, most of the milk supply is now imported. The $500 million lawsuit against the producers brought by consumers is still pending (van Ravenswaay and Smith, 1986).

When the costs attributable to domestic animal poisonings and contaminated meat, milk, and eggs are combined, the economic value of all livestock products in the United States lost to pesticide contamination is estimated to be at least $29.6 million annually (Table 2).

Similarly, other nations lose significant numbers of livestock and large amounts of animal products each year due to pesticide-induced illness or death. Exact data concerning livestock losses do not exist and the available information comes only from reports of the incidence of mass destruction of livestock. For example, when the pesticide leptophos was used by Egyptian farmers on rice and other crops, 1,300 draft animals were poisoned and lost (Sebae, 1977, in Bull, 1982). The estimated economic losses were significant but exact figures are not available.

In addition, countries exporting meat to the United States can experience tremendous economic losses if the meat is found contaminated with pesticides. In a 15-year period, the beef industries in Guatemala, Honduras, and Nicaragua lost more than $1.7 million due to pesticide contamination of exported meat (ICAITI, 1977). In these countries, meat which is too contaminated for export is sold in local markets. Obviously such policies contribute to public health problems.

Destruction of Beneficial Natural Predators and Parasites

In both natural and agroecosystems, many species, especially predators and parasites, control or help herbivorous populations. Indeed, these natural beneficial species make it possible for ecosystems to remain "green." With the parasites and predators keeping herbivore populations at low levels, only a relatively small amount of plant biomass is removed each growing season (Hairston et al., 1960). Natural enemies play a major role in keeping populations of many insect and mite pests under control (DeBach, 1964; Huffaker, 1977; Pimentel, 1988).

Like pest populations, beneficial natural enemies are adversely affected by pesticides (van den Bosch and Messenger, 1973; Adkisson, 1977; Ferro, 1987; Croft, 1990). For example, the following pests have reached outbreak levels in cotton and apple crops following the destruction of natural enemies by pesticides: *cotton*—cotton bollworm, tobacco budworm, cotton aphid, spider mites, and cotton loopers (Adkisson, 1977; OTA, 1979); and *apple*—European red mite, red-banded leafroller, San Jose scale, oystershell scale, rosy apple aphid, wooly apple aphid, white apple aphid, two-spotted spider mite, and apple rust mite (Tabashnik and Croft, 1985; Messing et al., 1989; Croft, 1990; Kovach and Agnello, 1991). Significant pest outbreaks also have occurred in other crops (Huffaker and Kennett, 1956; Huffaker, 1977; OTA, 1979; Croft, 1990; Pimentel, 1991). Also, because parasitic and predaceous insects often have complex searching and attack behaviors, sublethal insecticide dosages may alter this behavior and in this way disrupt effective biological controls (L. E. Ehler, University of California, PC, 1991).

Fungicides also can contribute to pest outbreaks when they reduce fungal pathogens that are naturally parasitic on many insects. For example, the use of benomyl reduces populations of entomopathogenic fungi, resulting in increased survival of velvet bean catepillars and cabbage loopers in soybeans. This eventually leads to reduced soybean yields (Ignoffo et al., 1975; Johnson et al., 1976).

When outbreaks of secondary pests occur because their natural enemies are destroyed by pesticides, additional and sometimes more expensive pesticide treatments have to be applied in efforts to sustain crop yields. This raises overall costs and contributes to pesticide-related problems.

An estimated $520 million can be attributed to costs of additional pesticide applications and increased crop losses, both of which follow the destruction of natural enemies by pesticides applied to crops (Table 3).

As in the United States, natural enemies are being adversely affected by pesticides worldwide. Although no reliable estimate is available concering the impact of this in terms of increased pesticide use and/or reduced yields, general observations by entomologists indicate the impact of loss of natural enemies is severe in many parts of the world.

Table 3. Losses due to the destruction of beneficial natural enemies in U.S. crops ($ millions).

Crops	Total expenditures for insect control with pesticides ($)[a]	Amount of added control costs ($)
Cotton	320	160
Tobacco	5	1
Potatoes	31	8
Peanuts	18	2
Tomatoes	11	2
Onions	1	0.2
Apples	43	11
Cherries	2	1
Peaches	12	2
Grapes	3	1
Oranges	8	2
Grapefruit	5	1
Lemons	1	0.2
Nuts	160	16
Other	500	50
TOTAL		$257.4 ($520)[b]

[a]Pimentel et al. (1991).

[b]Because the added pesticide treatments do not provide as effective control as the natural enemies, we estimate that at least an additional $260 million in crops are lost to pests. Thus, the total loss due to the destruction of natural enemies is estimated to be at least $520 million/year.

For example, from 1980 to 1985 insecticide use in rice production in Indonesia drastically increased (Oka, 1991). This caused the destruction of beneficial natural enemies of the brown planthopper and the pest populations exploded. Rice yields dropped to the extent that rice had to be imported into Indonesia. The estimated loss in rice in just a 2-year period was $1.5 billion (FAO, 1988).

Following that incident, Dr. I. N. Oka and his associates, who previously had developed a successful low-insecticide program for rice pests in Indonesia, were consulted by Indonesian President Suharto's staff to determine what should be done to rectify the situation (I. N. Oka, Bogor Food Research Institute, Indonesia, PC, 1990). Their advice was to substantially reduce insecticide use and return to a sound "treat-when-necessary" program that protected the natural enemies. Following Oka's advice, President Suharto mandated in 1986 that 57 of 64 pesticides would be withdrawn from use on rice and pest management practices would be improved. Pesticide subsidies also were reduced to zero. Subsequently, rice yields increased to levels well above those recorded during the period of heavy pesticide use (FAO, 1988).

D. Rosen (Hebrew University of Jerusalem, PC, 1991) estimates that natural enemies account for up to 90% of the control of pest species achieved in agroecosystems and natural systems; we estimate that at least 50% of control of pest species is due to natural enemies. Pesticides give an additional control of 10% (Pimentel et al., 1978), while the remaining 40% is due to host-plant resistance and other limiting factors present in the agroecosystem (Pimentel, 1988).

Parasites, predators, and host-plant resistance are estimated to account for about 80% of the nonchemical control of pest insects and plant pathogens in crops (Pimentel et al., 1991). Many cultural controls such as crop rotations, soil and water management, fertilizer management, planting time, crop-plant density, trap crops, polyculture, and others provide additional pest control. Together these nonchemical controls can be used effectively to reduce U.S. pesticide use by as much as one-half without any reduction in crop yields (Pimentel et al., 1991).

Pesticide Resistance in Pests

In addition to destroying natural enemy populations, the extensive use of pesticides has often resulted in the development of pesticide resistance in insect pests, plant pathogens, and weeds. In a report of the United Nations Environment Programme pesticide resistance was ranked as one of the top four environmental problems in the world (UNEP, 1979). About 504 insect and mite species (Georghiou, 1990), a total of nearly 150 plant pathogen species (Georghiou, 1986; Eckert, 1988), and about 273 weed species are now resistant to pesticides (LeBaron and McFarland, 1990).

Increased pesticide resistance in pest populations frequently results in the need for several additional applications of the commonly used pesticides to maintain expected crop yields. These additional pesticide applications compound the problem by increasing environmental selection of resistance traits. Despite attempts to deal with it, pesticide resistance continues to develop (Dennehy et al., 1987).

The impact of pesticide resistance, which develops gradually over time, is felt in the economics of agricultural production. A striking example of this occurred in northeastern Mexico and the Lower Rio Grande of Texas (Adkisson, 1972; NAS, 1975). Over time extremely high pesticide resistance had developed in the tobacco budworm population on cotton. Finally in early 1970 approximately 285,000 ha of cotton had to be abandoned because pesticides were ineffective and there was no way to protect the crop from the budworm. The economic and social impact on these Texan and Mexican farming communities dependent upon cotton was devastating.

The study by Carrasco-Tauber (1989) indicates the extent of costs attributed to pesticide resistance. They reported a yearly loss of $45 to $120/ha to pesticide resistance in California cotton. A total of 4.2 million hectares of cotton were harvested in 1984; thus, assuming a loss of $82.50/ha, approximately $348 million of the California cotton crop was lost to resistance. Since $3.6 billion of U.S. cotton were harvested in 1984 (USBC, 1990), the loss due to resistance for that year was approximately 10%. Assuming a 10% loss in other major crops that receive heavy pesticide treatments in the United States, crop losses due to pesticide resistance are estimated to be $1.4 billion/year.

A detailed study by Archibald (1984) further demonstrated the hidden costs of pesticide resistance in California cotton. She reported that 74% more organophosphorus insecticides were required in 1981 to achieve the same kill of pests, like *Heliothis* spp., than in 1979. Her analysis demonstrated that the diminishing effect of pesticides plus the intensified pest control reduced the economic return per dollar of pesticide invested to only $1.14.

Furthermore, efforts to control resistant *Heliothis* spp. exact a cost on other crops when large, uncontrolled populations of *Heliothis* and other pests disperse onto other crops. In addition, the cotton aphid and the whitefly exploded as secondary cotton pests because of their resistance and their natural enemies' exposure to the high concentrations of insecticides.

The total external cost attributed to the development of pesticide resistance is estimated to range between 10% to 25% of current pesticide treatment costs (La Farge, 1985; Harper and Zilberman, 1990), or approximately $400 million each year in the United States alone. In other words, at least 10% of pesticides used in the United States is applied just to combat increased resistance that has developed in various pest species.

In addition to plant pests, a large number of insect and mite pests of both livestock and humans have become resistant to pesticides (Georghiou, 1986). Although a relatively small quantity of pesticide is applied for control of livestock and human pests, the cost of resistance has become significant. Based on available data, the yearly cost of resistance in insect and mite pests of livestock and humans we estimated to be about $30 million for the United States.

Although the costs of pesticide resistance are high in the United States, the costs in tropical developing countries are significantly greater, because pesticides are not only used to control agricultural pests but are also vital for the control of disease vectors. One of the major costs of resistance in tropical countries is associated with malaria control. By 1961, the incidence of malaria in India after early pesticide use declined to only 41,000 cases. However, because mosquitoes developed resistance to pesticides, as did malarial parasites to drugs, the incidence of malaria in India now has exploded to about 59 million cases per year (Reuben, 1989). Similar problems are occurring not only in India but also in the rest of Asia, Africa, and South America: the total incidence of malaria is estimated to be 270 million cases (WHO, 1990; NAS, 1991).

Honeybee and Wild Bee Poisonings and Reduced Pollination

Honeybees and wild bees are absolutely vital for pollination of fruits, vegetables, and other crops. Their direct and indirect benefits to agricultural production range from $10 to $33 billion each year in the United States (Robinson et al., 1989; E. L. Atkins, University of California, PC, 1990). Because most insecticides used in agriculture are toxic to bees, pesticides have a major impact on both honeybee and wild bee populations. D. Mayer (Washington State University, PC, 1990) estimates that approximately 20% of all honeybee colonies are adversely affected by pesticides. He includes the approximately 5% of U.S. bee colonies that are killed outright or die during winter because of pesticide exposure. Mayer calculates that the direct annual loss reaches $13.3 million (Table 4). Another 15% of the bee colonies are seriously weakened by pesticides or suffer losses when apiculturists have to move colonies to avoid pesticide damage.

According to Mayer, the yearly estimated loss from partial bee kills, reduced honey production, plus the cost of moving colonies totals about $25.3 million. Also, as a result of heavy pesticide use on certain

Table 4. Estimated honeybee losses and pollination losses from honeybees and wild bees.

Colony losses from pesticides	$13.3 million/year
Honey and wax losses	25.3 million/year
Loss of potential honey production	27.0 million/year
Bee rental for pollination	4 million/year
Pollination losses	200 million/year
TOTAL	$319.6 million/year

crops, beekeepers are excluded from 4 to 6 million ha of otherwise suitable apiary locations (D. Mayer, Washington State University, PC, 1990). He estimates the yearly loss in potential honey production in these regions is about $27 million.

In addition to these direct losses caused by the damage to bees and honey production, many crops are lost because of the lack of pollination. In California, for example, approximately 1 million colonies of honeybees are rented annually at $20 per colony to augment the natural pollination of almonds, alfalfa, melons, and other fruits and vegetables (R. A. Morse, Cornell University, PC, 1990). Since California produces nearly 50% of our bee-pollinated crops, the total cost for bee rental for the entire country is estimated at $40 million. Of this cost, we estimate at least one-tenth or $4 million is attributed to the effects of pesticides (Table 4).

Estimates of annual agricultural losses due to the reduction in pollination by pesticides may range as high as $4 billion/year (J. Lockwood, University of Wyoming, PC, 1990). For most crops both crop yield and quality are enhanced by effective pollination. For example, McGregor et al. (1955) and Mahadevan and Chandy (1959) demonstrated that for several cotton varieties, effective pollination by bees resulted in yield increases from 20% to 30%. Assuming that a conservative 10% increase in cotton yield would result from more efficient pollination, and subtracting charges for bee rental, the net annual gain for cotton alone could be as high as $400 million. However, using bees to enhance cotton pollination is impossible at present because of the intensive use of insecticides on cotton (McGregor, 1976).

Mussen (1990) emphasizes that poor pollination will not only reduce crop yields, but more importantly, it will reduce the quality of crops such as melons and other fruits. In experiments with melons, E. L. Atkins (University of California, PC, 1990) reported that with adequate pollination melon yields were increased 10% and quality was raised 25% as measured by the dollar value of the crop.

Based on the analysis of honeybee and related pollination losses caused by pesticides, pollination losses attributed to pesticides are estimated to represent about 10% of pollinated crops and have a yearly cost of about $200 million. Adding the cost of reduced pollination to the other environmental costs of pesticides on honeybees and wild bees, the total annual loss is calculated to be about $320 million (Table 4). Clearly, the available evidence confirms that the yearly cost of direct honeybee losses, together with reduced yields resulting from poor pollination, are significant.

Crop and Crop Product Losses

Basically, pesticides are applied to protect crops from pests in order to increase yields, but sometimes the crops are damaged by pesticide treatments. This occurs when (1) the recommended dosages suppress crop growth, development, and yield; (2) pesticides drift from the targeted crop to damage adjacent nearby crops; (3) residual herbicides either prevent chemical-sensitive crops from being planted in rotation or inhibit the growth of crops that are planted; and/or (4) excessive pesticide residues accumulate on crops, necessitating the destruction of the harvest. Crop losses translate into financial losses for growers, distributors, wholesalers, transporters, retailers, food processors, and others. Potential profits as well as investments are lost. The costs of crop losses increase when the related costs of investigations, regulation, insurance, and litigation are added to the equation. Ultimately the consumer pays for these losses in higher marketplace prices.

Data on crop losses due to pesticide use are difficult to obtain. Many losses are never reported to the state and federal agencies because the parties often settle privately (B. D. Berver, Office of Agronomy Services, South Dakota, PC, 1990; R. Batteese, Board of Pesticide Control, Maine Dept. of Agriculture, PC, 1990; J. Peterson, Pesticide/Noxious Weed Division, Dept. of Agr., North Dakota, PC, 1990; E. Streams, EPA, region VII, PC, 1990). For example, in North Dakota, only an estimated one-third of the pesticide-induced crop losses are reported to the State Department of Agriculture (Peterson, PC, 1990). Furthermore, according to the Federal Crop Insurance Corporation, losses due to pesticide use are not insurable because of the difficulty of determining pesticide damage (E. Edgeton, Federal Crop Insurance Corp., Washington, D.C., PC, 1990).

Damage to crops may occur even when recommended dosages of herbicides and insecticides are applied to crops under normal environmental conditions (Chang, 1965; J. Neal, Chemical Pesticides Program, Cornell University, PC, 1990). Heavy dosages of insecticides used on crops have been reported to suppress growth and yield in both cotton and strawberry crops (ICAITI, 1977; Reddy et al., 1987; Trumbel et al., 1988). The increased susceptibility of some crops to insects and diseases following normal use of 2,4-D and other herbicides was demonstrated by Oka and Pimentel (1976), Altman (1985), and Rovira and McDonald (1986). Furthermore, when weather and/or soil conditions are inappropriate for pesticide application, herbicide treatments may cause yield reductions ranging from 2% to 50% (von Rumker and Horay, 1974; Elliot et al., 1975; Akins et al., 1976).

Crops are lost when pesticides drift from the target crops to non–target crops located as much as several miles downwind (Henderson, 1968; Barnes et al., 1987). Drift occurs with almost all methods of pesticide application including both ground and aerial equipment; the potential problem is greatest when pesticides are applied by aircraft (Ware et al., 1969). With aircraft 50% to 75% of pesticides applied miss the target area (Ware et al., 1970; ICAITI, 1977; Ware, 1983; Akesson and Yates, 1984; Mazariegos, 1985). In contrast, 10% to 35% of the pesticide applied with ground application equipment misses the target area (Ware et al., 1975; Hall, 1991). The most serious drift problems are caused by "speed sprayers" and "mist-blower sprayers," because large amounts of pesticide are applied by these sprayers and with these application technologies about 35% of the pesticide drifts away from the target area (E. L. Atkins, University of California, PC, 1990).

Crop injury and subsequent loss due to drift are particularly common in areas planted with diverse crops. For example, in southwest Texas in 1983 and 1984, nearly $20 million of cotton was destroyed from drifting 2,4-D herbicide when adjacent wheat fields were aerially sprayed with the herbicide (Hanner, 1984).

Clearly, drift damage, human exposure, and widespread environmental contamination are inherent in the process of pesticide application and add to the cost of using pesticides. As a result, commercial applicators are frequently sued for damage inflicted during or after treatment. Therefore, most U.S. applicators now carry liability insurance at an estimated cost of about $245 million/year (FAA, 1988; D. Witzman, U.S. Aviation Underwriters, Tennessee, PC, 1990; H. Collins, Nat. Agr. Aviation Assoc., Washington, D.C., PC, 1990).

When residues of some herbicides persist in the soil, crops planted in rotation may be injured (Nanjappa and Hosmani, 1983; Rogers, 1985; Keeling et al., 1989). In 1988/1989, an estimated $25 to $30 million of Iowa's soybean crop was lost due to the persistence of the herbicide Sceptor in the soil (R. G. Hartzler, Cooperative Extension Serv., Iowa State University, PC, 1990).

Herbicide persistence can sometimes prevent growers from rotating their crops and this situation may force them to continue planting the same crop (Altman, 1985; T. Tomas, Nebraska Sustainable Agriculture Society, Hartington, NE, PC, 1990). For example, the use of Sceptor in Iowa, as mentioned, has prevented farmers from implementing their plan to plant soybeans after corn (R. G. Hartzler, PC, 1990). Unfortunately, the continuous planting of some crops in the same field often intensifies insect, weed, and pathogen problems (PSAC, 1965; NAS, 1975; Pimentel et al., 1991). Such pest problems not only reduce crop yields but often require added pesticide applications.

Although crop losses caused by pesticides seem to be a small percentage of total U.S. crop production, their total value is significant. For instance, an average of 0.14% of San Joaquin County's (California) total crop production was lost to pesticides from 1986 to 1987 (OACSJC, 1990; OACSJC, Agricultural Commissioner, San Joaquin County, CA, PC, 1990). Similarly, in Yolo County, CA, approximately 0.18% of its total crop production was lost in 1989 (OACYC, Agricultural Commissioner, Yolo County, CA, PC, 1990; OACYC, 1990). Estimates from Iowa indicate that less than 0.05% of its annual soybean crop is lost to pesticides (R. G. Hartzler, PC, 1990).

An average 0.1% loss in the annual U.S. production of corn, soybeans, cotton, and wheat, which together account for about 90% of the herbicides and insecticides used in U.S. agriculture, was valued at $35.3 million in 1987 (USDA, 1989a; NAS, 1989). Assuming that only one-third of the incidents involving crop losses due to pesticides are reported to authorities, the total value of all crops lost because of pesticides could be as high as three times this amount, or $106 million annually.

However, this $106 million does not take into account other crop losses, nor does it include major but recurrent events such as the large-scale losses that occurred in Iowa in 1988–1989 ($25–$30 million), Texas in 1983–1984 ($20 million), and in California's aldicarb/watermelon crisis in 1985 ($8 million, see below). These recurrent losses alone represent an average of $30 million each year, raising the estimated average crop loss value from the use of pesticides to approximately $136 million.

Additional losses are incurred when food crops are disposed of because they exceed the EPA regulatory tolerances for pesticide residue levels. Assuming that all the crops and crop products that exceed the EPA regulatory tolerances (reported to be at least 1%) were disposed of as required by law, then about $550 million in crops annually would be destroyed because of excessive pesticide contamination (calculated based on data from FDA [1990] and USDA [1989a]). Because most of the crops with pesticides above the tolerance levels are neither detected nor destroyed, they are consumed by the public, avoiding financial loss but creating public health risks.

A well-publicized incident in California during 1985 illustrates this problem. In general, excess pesticides in the food go undetected unless a large number of people become ill after the food is consumed. Thus when more than 1,000 persons became ill from eating the contaminated watermelons, approximately $1.5 million dollars' worth of watermelons were ordered destroyed (R. Magee, State of California Dept. of Food and Agriculture, Sacramento, PC, 1990). After the public became ill, it was learned that several California farmers had treated watermelons with the insecticide aldicarb (Temik), which is not registered for use on watermelons (Taylor, 1986; Kizer, 1986). Following this crisis the California State Assembly appropriated $6.2 million to be awarded to claimants affected by state seizure and freeze orders (*Legislative Counsel's Digest*, 1986). According to the California Department of Food and Agriculture an estimated $800,000 in investigative costs and litigation fees resulted from this one incident (R. Magee, CDFA, PC, 1990). The California Department of Health Services was assumed to have incurred similar expenses, putting the total cost of the incident at nearly $8 million.

To avoid other dangerous and costly incidents like the California watermelon crisis, many private distributors and grocers are testing their produce for the presence of pesticides to reassure themselves and consumers of the safety of the food they handle (C. Merrilees, Consumer Pesticide Project, Nat. Toxics Campaign, San Francisco, PC 1990). Nationally, this testing is presently estimated to cost $1 million per year (C. Merrilees, PC, 1990), but if all the retail grocers nationwide were to undertake such testing, the calculated cost would be approximately $66 million per year based on data from California.

Special investigations of crop losses due to pesticide use, conducted by state and federal agencies, are also costly. From 1987 through 1989, the State of Montana Department of Agriculture conducted an average of 80 pesticide-related investigations per year at an average cost of $3,500 per investigation (S. F. Baril, State of Montana, Dept. of Agr., PC, 1990). Also, the State of Hawaii conducts approximately five investigations a year and these cost nearly $10,000 each (R. Boesch, Pesticide Programs, Dept. of Agriculture, State of Hawaii, PC, 1990). Averaging the number of investigations from seven states (Arkansas, Hawaii, Idaho, Iowa, Louisiana, Mississippi, and Texas) and using the low Montana figure of $3,500 per investigation, the average state conducts 70 investigations a year at a cost of $246,000 annually. Using these data, the investigations are estimated to total $10 million annually. This figure does not include investigation costs at the federal level.

When crop seizures, insurance, and investigation costs are added to the costs of direct crop losses due to the use of pesticides in commercial crop production, the total monetary loss is estimated to be about $942 million annually in the United States (Table 5).

Ground- and Surface Water Contamination

Certain pesticides applied to crops eventually end up in ground- and surface water. The three most common pesticides found in groundwa-

Table 5. Estimated loss of crops and trees due to the use of pesticides.

Impacts	Total costs (in millions of $)
Crop losses	136
Crop applicator insurance	245
Crops destroyed because of excess pesticide contamination	550
Investigations and testing	
Governmental	10
Private	1
TOTAL	$942 million

ter are aldicarb (an insecticide), alachlor, and atrazine (two herbicides) (Osteen and Szmedra, 1989). Estimates are that nearly one-half of the groundwater and well water in the United States is or has the potential to be contaminated (Holmes et al., 1988). The EPA (1990a) reported that 10.4% of community wells and 4.2% of rural domestic wells have detectable levels of at least one pesticide of the 127 pesticides tested in a national survey. It would cost an estimated $1.3 billion annually in the United States if well and groundwater were monitored for pesticide residues (Nielsen and Lee, 1987).

Two major concerns about groundwater contamination with pesticides are that about one-half of the population obtains its water from wells and that once groundwater is contaminated, the pesticide residues remain for long periods of time. Not only are there just a few microorganisms that have the potential to degrade pesticides (Larson and Ventullo, 1983; Pye and Kelley, 1984) but the groundwater recharge rate averages less than 1% per year (CEQ, 1980).

Monitoring pesticides in groundwater is only a portion of the total cost of U.S. groundwater contamination. There is also the high cost of cleanup. For instance, at the Rocky Mountain Arsenal near Denver, Colorado, the removal of pesticides from the groundwater and soil was estimated to cost approximately $2 billion (NYT, 1988). If all pesticide-contaminated groundwater were cleared of pesticides before human consumption, the cost would be about $500 million (based on the costs of cleaning water [Clark, 1979; van der Leeden et al., 1990]). Note the cleanup process requires a water survey to target the contaminated water for cleanup. Thus, adding monitoring and cleaning costs, the total cost regarding pesticide-polluted groundwater is estimated to be about $1.8 billion annually.

Fishery Losses

Pesticides are washed into aquatic ecosystems by water runoff and soil erosion. About 18 tons/ha/year of soil are washed and/or blown from pesticide-treated cropland into adjacent locations including streams and lakes (USDA, 1989b). Pesticides also drift into streams and lakes and contaminate these aquatic systems (Clark, 1989). Some soluble pesticides are easily leached into streams and lakes (Nielsen and Lee, 1987).

Once in aquatic systems, pesticides cause fishery losses in several ways. These include high pesticide concentrations in water that directly kill fish; low-level doses that may kill highly susceptible fish fry; or the elimination of essential fish foods like insects and other invertebrates. In addition, because government safety restrictions ban the catching or sale of fish contaminated with pesticide residues, such unmarketable fish are considered an economic loss (EPA, 1990b; Knuth, 1989; ME & MNR, 1990).

Each year large numbers of fish are killed by pesticides. Based on EPA (1990b) data we calculate that from 1977 to 1987 the cost of fish kills due to all factors has been 141 million fish/year; from 6 to 14 million fish/year are killed by pesticides. These estimates of fish kills are considered to be low for the following reasons. First, 20% of the reported fish kills do not estimate the number of fish killed, and second, fish kills frequently cannot be investigated quickly enough to determine accurately the primary cause (Pimentel et al., 1980). In addition, fast-moving waters in rivers dilute pollutants so that these causes of kills frequently cannot be identified, and also wash away the poisoned fish, while other poisoned fish sink to the bottom and cannot be counted (EPA, 1990b). Perhaps most important is the fact that, unlike direct kills, few, if any, of the widespread and more frequent low-level pesticide poisonings are dramatic enough to be observed and therefore go unrecognized and unreported (EPA, 1990b).

The average value of a fish has been estimated to be about $1.70, using the guidelines of the American Fisheries Society (AFS, 1982); however, it was reported that Coors Beer might be "fined up to $10 per dead fish, plus other penalties" for an accidental beer spill in a creek (Barometer, 1991). At $1.70, the value of the estimated 6–14 million fish killed per year ranges from $10 to $24 million. For reasons mentioned earlier, this is considered an extremely low estimate; the actual loss is probably several times this amount.

Wild Birds and Mammals

Wild birds and mammals are also damaged by pesticides, but these animals make excellent "indicator species." Deleterious effects on wildlife include death from direct exposure to pesticides or secondary poisonings from consuming contaminated prey; reduced survival, growth, and reproductive rates from exposure to sublethal dosages; and habitat reduction through elimination of food sources and refuges (McEwen and Stephenson, 1979; Grue et al., 1983; Risebrough, 1986; Smith, 1987). In the United States, approximately 3 kg of pesticide per ha is applied on about 160 million ha/year of land (Pimentel et al., 1991). With such a large portion of the land area treated with heavy dosages of pesticide, it is to be expected that the impact on wildlife is significant.

The full extent of bird and mammal destruction is difficult to determine because these animals are often secretive, camouflaged, highly mobile, and live in dense grass, shrubs, and trees. Typical field studies of the effects of pesticides often obtain extremely low estimates of bird and mammal mortality (Mineau and Collins, 1988). This is because bird carcasses disappear quickly, well before the dead birds can be found and counted. Studies show only 50% of birds are recovered even when the bird's location is known (Mineau, 1988). Furthermore, where known numbers of bird carcasses were placed in identified locations in the field, 62% to 92% disappeared overnight due to vertebrate and invertebrate scavengers (Balcomb, 1986). Then, too, field studies seldom account for birds that die a distance from the treated areas. Finally, birds often hide and die in inconspicuous locations.

Nevertheless, many bird casualties caused by pesticides have been reported. For instance, White et al. (1982) reported that 1,200 Canada geese were killed in one wheat field that was sprayed with a 2:1 mixture of parathion and methyl parathion at a rate of 0.8 kg/ha. Carbofuran applied to alfalfa killed more than 5,000 ducks and geese in five incidents, while the same chemical applied to vegetable crops killed 1,400 ducks in a single incident (Flickinger et al., 1980, 1991). Carbofuran is estimated to kill 1–2 million birds each year (EPA, 1989). Another pesticide, diazinon, applied on just three golf courses killed 700 Atlantic brant geese of the wintering population of 2,500 geese (Stone and Gradoni, 1985).

Several studies report that the use of herbicides in crop production result in the total elimination of weeds that harbor some insects (Potts, 1986; R. Beiswenger, University of Wyoming, PC, 1990). This has led to significant reductions in the grey partridge in the United Kingdom and in the common pheasant in the United States. In the case of the partridge, population levels have decreased more than 77%, because partridge chicks (also pheasant chicks) depend on insects to supply them with needed protein for their development and survival (Potts, 1986; R. Beiswenger, University of Wyoming, PC, 1990).

Frequently the form of a pesticide influences its toxicity to wildlife

(Hardy, 1990). For example, treated seed and insecticide granules, including carbofuran, fensulfothion, fonofos, and phorate, are particularly toxic to birds when consumed. Estimates are that from 0.23 to 1.5 birds/ha were killed in Canada, while in the United States the estimates ranged from 0.25 to 8.9 birds/ha killed per year by the pesticides (Mineau, 1988).

Pesticides also adversely affect the reproductive potential of many birds and mammals. Exposure to birds, especially predatory birds, to chlorinated insecticides has caused reproductive failure, sometimes attributed to eggshell thinning (Stickel et al., 1984; Risebrough, 1986; Gonzalez and Hiraldo, 1988; Elliot et al., 1988). Most of the affected population have recovered after the ban of DDT in the United States (Bednarz et al., 1990). However, DDT and its metabolite DDE remain a concern, because DDT continues to be used in some South American countries, which are the wintering areas for numerous bird species (Stickel et al., 1984).

Several pesticides, especially DBCP, dimethoate, and deltamethrinare, have been reported to reduce sperm production in certain mammals (Salem et al., 1988; Foote et al., 1986). Clearly, when this occurs the capacity of certain wild mammals to survive is reduced.

Habitat alteration and destruction can be expected to reduce mammal populations. For example, when glyphosphate was applied to forest clear-cuts to eliminate low-growing vegetation, the southern red-backed vole population was greatly reduced because its food source and cover were practically eliminated (D'Anieri et al., 1987). Similar effects from herbicides on other mammals have been reported (Pimentel, 1971). However, overall, the impacts of pesticides on mammals have been inadequately investigated.

Although the gross values for wildlife are not available, expenditures involving wildlife made by humans are one measure of the monetary value. Nonconsumptive users of wildlife spent an estimated $14.3 billion on their sport in 1985 (USFWS, 1988). Yearly, U.S. bird-watchers spend an estimated $600 million on their sport and an additional $500 million on birdseed, or a total of $1.1 billion (USFWS, 1988). The money spent by bird hunters to harvest 5 million game birds was $1.1 billion, or approximately $216/bird (USFWS, 1988). In addition, estimates of the value of all types of birds ranged from $0.40 to more than $800/bird. The $0.40/bird was based on the costs of bird-watching and the $800/bird was based on the replacement costs of the affected species (Walgenbach, 1979; Tinney, 1981; Dobbins, 1986).

If it is assumed that the damages pesticides inflict on birds occur primarily on the 160 million ha of cropland that receives most of the pesticide, and the bird population is estimated to be 4.2 birds per ha of cropland (Blew, 1990), then 672 million birds are directly exposed to pesticides. If it is conservatively estimated that only 10% of the bird population is killed, then the total number killed is 67 million birds. Note this estimate is at the lower end of the range of 0.25 to 8.9 birds/ ha killed per year by pesticides mentioned earlier in this section. Also, this is considered a conservative estimate because secondary losses due to reductions in invertebrate prey were not included in the assessment. Assuming the average value of a bird is $30, an estimated $2 billion in birds are destroyed annually.

Also, a total of $102 million is spent yearly by the U.S. Fish and Wildlife Service on their Endangered Species Program, which aims to reestablish species such as the bald eagle, peregrine falcon, osprey, and brown pelican that in some cases were reduced by pesticides (USFWS, 1991).

When all the above costs are combined, we estimate that the U.S. bird losses associated with pesticide use represent a cost of about $2.1 billion/year.

Microorganisms and Invertebrates

Pesticides easily find their way into soils, where they may be toxic to the arthropods, earthworms, fungi, bacteria, and protozoa found there. Small organisms are vital to ecosystems because they dominate both the structure and function of natural systems.

For example, an estimated 4.5 tons/ha of fungi and bacteria exist in the upper 15 cm of soil (Stanier et al., 1970). They with the arthropods make up 95% of all species and 98% of the biomass (excluding vascular plants). The microorganisms are essential to proper functioning in the ecosystem, because they break down organic matter, enabling the vital chemical elements to be recycled (Atlas and Bartha, 1987). Equally important is their ability to "fix" nitrogen, making it available for plants (Brock and Madigan, 1988). The role of microorganisms cannot be overemphasized, because in nature, agriculture, and forestry they are essential agents in biogeochemical recycling of the vital elements in all ecosystems (Brock and Madigan, 1988).

Earthworms and insects aid in bringing new soil to the surface at a rate of up to 200 tons/ha/year (Kevan, 1962; Satchel, 1967). This action improves soil formation and structure for plant growth and makes various nutrients more available for absorption by plants. The holes (up to 10,000 holes per square meter) in the soil made by earthworms and insects also facilitate the percolation of water into the soil (Hole, 1981;

Edwards and Lofty, 1982), thereby slowing rapid water runoff from the land and preventing soil erosion.

Insecticides, fungicides, and herbicides reduce species diversity in the soil as well as the total biomass of these biota (Pimentel, 1971). Stringer and Lyons (1974) reported that where earthworms had been killed by pesticides, the leaves of apple trees accumulated on the surface of the soil. Apple scab, a disease carried over from season to season on fallen leaves, is commonly treated with fungicides. Some fungicides, insecticides, and herbicides can be toxic to earthworms, which would otherwise remove and recycle the surface leaves (Edwards and Lofty, 1977).

On golf courses and other lawns the destruction of earthworms by pesticides results in the accumulation of dead grass or thatch in the turf (Potter and Braman, 1991). To remove this thatch special equipment must be used at considerable expense.

Although these invertebrates and microorganisms are essential to the vital structure and function of all ecosystems, it is impossible to place a dollar value on the damage caused by pesticides to this large group of organisms. To date no relevant quantitative data on the value of microorganism destruction by pesticides has been collected.

Government Funds for Pesticide Pollution Control

A major environmental cost associated with all pesticide use is the cost of carrying out state and federal regulatory actions, as well as pesticide-monitoring programs needed to control pesticide pollution. Specifically, these funds are spent to reduce the hazards of pesticides and to protect the integrity of the environment and public health.

At least $1 million is spent each year by the state and federal government to train and register pesticide applicators (D. Rutz, Cornell University, PC, 1991). Also, more than $40 million is spent each year by the EPA just to register and reregister pesticides (GAO, 1986). Based on these known expenditures, estimates are that the federal and state governments spend approximately $200 million/year for pesticide pollution control (USBC, 1990) (Table 6).

Although enormous amounts of government money are being spent to reduce pesticide pollution, many costs of pesticide are not taken into account. Also, many serious environmental and social problems remain to be corrected by improved government policies.

Ethical and Moral Issues

Although pesticides provide about $16 billion/year in saved U.S. crops, the data of this analysis suggest that the environmental and so-

Table 6. Total estimated environmental and social costs from pesticides in the United States.

Costs	Millions of $/year
Public health impacts	787
Domestic animals deaths and contamination	30
Loss of natural enemies	520
Cost of pesticide resistance	1,400
Honeybee and pollination losses	320
Crop losses	942
Fishery losses	24
Bird losses	2,100
Groundwater contamination	1,800
Government regulations to prevent damage	200
TOTAL	8,123

cial costs of pesticides to the nation total approximately $8 billion. From a strictly cost/benefit approach, it appears that pesticide use is beneficial. However, the nature of environmental costs of pesticides has other trade-offs involving environmental quality and human health.

One of these issues concerns the importance of public health vs. pest control. For example, assuming that pesticide-induced cancers number 10,000 cases per year and that pesticides return a net agricultural benefit of $12 billion/year, each case of cancer is "worth" $1.2 million in pest control. In other words, for every $1.2 million in pesticide benefits, one person falls victim to cancer. Social mechanisms and market economics provide these ratios, but they ignore basic ethics and values.

In addition, pesticide pollution of the global environment raises numerous other ethical questions. The environmental insult of pesticides has the potential to demonstrably disrupt entire ecosystems. All through history, humans have felt justified in removing forests, draining wetlands, and constructing highways and housing everywhere. L. White (1967) has blamed the environmental crisis on religious teachings of mastery over nature. Whatever the origin, pesticides exemplify this attempt at mastery, and even a noneconomic analysis would question its justification. There is a clear need for a careful and comprehensive assessment of the environmental impacts of pesticides on agriculture and the natural ecosystem.

In addition to the ethical status of ecological concerns are questions of economic distribution of costs. Although farmers spend about $4 billion/year for pesticides, little of the pollution costs that result are borne by them or the pesticide chemical companies. Rather, most of the costs are borne off-site by public illnesses and environmental destruction. Standards of social justice suggest that a more equitable allocation of responsibility is desirable.

These ethical issues do not have easy answers. Strong arguments can be made to support pesticide use based on its definite social and economic benefits. However, evidence of these benefits should not cover up the public health and environmental problems. One goal should be to maximize the benefits while at the same time minimizing the health, environmental, and social costs. A recent investigation pointed out that U.S. pesticide use could be reduced by one-half without any reduction in crop yields or cosmetic standards and would increase food costs less than 0.6% (Pimentel et al., 1991). Judicious use of pesticides could reduce the environmental and social costs, while benefiting farmers economically in the short-term and supporting sustainability of agriculture in the long-term. That pesticide use be discontinued is not suggested, but that current pesticide policies be reevaluated to determine safer ways to employ them in pest control is suggested.

The major environmental and public health problems associated with pesticides are in large measure responsible for the loss of public confidence in state and federal regulatory agencies as well as in institutions that conduct agricultural research. A recent survey by Sachs et al. (1987) confirmed that confidence in the ability of the U.S. government to regulate pesticides declined from 98% in 1965 to only 46% in 1985. Another survey conducted by the FDA (1989) found that 97% of the public were genuinely concerned that pesticides contaminate their food.

Public concern over pesticide pollution confirms a national trend in the country toward environmental values. Media emphasis on the issues and problems caused by pesticides has contributed to a heightened public awareness of ecological concerns. This awareness is encouraging research in sustainable agriculture and nonchemical pest management.

Granted, substituting nonchemical pest controls in U.S. agriculture would be a major undertaking and would not be without its costs. The direct and indirect benefits and costs of implementation of a policy to reduce pesticide use should be researched in detail. Ideally, such a program would both enhance social equitability and promote public understanding of how to better protect human health and the environment while abundant, safe food is supplied. Clearly, it is essential that the environmental and social costs and benefits of pesticide use be considered when future pest control programs are being developed and evaluated. Such costs and benefits should be given ethical and moral scrutiny before policies are implemented, so that sound, sustainable pest management practices are available to benefit farmers, society, and the environment.

Conclusion

An investment of about $4 billion dollars in pesticide control saves approximately $16 billion in U.S. crops, based on direct costs and benefits (Pimentel et al., 1991). However, the indirect costs of pesticide use to the environment and public health need to be balanced against these benefits. Based on the available data, the environmental and social costs of pesticide use total approximately $8 billion each year (Table 6). Users of pesticides in agriculture pay directly for only about $3 billion of this cost, which includes problems arising from pesticide resistance and destruction of natural enemies. Society eventually pays this $3 billion plus the remaining $5 billion in environmental and public health costs (Table 6).

Our assessment of the environmental and health problems associated with pesticides faced problems of scarce data that made this assessment of the complex pesticide situation incomplete. For example, what is an acceptable monetary value for a human life lost or a cancer illness due to pesticides? Also, equally difficult is placing a monetary value on killed wild birds and other wildlife; on the death of invertebrates, or microbes; or on the price of contaminated food and groundwater.

In addition to the costs that cannot be accurately measured, there are additional costs that have not been included in the $8 billion/year figure. A complete accounting of the indirect costs should include accidental poisonings like the "aldicarb/watermelon" crisis; domestic animal poisonings; unrecorded losses of fish, wildlife, crops, trees, and other plants; losses resulting from the destruction of soil invertebrates, microflora, and microfauna; true costs of human pesticide poisonings; water and soil pollution; and human health effects like cancer and sterility. If the full environmental and social costs could be measured as a whole, the total cost would be significantly greater than the estimate of $8 billion/year. Such a complete long-term cost/benefit analysis of pesticide use would reduce the perceived profitability of pesticides.

Human pesticide poisonings, reduced natural enemy populations, increased pesticide resistance, and honeybee poisonings account for a substantial portion of the calculated environmental and social costs of pesticide use in the United States. Fortunately some losses of natural enemies and pesticide resistance problems are being alleviated through carefully planned use of integrated pest management (IPM) practices. But a great deal remains to be done to reduce these important environmental costs (Pimentel et al., 1991).

This investigation not only underscores the serious nature of the environmental and social costs of pesticides but emphasizes the great need for more detailed investigation of the environmental and economic impacts of pesticides. Pesticides are and will continue to be a valuable pest control tool. Meanwhile, with more accurate, realistic cost/benefit analyses, we will be able to work to minimize the risks and develop and increase the use of nonchemical pest controls to maximize the benefits of pest control strategies for all society.

Acknowledgments

We thank the following people for reading an earlier draft of this manuscript, for their many helpful suggestions, and, in some cases, for providing additional information: A. Blair, National Institutes of Health; J. Blondell, U.S. Environmental Protection Agency; S. A. Briggs, Rachel Carson Council; L. E. Ehler, University of California (Davis); E. L. Flickinger, U.S. Fish and Wildlife Service; T. Frisch, NYCAP; E. L. Gunderson, U.S. Food and Drug Administration; R. G. Hartzler, Iowa State University; H. Lehman and G. A. Surgeoner, University of Guelph; P. Mineau, Environment Canada; I. N. Oka, Bogor Food and Agriculture Institute, Indonesia; C. Osteen, U.S. Department of Agriculture; O. Petersson, The Swedish University of Agricultural Sciences; D. Rosen, Hebrew University of Jerusalem; J. Q. Rowley, Oxfam, United Kingdom; P. A. Thomson and J. J. Jenkins, Oregon State University; C. Walters, Acres, U.S.A.; G. W. Ware, University of Arizona; and D. H. Beerman, T. Brown, E. L. Madsen, R. Roush, C. R. Smith, Cornell University.

References

Adkisson, P. L. 1972. The integrated control of the insect pests of cotton. Tall Timbers Conf. Ecol. Control Habitat Mgmt. 4:175–188.

Adkisson, P. L. 1977. Alternatives to the unilateral use of insecticides for insect pest control in certain field crops. Edited by L. F. Seatz. Symposium on Ecology and Agricultural Production. Knoxville: Univ. of Tennessee, pp. 129–144.

AFS. 1982. Monetary values of freshwater fish and fish-kill counting guidelines. Bethesda, MD: Amer. Fisheries Soc. Special Pub. No. 13.

Akesson, N. B., and W. E. Yates. 1984. Physical parameters affecting aircraft spray application. Edited by W. Y. Garner, and J. Harvey. Chemical and Biological Controls in Forestry. Washington, D.C.: Amer. Chem. Soc. Ser. 238, pp. 95–111.

Akins, M. B., L. S. Jeffery, J. R. Overton, and T. H. Morgan. 1976. Soybean response to preemergence herbicides. Proc. S. Weed Sci. Soc. 29:50.

Altman, J. 1985. Impact of herbicides on plant diseases. Edited by C. A. Parker, A. D. Rovia, K. J. Moore, and P. T. W. Wong. Ecology and Management of Soilborne Plant Pathogens. St. Paul: Amer. Phytopathological Soc., pp. 227–231.

Archibald, S. O. 1984. A Dynamic Analysis of Production Externalities: Pesticide Resistance in California. Davis, CA: Ph.D. Thesis. University of California (Davis).

Atlas, R. M., and R. Bartha. 1987. Microbial Ecology: Fundamentals and Applications. 2nd ed. Menlo Park, CA: Benjamin Cummings Co.

Balcomb, R. 1986. Songbird carcasses disappear rapidly from agricultural fields. Auk 103:817–821.

Barnes, C. J., T. L. Lavy, and J. D. Mattice. 1987. Exposure of non-applicator personnel and adjacent areas to aerially applied propanil. Bul. Environ. Contam. and Tox. 39:126–133.

Barometer. 1991. Too Much Beer Kills Thousands. Oregon State University Barometer, May 14, 1991.

Barton, J., and F. Oehme. 1981. Incidence and characteristics of animal poisonings seen at Kansas State University from 1975 to 1980. Vet. and Human Tox. 23:101–102.

Beasley, V. R., and H. Trammel. 1989. Insecticide. Edited by R. W. Kirk. Current Veterinary Therapy: Small Animal Practice. Philadelphia: W. B. Saunders, pp. 97–107.

Bednarz, J. C., D. Klem, L. J. Goodrich, and S. E. Senner. 1990. Migration counts of raptors at Hawk Mountain, Pennsylvania, as indicators of population trends, 1934–1986. Auk 107:96–109.

Blair, A., O. Axelson, C. Franklin, O. E. Paynter, N. Pearce, D. Stevenson, J. E. Trosko, H. Vaubui, G. Williams, J. Woods, and S. H. Zahm. 1990. Carcinogenic effects of pesticides. Edited by C. F. Wilkinson and S. R. Baker. The Effect of Pesticides on Human Health. Princeton, N.J.: Princeton Scientific Pub. Co., Inc., pp. 201–260.

Blair, A., H. Malker, K. P. Cantor, L. Burmeister, and K. Wirklund. 1985. Cancer among farmers: a review. Scand. Jour. Work. Environ. Health 11:397–407.

Blair, A., and H. Zahm. 1990. Methodologic issues in exposure assessment for case-control studies of cancer and herbicides. Amer. J. of Industrial Med. 18:285–293.

Blew, J. H. 1990. Breeding Bird Census. 92 Conventional Cash Crop Farm. Jour. Field Ornithology. 61 (Suppl.) 1990:80–81.

Brock, T., and M. Madigan. 1988. Biology of Microorganisms. London: Prentice Hall.

Brown, L. M., A. Blair, R. Gibson, G. D. Everett, K. P. Cantor, L. M. Schuman, L. F. Burmeister, S. F. Van Lier, and F. Dick. 1990. Pesticide exposures and other agricultural risk factors for leukemia among men in Iowa and Minnesota. Cancer Res. 50:6585–6591.

Bull, D. 1982. A Growing Problem: Pesticides and the Third World Poor. Oxford: Oxfam.

Carrasco-Tauber, C. 1989. Pesticide Productivity Revisited. Amherst: M. S. Thesis. University of Massachusetts.

CEQ. 1980. The Global 2000 Report to the President. Washington, D.C.: Council on Environmental Quality and the U.S. Department of State.

Chang, W. L. 1965. Comparative study of weed control methods in rice. Jour. Taiwan Agr. Res. 14:1–13.

Clark, R. B. 1989. Marine Pollution. Oxford: Clarendon Press.

Clark, R. M. 1979. Water supply regionalization: a critical evaluation. Proc. Am. Soc. Civil Eng. 105:279–294.

Colvin, B. M. 1987. Pesticide uses and animal toxicoses. Vet. and Human Tox. 29 (suppl. 2):15 pp.

Committee, House of Commons Agriculture. 1987. The Effects of Pesticides on Human Health. London: Report and Proceedings of the Committee. 2nd Special Report. Vol. I. Her Majesty's Stationery Office.

Croft, B. A. 1990. Arthropod Biological Control Agents and Pesticides. New York: Wiley.

D'Anieri, P., D. M. Leslie, and M. L. McCormack. 1987. Small mammals in glyphosphate-treated clearcuts in Northern Maine. Canadian Field Nat. 101:547–550.

DeBach, P. 1964. Biological Control of Insect Pests and Weeds. New York: Rheinhold.

Dennehy, T. J., J. P. Nyrop, R. T. Roush, J. P. Sanderson, and J. G. Scott. 1987. Managing pest resistance to pesticides: A challenge to New York's agriculture. New York's Food and Life Sciences Quarterly 17:4, 13–17.

Devens, B. H., M. H. Grayson, T. Imamura, and K. E. Rodgers. 1985. O,O,S-trimethyl phosphorothioate effects on immunocompetence. Pestic. Biochem. Physiol. 24:251–259.

Dobbins, J. 1986. Resources Damage Assessment of the T/V Puerto Rican Oil Spill Incident. Washington, D.C.: James Dobbins Associates, Inc., Report to NOAA, Sanctuary Program Division.

Dogheim, S. M., E. N. Nasr, M. M. Almaz, and M. M. El-Tohamy. 1990. Pesticide residues in milk and fish samples collected from two Egyptian Governorato. Jour. Assoc. Off. Anal. Chem. 73:19–21.

Eckert, J. W. 1988. Historical development of fungicide resistance in plant pathogens. Edited by C. J. Delp. Fungicide Resistance in North America. St. Paul: APS Press, pp. 1–3.

Ecobichon, D. J., J. E. Davies, J. Doull, M. Ehrich, R. Joy, D. McMillan, R. MacPhail, L. W. Reiter, W. Slikker, and H. Tilson. 1990. Neurotoxic effects of pesticides. Edited by C. F. Wilkinson and S. R. Baker. The Effect of Pesticides on Human Health. Princeton, NJ: Princeton Scientific Pub. Co., Inc., pp. 131–199.

Edwards, C., and J. Lofty. 1977. Biology of Earthworms. London: Chapman and Hall.

Edwards, C. A., and J. R. Lofty. 1982. Nitrogenous fertilizers and earthworm populations in agricultural soils. Soil Biol. Biochem. 14:515–521.

Elliot, B. R., J. M. Lumb, T. G. Reeves, and T. E. Telford. 1975. Yield losses in weed-free wheat and barley due to post-emergence herbicides. Weed Res. 15:107–111.

Elliot, J. E., R. J. Norstrom, and J. A. Keith. 1988. Organochlorines and eggshell thinning in Northern Gannets (Sula bassanus) from Eastern Canada 1968–1984. Environ. Poll. 52:81–102.

EPA. 1977. Minutes of Administrator's Pesticide Policy Advisory Committee. March. Washington, D.C.: U.S. Environmental Protection Agency.

EPA. 1989. Carbofuran: A Special Review Technical Support Document. Washington, D.C.: U.S. Environmental Protection Agency, Office of Pesticides and Toxic Substances.

EPA. 1990a. National Pesticide Survey—Summary. Washington, D.C.: U.S. Environmental Protection Agency.

EPA. 1990b. Fish Kills Caused by Pollution. 1977–1987. Washington, D.C.: Draft Report of U.S. Environmental Protection Agency. Office of Water Regulations and Standards.

FAA. 1988. Census of Civil Aircraft. Washington, D.C.: U.S. Federal Aviation Administration.

FAO. 1986. Production Yearbook. Rome: Food and Agriculture Organization. United Nations. Vol. 40.

FAO. 1988. Integrated Pest Management in Rice in Indonesia. Jakarta: Food and Agriculture Organization. United Nations. May.

FDA. 1989. Food and Drug Administration Pesticide Program Residues in Foods—1988. Jour. Assoc. Off. Anal. Chem. 72:133A–152A.

FDA 1990. Food and Drug Administration Pesticide Program Residues in Foods—1989. Jour. Assoc. Off. Anal. Chem. 73:127A–146A.

Ferro, D. N. 1987. Insect pest outbreaks in agroecosystems. Edited by P. Barbosa and J. C. Schultz. Insect Outbreaks. San Diego: Academic Press, pp. 195–215.

Fiore, M. C., H. A. Anderson, R. Hong, R. Golubjatnikov, J. E. Seiser, D. Nordstrom, L. Hanrahan, and D. Belluck. 1986. Chronic exposure to aldicarb-contaminated groundwater and human immune function. Environ. Res. 41:633–645.

Fisher, A., L. G. Chestnut, and D. M. Violette. 1989. The value of reducing risks of death: A note on new evidence. Jour. Policy Anal. and Mgt. 8:88–100.

Flickinger, E. L., K. A. King, W. F. Stout, and M. M. Mohn. 1980. Wildlife hazards from furadan 3G applications to rice in Texas. Jour. Wildlife Mgt. 44:190–197.

Flickinger, E. L., G. Juenger, T. J. Roffe, M. R. Smith, and R. J. Irwin. 1991. Poisoning of Canada geese in Texas by parathion sprayed for control of Russian wheat aphid. Jour. Wildlife Diseases 27:265–268.

Foote, R. H., E. C. Schermerhorn, and M. E. Simkin. 1986. Measurement of semen quality, fertility, and reproductive hormones to assess dibromochloropropane (DBCP) effects in live rabbits. Fund. and Appl. Tox. 6:628–637.

Gaafar, S. M., W. E. Howard, and R. Marsh. 1985. World Animal Science B: Parasites, Pests, and Predators. Amsterdam: Elsevier.

GAO. 1986. Pesticides: EPA's Formidable Task to Assess and Regulate Their Risks. Washington, D.C.: General Accounting Office.

Georghiou, G. P. 1986. The magnitude of the resistance problem. Pesticide Resistance, Strategies and Tactics for Management. Washington, D.C.: National Academy of Sciences, pp. 18–41.

Georghiou, G. P. 1990. Overview of insecticide resistance. Edited by M. B. Green, H. M. LeBaron, and W. K. Moberg. Managing Resistance to Agrochemicals: From Fundamental Research to Practical Strategies. Washington, D.C.: Amer. Chem. Soc. pp. 18–41.

Gonzalez, L. M., and F. Hiraldo. 1988. Organochlorines and heavy metal contamination in the eggs of the Spanish Imperial Eagle (Aquila [heliaca] adaberti) and accompanying changes in eggshell morphology and chemistry. Environ. Poll. 51:241–258.

Grue, C. E., W. J. Fleming, D. G. Busby, and E. F. Hill. 1983. Assessing hazards of organophosphate pesticides to wildlife. Transactions of the North American Wildlife and Natural Resources Conference (48th). Washington, D.C.: Wildlife Management Institute.

Hairston, N. G., F. E. Smith, and L. B. Slobodkin. 1960. Community structure, population control and competition. Amer. Nat. 94:421–425.

Hall, F. R. 1991. Pesticide application technology and integrated pest management (IPM). Edited by D. Pimentel. Handbook of Pest Management in Agriculture. Boca Raton, FL: CRC Press. pp. 135–167.

Hanner, D. 1984. Herbicide drift prompts state inquiry. Dallas (Texas) Morning News July 25.

Hardy, A. R. 1990. Estimating exposure: the identification of species at risk and routes of exposure. Edited by L. Somerville and C. H. Walker. Pesticide Effects on Terrestrial Wildlife. London: Taylor & Francis, 81–97.

Harper, C. R., and D. Zilberman. 1990. Pesticide regulation: problems in trading off economic benefits against health risks. Edited by D. Zilberman and J. B. Siebert. Economic Perspectives on Pesticide Use in California. Berkeley: University of California. October. pp. 181–208.

Henderson, J. 1968. Legal aspects of crop spraying. Univ. Ill. Agr. Exp. Sta. Circ. 99.

Hoar, S. K., A. Blair, F. F. Holmes, C. D. Boysen, R. J. Robel, R. Hoover, and J. F. Fraumeni. 1986. Agricultural herbicide use and risk of lymphoma and soft-tissue sarcoma. J. Amer. Med. Assoc. 256:1141–1147.

Hole, F. D. 1981. Effects of animals on soil. Geoderma 25:75–112.

Holmes, T., E. Nielsen, and L. Lee. 1988. Managing groundwater contamination in rural areas. Rural Development Perspectives. Washington, D.C.: U.S. Dept. of Agr., Econ. Res. Ser. vol. 5 (1). pp. 35–39.

Huffaker, C. B. 1977. Biological Control. New York: Plenum.

Huffaker, C. B., and C. E. Kennett. 1956. Experimental studies on predation: predation and cyclamen mite populations on strawberries in California. Hilgardia 26:191–222.

Hundley, H. K., T. Cairns, M. A. Luke, and H. T. Masumoto. 1988. Pesticide residue findings by the Luke method in domestic and imported foods and animal feeds for fiscal years 1982–1986. Jour. Assoc. Off. Anal. Chem. 71:875–77.

ICAITI. 1977. An Environmental and Economic Study of the Consequences of Pesticide Use in Central American Cotton Production. Guatemala City, Guatemala: Final Report, Central American Research Institute for Industry, United Nations Environment Programme.

Ignoffo, C. M., D. L. Hostetter, C. Garcia, and R. E. Pinnelle. 1975. Sensitivity of the entomopathogenic fungus *Nomuraea rileyi* to chemical pesticides used on soybeans. Environ. Ent. 4:765–768.

Johnson, D. W., L. P. Kish, and G. E. Allen. 1976. Field evaluation of selected pesticides on the natural development of the entomopathogen, *Nomuraea rileyi*, on the velvetbean caterpillar on soybean. Environ. Ent. 5:964–966.

Keefe, T. J., E. P. Savage, S. Munn, and H. W. Wheeler. 1990. Evaluation of epidemiological factors from two national studies of hospitalized pesticide poisonings, U.S.A. Washington, D.C.: Exposure Assessment Branch, Hazard Evaluation Division, Office of Pesticides and Toxic Substances, U.S. Environmental Protection Agency.

Keeling, J. W., R. W. Lloyd, and J. R. Abernathy. 1989. Rotational crop response to repeated applications of korflurazon. Weed Tech. 3:122–125.

Kevan, D. K. McE. 1962. Soil Animals. New York: Philosophical Library.

Kizer, K. 1986. California's Fourth of July Food Poisoning Epidemic from Aldicarb-Contaminated Watermelons. Sacramento: State of California Department of Health Services.

Knuth, B. A. 1989. Implementing chemical contamination policies in sportfisheries. Agency partnerships and constitutency influence. Jour. Mgt. Sci. and Policy Anal. 6:69–81.

Kovach, J., and A. M. Agnello. 1991. Apple pests—pest management system for insects. Edited by D. Pimentel. Handbook of Pest Management in Agriculture. Boca Raton, FL: CRC Press, pp. 107–116.

LaFarge, A. M. 1985. The presistence of resistance. Agrichemical Age 29:10–12.

Larson, R. J., and R. M. Ventullo. 1983. Biodegradation potential of groundwater bacteria. Edited by D. M. Nielsen. Proceedings of the Third National Symposium on Aquifer Restoration and Groundwater Monitoring. May 25–27. Worthington, Ohio: National Water Well Association, pp. 402–409.

LeBaron, H. M., and J. McFarland. 1990. Herbicide Resistance in Weeds and Crops. Edited by M. B. Green, H. M. LeBaron, and W. K. Moberg. Managing Resistance From Fundamental Research to Practical Strategies. Washington, D.C.: Amer. Chem. Soc., pp. 336–52.

Legislative Counsel's Digest (State of California). 1986. Legislative Assembly Bill No. 2755.

Lijinsky, W. 1989. Statement by D. William Lijinsky before the Committee on Labor and Human Resources. Washington, D.C.: U.S. Senate, June 6, 1989.

Litovitz, T. L., B. F. Schmitz, and K. M. Bailey. 1990. 1989 Annual report of the American Association of Poison Control Centers for National Data Collection System. Amer. Jour. Emergency Med. 8:394–442.

Lotti, M. 1984. The delayed polyneuropathy caused by some organophospho-rus esters. Edited by C. L. Galli, L. Manzo and P. S. Spencer. Recent Advances in Nervous Systems Toxicology. Proceedings of a NATO Advanced Study Institute on Toxicology. Proceedings of a NATO Advanced Study Institute on Toxicology of the Nervous System, Dec. 10–20, 1984, Belgirate, Italy, pp. 247–257.

Luster, M. I., J. A. Blank, and J. H. Dean. 1987. Molecular and cellular basis of chemically induced immunotoxicity. Ann. Rev. Pharmacol. Tox. 27:23–49.

Mahadevan, V., and K. C. Chandy. 1959. Preliminary studies on the increase in cotton yield due to honeybee pollination. Madras Agr. Jour. 46:23–26.

Mazariegos, F. 1985. The Use of Pesticides in the Cultivation of Cotton in Cen-tral America. Guatemala: United Nations Environment Programme, Indus-try and Environment. July/August/September.

McEwen, F. L., and G. R. Stephenson. 1979. The Use and Significance of Pesti-cides in the Environment. New York: John Wiley and Sons, Inc.

McGregor, S. E. 1976. Insect pollination of cultivated crop plants. Washington, D.C.: U.S. Dept. of Agr., Agr. Res. Ser., Agricultural Handbook No. 496.

McGregor, S. E., C. Rhyne, S. Worley, and F. E. Todd. 1955. The role of honey-bees in cotton pollination. Agron. Jour. 47:23–25.

ME & MNR. 1990. Guide to Eating Ontario Sport Fish. Ontario: Ministry of Environment and Ministry of Natural Resources.

Messing, R. H., B. A. Croft, and K. Currans. 1989. Assessing pesticide risk to arthropod natural enemies using expert system technology. Appl. Nat. Re-sour. Manage., Moscow, Idaho 3:1–11.

Mineau, P. 1988. Avian mortality in agroecosystems: 1. The case against gran-ule insecticides in Canada. In Field methods for the study of environmental effects of pesticides. Edited by M. P. Greaves, B. D. Smith, and P. W. Greig-Smith. London: British Crop Protection Council (BPCP) Monograph 40, BPCP, Thornton Heath, pp. 3–12.

Mineau, P., and B. T. Collins. 1988. Avian mortality in agro-ecosystems: 2. Meth-ods of detection. In Field methods for the study of environmental effects of pesticides. Edited by M. P. Greaves, G. Smith, and B. D. Smith. Cambridge, UK: Proceedings of a Symposium Organized by the British Crop Protection Council, held at Churchill College, pp. 13–27.

Mussen, E. 1990. California crop pollination. Gleanings in Bee Culture 118:636–647.

Nanjappa, H. V., and N. M. Hosmani. 1983. Residual effect of herbicides ap-plied in transplanted fingermillet (Eleusive coracana Gaertu.) on the suc-ceeding crops. Indian Jour. of Agron. 28:42–45.

NAS. 1975. Pest Control: An Assessment of Present and Alternative Technolo-gies. 4 volumes. Washington, D.C.: National Academy of Sciences.

NAS. 1985. Meat and Poultry Inspection. The Scientific Basis of the Nation's Program. Washington, D.C.: National Academy of Sciences.

NAS. 1987. Regulating Pesticides in Food. Washington, D.C.: National Acad-emy of Sciences.

NAS. 1989. Alternative Agriculture. Washington, D.C.: National Academy of Sciences.

NAS. 1991. Malaria Prevention and Control. Washington, D.C.: National Academy of Sciences.

Nielsen, E. G., and L. K. Lee. 1987. The Magnitude and Costs of Groundwater Contamination from Agricultural Chemicals. A National Perspective. Washington, D.C.: U.S. Dept. of Agr., Econ. Res. Ser., Natural Resour. Econ. Div., ERS Staff Report, AGES870318.

NYT. 1988. Shell Loses Suit on Cleanup Cost. New York, NY: New York Times.

OACSJC. 1990. San Joaquin County Agricultural Report 1989. San Joaquin County, CA: San Joaquin County Department of Agriculture.

OACYC. 1990. Yolo County 1989 Agricultural Report. Yolo County, CA: Office of the Agricultural Commissioner, Yolo County.

Oka, I. N. 1991. Success and challenges of the Indonesian national integrated pest management program in the rice-based cropping system. Crop Protection 10:163–165.

Oka, I. N., and D. Pimentel. 1976. Herbicide (2,4-D) increases insect and pathogen pests on corn. Science 193:239–240.

Olson, L. J., B. J. Erikson, R. D. Hinsdill, J. A. Wyman, W. P. Porter, R. C. Bidgood, and E. V. Norheim. 1987. Aldicarb immunomodulation in mice: an inverse dose-response to parts per billion in drinking water. Arch. Environ. Contam. Tox. 16:433–439.

Osteen, C. D., and P. I. Szmedra. 1989. Agricultural Pesticide Use Trends and Policy Issues. Washington, D.C.: U.S. Dept. of Agr., Econ. Res. Serv., Agr. Econ. Report No. 622.

OTA. 1979. Pest Management Strategies. Washington, D.C.: Office of Technology Assessment, Congress of the United States. 2 volumes.

OTA. 1988. Pesticide Residues in Food: Technologies for Detection. Washington, D.C.: Office of Technology Assessment. U.S. Congress.

Pimentel, D. 1971. Ecological Effects of Pesticides on Non-Target Species. Washington, D.C.: U.S. Government Printing Office.

Pimentel, D. 1988. Herbivore population feeding pressure on plant host: Feedback evolution and host conservation. Oikos 53:185–238.

Pimentel, D. 1990. Estimated annual world pesticide use. Edited by Ford Foundation. Facts and Figures. New York: Ford Foundation.

Pimentel, D., D. Andow, R. Dyson-Hudson, D. Gallahan, S. Jacobson, M. Irish, S. Kroop, A. Moss, I. Schreiner, M. Shepard, T. Thompson, and B. Vinzant. 1980. Environmental and social costs of pesticides: a preliminary assessment. Oikos 34:127–140.

Pimentel, D., J. Krummel, D. Gallahan, J. Hough, A. Merrill, I. Schreiner, P. Vittum, F. Koziol, E. Back, D. Yen, and S. France. 1978. Benefits and costs of pesticide use in U.S. food production. BioScience 28:778–784.

Pimentel, D., L. McLaughlin, A. Zepp, B. Lakitan, T. Kraus, P. Kleinman, F. Vancini, W. J. Roach, E. Graap, W. S. Keeton, and G. Selig. 1991. In Environmental and economic impacts of reducing U.S. agricultural pesticide use. Edited by D. Pimentel. Handbook on Pest Management in Agriculture. Boca Raton, FL: CRC Press, pp. 679–718.

PN. 1990. Towards a reduction in pesticide use. Pesticide News (March).

Potashnik, G., and I. Yanai-Inbar. 1987. Dibromochloropropane (DBCP): an 8-year reevaluation of testicular function and reproductive performance. Fertility and Sterility 47:317–323.

Potter, D. A., and S. K. Braman. 1991. Ecology and management of turfgrass insects. Ann. Rev. Entom. 36:383–406.

Potts, G. R. 1986. The Partridge: Pesticides, Predation and Conservation. London: Collins.

PSAC. 1965. Restoring the Quality of Our Environment. Washington, D.C.: President's Science Advisory Committee. The White House.

Pye, V., and J. Kelley. 1984. The extent of groundwater contamination in the United States. Edited by NAS. Groundwater Contamination. Washington, D.C.: National Academy of Sciences.

Reddy, V. R., D. N. Baker, F. D. Whisler, and R. E. Fye. 1987. Application of GOSSYM to yield decline in cotton, I. Systems analysis of effects of herbicides on growth, development, and yield. Agron. Jour. 79:42–47.

Reuben, R. 1989. Obstacles to malaria control in India—the human factor. In Demography and vector-borne diseases. Edited by W. W. Service. Boca Raton, FL: CRC Press. pp. 143–154.

Risebrough, R. W. 1986. Pesticides and bird populations. In Current ornithology. Edited by R. F. Johnston. New York: Plenum Press, pp. 397–427.

Robinson, W. E., R. Nowogrodzki, and R. A. Morse. 1989. The value of honey bees as pollinators of U.S. crops. Amer. Bee Jour. 129:477–487.

Rogers, C. B. 1985. Fluometuron carryover and damage to subsequent crops. 45:8,2375B: Dissertation Abstracts International.

Rovira, A. D., and H. J. McDonald. 1986. Effects of the herbicide chlorsulfuron on Rhizoctonia Bare Patch and Take-all of barley and wheat. Plant Disease 70:879–882.

Sachs, C., D. Blair, and C. Richter. 1987. Consumer pesticide concerns: A 1965 and 1984 comparison. Jour. Consu. Aff. 21:96–107.

Salem, M. H., Z. Abo-Elezz, G. A. Abd-Allah, G. A. Hassan, and N. Shakes. 1988. Effect of organophophorus (dimethoate) and pyrethroid (deltamethrin) pesticides on semen characteristics in rabbits. Jour. Environ. Sci. Health B23:279–290.

Satchel, J. E. 1967. Lumbricidae. In Soil biology. Edited by A. Burges and F. Raw. New York: Academic Press, pp. 259–322.

Savage, E. P., T. J. Keefe, L. W. Mounce, R. K. Heaton, A. Lewis, and P. J. Burcar. 1988. Chronic neurological sequelae of acute organophosphate pesticide poisoning. Arch. Environ. Health 43:38–45.

Sebae, A. H. 1977. Incidents of local pesticide hazards and their toxicological interpretation. Alexandria, Egypt: Proceedings of UC/AID University of Alexandria Seminar Workshop in Pesticide Management.

Shaked, I., U. A. Sod-Moriah, J. Kaplanski, G. Potashnik, and O. Buckman. 1988. Reproductive performance of dibromochloropropane-treated female rats. Int. Jour. Fert. 33:129–133.

Sharp, D. S., B. Eskenazi, R. Harrison, P. Callas, and A. H. Smith. 1986. Delayed health hazards of pesticide exposure. Ann. Rev. Public Health 7:441–471.

Smith, G. J. 1987. Pesticide Use and Toxicology in Relation to Wildlife: Organophosphorus and Carbamate Compounds. Washington, D.C.: Resource Publication 170, U.S. Dept. of Interior, Fish and Wildlife Service.

Stanier, R., M. Doudoroff, and E. Adelberg. 1970. The microbial world. London: Prentice Hall.

Stickel, W. H., L. F. Stickel, R. A. Dyrland, and D. L. Hughes. 1984. DDE in birds: Lethal residues and loss rates. Arch. Environ. Contam. and Tox. 13:1–6.

Stone, W. B., and P. B. Gradoni. 1985. Wildlife mortality related to the use of the pesticide diazinon. Northeastern Environ. Sci. 4:30–38.

Stringer, A., and C. Lyons. 1974. The effect of benomyl and thiophanate-methyl on earthworm populations in apple orchards. Pesticide Sci. 5:189–196.

Tabashnik, B. E., and B. A. Croft. 1985. Evolution of pesticide resistance in apple pests and their natural enemies. Entomophaga 30:37–49.

Taylor, R. B. 1986. State sues three farmers over pesticide use on watermelons. Los Angeles Times, I (CC)–3–4.

Thomas, P. T., and R. V. House. 1989. Pesticide-induced modulation of immune system. In Carcinogenicity and pesticides: Principles, issues, and relationships. Edited by N. N. Ragsdale and R. E. Menzer. Washington, D.C.: Amer. Chem. Soc., ACS Symp. Ser. 414, pp. 94–106.

Tinney, R. T. 1981. The oil drilling prohibitions at the Channel Islands and Pt. Reyes-Fallallon Islands National Marine Sanctuaries: Some costs and benefits. Washington, D.C.: Report to Center for Environmental Educations.

Trumbel, J. T., W. Carson, H. Nakakihara, and V. Voth. 1988. Impact of pesticides for tomato fruitworm (Lepidoptera:Noctuidae) suppression on photosynthesis, yield, and nontarget arthropods in strawberries. Jour. Econ. Ent. 81:608–614.

UNEP. 1979. The State of the Environment: Selected topics—1979. Nairobi: United Nations Environment Programme, Governing Council, Seventh Session.

USBC. 1990. Statistical Abstract of the United States. Washington, D.C.: U.S. Bureau of the Census, U.S. Dept. of Commerce.

USDA. 1989a. Agricultural Statistics. Washington, D.C.: U.S. Department of Agriculture.

USDA. 1989b. The Second RCA Appraisal. Soil, Water, and Related Resources on Non-federal Land in the United States. Analysis of Conditions and Trends. Washington, D.C.: U.S. Department of Agriculture.

USFWS. 1988. 1985 Survey of Fishing, Hunting, and Wildlife Associated Recreation. Washington, D.C.: U.S. Fish and Wildlife Service. U.S. Dept. of Interior.

USFWS. 1991. Federal and State Endangered Species Expenditures. Washington, D.C.: U.S. Fish and Wildlife Service.

van den Bosch, R., and P. S. Messenger. 1973. Biological Control. New York: Intext Educational.

van der Leeden, F., F. L. Troise, and D. K. Todd. 1990. The Water Encyclopedia (2nd ed.). Chelsea, MI: Lewis Pub.

van Ravenswaay, E., and E. Smith. 1986. Food contamination: Consumer reactions and producer losses. National Food Review (Spring):14–16.

von Rumker, R., and F. Horay. 1974. Farmers' Pesticide Use Decisions and Attitudes on Alternate Crop Protection Methods. Washington, D.C.: U.S. Environmental Protection Agency.

Walgenbach, F. E. 1979. Economic Damage Assessment of Flora and Fauna Resulting from Unlawful Environmental Degradation. Sacramento, CA: Manuscript, California Department of Fish and Game.

Ware, G. W. 1983. Reducing pesticide application drift-losses. Tucson: University of Arizona, College of Agriculture, Cooperative Extension.

Ware, G. W., W. P. Cahill, B. J. Estesen, W. C. Kronland, and N. A. Buck. 1975. Pesticide drift deposit efficiency from ground sprays on cotton. Jour. Econ. Entomol. 68:549–550.

Ware, G. W., W. P. Cahill, P. D. Gerhardt, and J. W. Witt. 1970. Pesticide drift IV. On-target deposits from aerial application of insecticides. Journ. Econ. Ent. 63:1982–1983.

Ware, G. W., B. J. Estesen, W. P. Cahill, P. D. Gerhardt, and K. R. Frost. 1969. Pesticide drift. I. High-clearance vs aerial application of sprays. Jour. Econ. Entomo. 62:840–843.

White, D. H., C. A. Mitchell, L. D. Wynn, E. L. Flickinger, and E. J. Kolbe. 1982. Organophosphate insecticide poisoning of Canada geese in the Texas Panhandle. Jour. Field Ornithology 53:22–27.

White, L. 1967. The historical roots of our ecological crisis. Science 155:1203–1207.

WHO. 1990 March 28. World Health Organization Press Release. Tropical Diseases News 31:3.

WHO/UNEP. 1989. Public Health Impact of Pesticides Used in Agriculture. Geneva: World Health Organization/United Nations Environment Programme.

Whorton, D., R. M. Krauss, S. Marshall, and T. H. Milby. 1977. Infertility in male pesticide workers. The Lancet, Dec. 17, 1977:1259–1261.

Wigle, D. T., R. M. Samenciw, K. Wilkins, D. Riedel, L. Ritter, H. I. Morrison, and Y. Mao. 1990. Mortality study of Canadian male farm operators: Non-Hodgkin's lymphoma mortality and agricultural practices in Saskatchewan. J. Nat. Canc. Instit. 82:575–582.

Wilkinson, C. F. 1990. Introduction and overview. In The effect of pesticides on human health. Edited by C. F. Wilkinson and S. R. Baker. Princeton, NJ: Princeton Scientific Pub. Co., Inc., pp. 1–33.

Zahm, S. H., D. D. Weisenburger, P. A. Babbitt, R. C. Saal, J. B. Vaught, K. P. Cantor, and A. Blair. 1990. A case-control study of non-Hodgkin's lymphoma and the herbicide 2,4-dichlorophenoxyacetic acid (2,4-D) in Eastern Nebraska. Epidemiology 1:349–356.

Study Questions

1. What is the main point of the article by Pimentel and his coauthors? Why do you think it might be correct? Why do you think it might be wrong? Explain and defend your answers.

2. What assumptions, if any, do Pimentel and his coauthors make about the chemical technology of pesticides? Do you think these assumptions are correct? Explain and defend your answers.

3. If society accepts the main conclusions of the article by Pimentel and colleagues, what consequences might follow? Would these consequences be desirable? Explain and defend your answers.

4. What are the ethical dangers and health risks associated with pesticides? Explain.

5. How can pesticides contribute to pest outbreaks? Explain.

6. Can pesticide use be uneconomical? Explain.

7. Why are honeybees and wild bees important to plants? How do pesticides affect bees? Explain.

8. Do pesticide benefits outweigh costs? Explain.

4.8

Ethical Issues in Human Genome Research

Thomas H. Murray

ABSTRACT *In addition to provocative questions about science policy, research on the human genome will generate important ethical questions in at least three categories. First, the possibility of greatly increased genetic information about individuals and populations will require choices to be made about what that information should be and about who should control the generation and dissemination of genetic information. Presymptomatic testing, carrier screening, workplace genetic screening, and testing by insurance companies pose significant ethical problems. Second, the burgeoning ability to manipulate human genotypes and phenotypes raises a number of important ethical questions. Third, increasing knowledge about genetic contributions to ethically and politically significant traits and behaviors will challenge our self-understanding and social institutions.—Murray, T. H. Ethical issues in human genome research. FASEB J. 5: 55–60; 1991.*

Key Words: ethics • gene therapy • insurance testing • genetic screening

Scientific research into human genetics has been a continuing source of intriguing, and at times formidable, ethical issues. The recent world-wide interest in a project to map and ultimately sequence the estimated three billion base pairs of the human genome has generated controversy over the effect such knowledge might have on us, as well as about the wisdom of investing so much research funding—an esti-

Originally published in *FASEB Journal*, vol. 5, no. 1 (January 1991), 55–60.

mated $3 billion over 15 years in the United States alone—on such a targeted effort.

The latter question is largely a matter of science policy rather than ethics. As an investment of scarce resources, the genome initiative may not be the wisest at this time. Respectable arguments have been made on both sides of the question. But even if it is not the wisest way to spend research resources, that would not make it unethical. The confusion may lie in the profligate manner in which epithets such as unethical or immoral are used to tarnish a person or enterprise we do not like.

It is difficult to define ethics comprehensively. In general, the study of ethics is understood as the attempt to understand good and bad, right and wrong. Some writers distinguish between ethics and morality, drawing on the association between morality and mores—the accepted ways of a people, both stemming from the same Latin root. For these authors, ethics connotes systematic study whereas morality suggests the more common, everyday, rough-and-ready practical effort to do good and avoid evil. Many contemporary scholars use ethics and morality interchangeably, except in special circumstances where the meaning of the two diverage. They will be used as synonyms in this article.

Field of Bioethics

A precise definition of ethics may be difficult to find, but there is considerable agreement on what is meant by bioethics. Bioethics is the study of ethical issues in medicine, health care, and the life sciences. As a formal field of study, bioethics is approximately 20 years old, although a few pioneering scholars were writing in the 1950s on issues that were later incorporated into bioethics. From its earliest stirrings, bioethics has paid considerable attention to human genetics, wrestling with such issues as prenatal genetic testing and abortion, genetic manipulation, and eugenics. In the past 5–10 years, the field of bioethics has grown rapidly, expanding the range of questions scholars ask, and even experiencing controversy over method.[1]

From the standpoint of bioethics, research on the human genome presents no completely novel ethical questions, at least for now. That is partly because of the nature of new ethical questions, which typically are variants of ethical questions that scholars and others have wrestled with before. This embeddedness of questions in experience with analogous questions means that we do not have to invent every response totally anew, but rather can draw on the history of scholarly analysis that has come before. The acceleration of knowledge about human genetics promised by genome research assures that the ethical questions presented will be plentiful and significant. They may be grouped into three categories: *1*) the possibility of greatly increased genetic information about individuals and populations; *2*) the manipulation of human genotypes and phenotypes; and *3*) challenges to our understanding of ourselves, individually and collectively.

Uses and Misuses of Genetic Information

Genome research will allow us to learn a great deal about the genetic makeup of individuals, especially their propensities toward diseases. In many instances, our powers to predict the likelihood (or in some cases certainty) of disease will come years or decades before any effective treatment for the disease is available. Nancy Wexler, known for her work on Huntington's disease, has dubbed this genetic prophecy.[2]

Huntington's Disease

Huntington's disease is inherited in an autosomal dominant manner. The first symptoms usually appear in the victims' 30s or 40s, after they have had an opportunity to have children. The disease is progressive and invariably fatal, causing uncontrolled movements and dementia. The gene has not yet been discovered, but markers close to it have been found, at a recombination frequency of as little as 1%, permitting an indirect test (linkage analysis) for the defective allele in families with adequate and informative DNA from multiple members. Presymptomatic testing has been made available for people at risk of Huntington's disease, but only a fraction of those eligible have chosen to be tested.[3]

As dramatically as any disease, presymptomatic testing for Huntington's disease illustrates the psychological dynamics and ethical difficulties likely to arise with testing for genetic predisposition.[4] Until recently, testing for Huntington's disease has taken place only under research protocols that included substantial counseling in advance of the actual genetic testing, as well as follow-up supportive counseling. Experience has shown that the process of explaining genetic risks is complex; understanding often comes only slowly and painfully, the psychological burdens of genetic disease can be massive, and not everyone wants to know his or her own risk. Using linkage analysis poses additional problems. It relies on DNA samples from affected and unaffected relatives of the person wishing to know. The tests may yield information about the risks of other family members who may not wish to know their own status; family members may not wish to participate, which yields conflict.

Huntington's disease is rare, with few if any new mutations. Individuals at risk are likely to know that they are. On the other hand, genetic testing for recessive diseases with high prevalence of carriers in identifiable populations is likely to affect many. Cystic fibrosis (CF) is one such example.

Cystic Fibrosis and Carrier Screening

The gene that causes CF has been identified, and the most frequent variant has been cloned.[5] It appears that CF can be caused by a variety of mutant alleles for the gene that sits on the long arm of chromosome 7. The most common mutant allele is known as ∆F508. It accounts for approximately 68% of known CF-inducing chromosomes. Other alleles can also cause CF, but are not yet well-characterized.

CF occurs in approximately 1 of every 2500 live births in white populations, which implies (for an autosomal recessive disease) a carrier frequency of roughly 1 in 25. The potential market in the U.S. alone for a screening test is enormous. The prospect of such widespread screening for CF carriers provokes ethical concerns.

First there are questions of accuracy. Would the test correctly identify those who carry a CF gene as well as those who do not? For a test that would pick up 75% of individuals carrying a CF gene, one-quarter of the carriers would be missed. For those planning to have a child, only 56% of couples at risk would have both partners identified as carriers.[6] A person whose test results are negative (with a test of the same sensitivity) would have his or her probability of being a carrier reduced from 1 in 25 to 1 in 99—better odds, but still not certain.

A second set of concerns is prompted by the history of population screening for such recessive disorders as sickle cell trait and phenylketone urea (PKU). Screening programs were often adopted without careful planning, and without provision for appropriate follow-up to assure that the programs' purposes would be achieved. Carrier screening programs sometimes have resulted in misunderstanding, stigmatization of carriers, and other problems. Properly designed screening programs would avoid such outcomes.

A crucial part of any acceptable screening program would be adequate explanatory and supportive counseling. Public misinformation about genetics is substantial. Many physicians do not have a good grasp of genetics.[7] One recent study estimated the counseling requirements of a population screening program for CF carriers. Using modest estimates of the counseling time needed by unaffected and affected couples, the authors concluded that 651,000 hours would be required. In light of the number of certified genetic counselors (450) and clinical geneticists (500) in the U.S., the supply appears grossly inadequate to meet the expected demand.[6]

At present, CF presents a volatile mixture pushing toward increased testing—enthusiasm for scientific breakthroughs, anxious prospective parents, and commercial enterprises eager to find profitable outlets for their biotechnologic skills. Just as we may hope that the development of CF carrier screening benefits from the history of other genetic screening efforts, so we may also hope that future screening programs for genetic diseases that will be discovered in genome research will benefit from our experience with CF screening. Professional bodies are beginning to speak out in favor of responsible positions. The American Society of Human Genetics, for example, has stated recently that routine carrier screening is not appropriate.[8]

Genetic testing for presymptomatic disease and carrier screening are only two of the uses to which knowledge gained in the Human Genome Initiative might be put. Prenatal screening for genetic disease is another; it stirs controversy because one of the choices upon finding that a fetus is afflicted with genetic disease is abortion. Much scholarship has been devoted to the ethics of abortion, with no resolution of the political battle.[9, 10]

These three forms of screening are done, at least purportedly, for the benefit of the individual being screened. New uses of genetic tests are evoking controversy. In these proposed uses, the test is being done not for the good of the person being tested, but rather for some organization—for example, a prospective employer or insurer.

Genetic Testing in the Workplace

In 1938, the geneticist J. B. S. Haldane observed that not all workers exposed to a particular occupational hazard became symptomatic. He postulated that the difference in response to toxic exposures was at least in part genetically determined. If we could assure that individuals who were genetically susceptible to the disease were steered to other occupations, Haldane reasoned, we could reduce the number of people who became ill.[11] Workplace genetic screening was justified by its consequences for public health. In Haldane's time, the technology to do genetic screening was not available. But his rationale for workplace genetic screening was revivified in the 1960s and 1970s with suggestions that new screening techniques might make it possible to put his idea into practice.[12]

The early proponents of workplace genetic screening did not foresee the political, economic, and ethical complexities their idea would later reveal. The first publicized case of such screening was by a U.S. corporation. A plant owned by this company began screening for sickle cell trait among its black workers. The company maintains that the screening was purely voluntary, initiated at the request of an organization of black employees, and that the results were not used in hiring or placement decisions. The journalist who reported the story claims otherwise.[13]

Debate about the ethics of workplace genetic testing has focused on the purposes for which such tests might be used. Four purposes have been identified: diagnosis, research, information, and exclusion.[14] Genetic tests, like many other procedures, can be helpful in diagnosis and research. Their use in those contexts are governed by the ethics of medical diagnosis and treatment and the ethics of research with human subjects. No novel moral dilemmas attend the use of genetic tests for those purposes.

Genetic tests may also be used before hiring or placing a particular worker to uncover a genetic susceptibility believed to put the individual at greater risk of occupational disease associated with hazards in that workplace. The crucial distinction here is between giving the tests voluntarily, with the information given to the worker who then decides whether to accept any additional risk, or compelling workers to take the tests and using the results to exclude workers who may have genetic susceptibilities.

A program of voluntary testing to inform workers of their risks is ethically defensible. In general, we believe that the individuals affected are the ones with the greatest ethical right to decide whether or not to accept risks. There are some risks that are so nearly certain to occur and cause grievous harm that we do limit choices for individuals. Except for such circumstances we usually allow competent adults to decide for themselves what risks are acceptable to them.

With presymptomatic genetic testing, a program of workplace genetic testing should include effective education and counseling to prevent misunderstanding of the results and to deal with emotional responses to learning that one has genetic risks. A carefully designed program of workplace genetic testing to inform workers, leaving the choice up to individual workers about whether to accept risks, does not confront any insurmountable ethical barriers. The same cannot be said for compelled genetic testing.

Compulsory genetic testing, leading to possible exclusion from jobs, has been controversial. Critics argue that it violates deeply held notions of individual authonomy and could be used in socially undesirable ways. For example, testing for sickle cell anemia followed by exclusion of those with the trait would effectively exclude one of every eight black job candidates in the U.S. Where workers are plentiful, employers might prefer to screen out susceptible workers rather than invest in equipment to reduce exposure to hazards. The prospect of such undesirable effects, along with our respect for individual liberty and choice, make compulsory workplace genetic testing ethically problematic.

The U.S. Congress Office of Technology Assessment (OTA) studied workplace genetic screening in the early 1980s.[15] The OTA report concluded that the present state of genetic testing and knowledge about genetic contributions to workplace disease did not justify screening employees for genetic susceptibility. A survey reported in the same study shows some use of genetic tests by U.S. employers, but ambiguities in the report make it impossible to know if the tests were used for legitimate and unproblematic purposes, such as diagnosing the illness of an employee or for the controversial purpose of excluding so-called hypersusceptibles.

From Reducing Illness to Reducing Cost

The most important movement in the ethics of workplace genetic testing has been away from the original vision of a public health measure to screening as a way of reducing illness-related costs with no effect on the overall incidence of disease.

Although there is still little evidence that workplace genetic screening could identify individuals with increased risks of workplace-related disease, there is increasing reason to believe that genetic screening for common diseases such as arterial disease (including coronary disease), stroke, and cancer may soon be possible. Other disabling diseases, including mental diseases such as depression and schizophrenia, might also become the targets of genetic screening. Employers might find such tests attractive ways to save money by screening prospective employees and hiring those without evidence of genetic susceptibilities to disease.

Employee illness, at least in the U.S., costs employers money. With health insurance costs becoming an increasingly larger proportion of employer's expenses, employers are looking for ways to diminish health-related costs including health insurance, disability insurance, the cost of lost productivity from ill workers, and the cost of training replacement workers for skilled positions.

The combination of increased employer concerns about the costs of illness and the prospect of genetic tests for predisposition to common, costly diseases are fertile ground for the use of such tests to screen workers.

The ethics of genetic screening for non-workplace-related disease differ in part from genetic screening for workplace-related disease. Working in the particular workplace does not put the worker at an increased risk; the disease to which the person may be susceptible is not related to the workplace. The information to be gained from such tests is not relevant to the individual's choice of whether to work in that environment, although it might be relevant to other life choices, such as diet, exercise, and other health-related behavior. Being denied a job because of predisposition to a non-workplace-related disease is as great an affront to an individual's liberty as a denial for predisposition to a workplace-related disease. But here there is no compensating reduction in risk to the individuals denied employment nor is there any public health benefit; those who will die from heart disease or cancer will still die from it—unemployed and possibly unemployable.

Employers are not the only ones likely to be interested in people's predisposition to common disabling and killing diseases. Companies selling life, disability, or health insurance are also interested in genetic tests.

Genetic Testing in Insurance

Insurance works on the principle of sharing risk. When the risk is equally uncertain to all, then all can be asked to contribute equally to the insurance pool. Not all individuals have the same risk of dying, for example. The older one is (beyond early childhood), the greater the risk of death. Older people are charged more for the same amount of life insurance than younger people. Some occupations are riskier than others. Test pilots pay more for life insurance than accountants. None of this seems unfair. The ethics of discriminating according to genetic predisposition, on the other hand, seems much more ethically complex.

Insurance companies have begun to think about what to do with tests for genetic predisposition to disease.[16] As the authors of an industry-sponsored report see it, two factors may force insurers to use genetic tests. First, once such tests become available in medical practice, individuals can be tested privately to learn whether they have enhanced risks of disease. People who learn they have higher risks are more likely to buy insurance and are more likely to buy larger amounts. In the insurance industry, this phenomenon is known as adverse selection—the tendency to purchase insurance when one expects to file a claim. Second, competition among insurance companies will tend to drive companies toward screening for predisposition. If one company begins using such tests, it would be able to offer lower rates to individuals who do not have genetic predisposition to disease and higher rates to those with such predispositions. Individuals offered the lower rates are more likely to purchase insurance from that company, whereas the ones with genetic preddisposition to disease will seek insurance from another company that does not do genetic testing. The latter company will either have to raise its rates (to avoid bankruptcy) or it will also have to use genetic tests.

Genetic Manipulation

Probably the most widely discussed and fear-inspiring use of genetic science is genetic manipulation, especially gene therapy. Other uses of genetics to manipulate human physique, physiology, or behavior may be equally significant and raise important ethical issues of their own.

Gene Therapy

Many human diseases are caused by abnormal alleles. The idea that the most effective way to correct such genetic deficiencies is by replacing, correcting, or supplementing the malfunctioning gene has been discussed for almost 2 decades, but has only recently become technically plausible.[17] The ethics of gene therapy have been discussed exhaustively. Indeed, there may be no other manifestation of modern genetics that has received such thorough ethical examination or that must pass through such extensive scientific and ethical review. In June of 1980 the general secretaries of the three largest religious bodies in the U.S. wrote to the President expressing concern about genetic manipulation. Partly in response to this concern, a U.S. Presidential Commission issued a report,[18] as did an office of the U.S. Congress.[19]

In the U.S., research on gene therapy funded by the National Institutes of Health must pass multiple levels of review, including a special subcommittee of the Recombinant DNA Advisory Committee. The subcommittee has issued several revisions of a "Points to Consider" document, outlining the many scientific and ethical considerations that must be satisfied before a research protocol could be approved.[20]

Scholarly discussion of the ethics of human gene therapy has reached a high level of sophistication, with both ethicists[21] and scientists[22] making substantive contributions. The most important question in the debate over the ethics of gene therapy is whether gene therapy is ethically distinctive from other forms of medical therapy. There are questions about such a novel and potentially risky therapy—for example, how to obtain genuine informed consent from patients in desperate circumstances. But these kinds of questions are not decisively different for gene therapy than for other highly innovative therapies for devastating diseases. The main ethically significant difference between gene therapy and other therapies is the potential for the changes wrought in an individual by gene therapy to be passed onto offspring of the treated person. Other therapies affect directly only the life of the particular individual being treated. The prospect of intentional genetic changes being passed on indefinitely struck most commentators as distinctive and significant for at least two reasons. First, the principle of informed consent that justifies many of the risks of medical experimentation cannot apply in any simple way to the descendants of subjects of gene therapy research. These persons were not yet conceived at the time of the research, and hence cannot be said to consent to risks to their own genome. Second, whatever harm might be caused by gene therapy might be magnified manyfold if it were passed onto future generations rather than dying out with the subjects of the current research.

Considerations such as these led to a crucial distinction in the ethics of gene therapy research: germ-line vs. somatic cell line gene therapy. Only genetic alterations in germ-line cells have the potential to be passed on to future persons. Genetic changes in a person's somatic cells will die with that individual. Somatic cell gene therapy, then, to the extent that it is reliably confined to somatic cells only, is ethically analogous to other novel medical therapies. The ethically distinctive element of gene therapy is only characteristic of germ-line manipulation.

Most scholars agree that research on somatic cell gene therapy is ethically defensible with the usual safeguards for research with human beings. There is less consensus about germ-line therapy. Some argue that there is no conclusive argument against germ-line gene therapy, and that the prospect of curing genetic disease not merely for the person under treatment but for all of his or her descendants could justify research. However, scholars and public bodies agree that it is better to exercise caution and not engage in any germ-line gene therapy research at this time. Further debate over the ethics of germ-line gene therapy is likely.

The distinctiveness of germ-line gene therapy can be questioned. Of course, other medical therapies or other human interventions can have genetic consequences for future generations. Mutagenic drugs, such as those often used in cancer treatment, and radiation therapy cause genetic changes, as do many environmental or occupational exposures.

Such changes, if they are incorporated stably into germ-line cells, can be passed on to future generations. Even standard, nonmutagenic therapies have genetic consequences for populations when they permit the survival and reproduction of individuals who would otherwise have perished or who did not have children. Future discussions will help to clarify the meaning and significance of the distinction between germ-line and somatic cell gene therapy.

Other Uses of Recombinant DNA in Humans

Ironically, the first approved experiment using recombinant DNA in humans was not gene therapy per se but involved labeling patients' tumor infiltrating lymphocyctes (TIL cells) with a gene that conferred resistance to neomycin. This biochemical flag allowed researchers to track the survival and migration of cells believed to fight the advanced melanoma from which these individuals suffered.[23] The first true approved gene therapy experiment is just now under way.[24] Genetic engineering techniques are also being developed that might enhance the cancer-fighting properties of cells such as TILs.

Modifying People with Biotechnology

Gene therapy may have attracted most of the public's attention, but other applications of new genetic knowledge that might be generated by the genome project could have more immediate and wider impact. The ability to clone a human gene, incorporate it into a microorganism or mammalian cell line, and produce and purify the gene product can have substantial ethical consequences. Human growth hormone (hGH) is now produced this way; before the introduction of biosynthetic hGH the only supply was from pituitaries of human cadavers. hGH was scarce; its only major therapeutic use was for hGH-deficient pituitary dwarfs.

Once biosynthetic hGH became available, the temptation to use it to make non-hGH-deficient children taller and to gain the competitive advantage that height gives in certain cultures (including the U.S.) became strong, and anecdotal reports of parents trying to obtain it for their normal children circulated. Thus far, misuse has been deterred by a restrictive policy that makes hGH available only through certain hospital pharmacies, even though it is not a drug whose sale is restricted by law. Recent research, however, indicating that hGH may mitigate or reverse some of the effects of aging in individuals who have low levels of endogenous hGH may presage more uses for hGH that will make the current method of control unworkable. Should hGH become widely available for indications other than short stature, we may see complex ethical problems as the knowledgeable, wealthy, and adventurous seek it for their children.[25]

In a similar manner, although for a more restricted group, another fruit of genetic research is having unintended social and ethical consequences. Biosynthetic erythropoietin (EPO) may offer enormous benefit to hundreds of thousands of people suffering from chronic anemia, such as those undergoing hemodialysis for treatment of end-stage renal disease. EPO induces the production of erythrocytes and can increase the oxygen-carrying capacity of the blood. Athletes engaged in sports requiring endurance such as distance running, skating, skiing, or cycling may gain a competitive advantage from EPO comparable to that obtained from blood doping—autologous or homologous blood infusion to raise the blood's ability to carry oxygen to rapidly metabolizing tissue. Using blood doping or performance-enhancing drugs to gain a competitive advantage is ethically suspect and prohibited by many sports authorities.[26] The organizations that oversee sports, such as the International Olympic Committee, have classified EPO as a performance-enhancing drug and are seeking ways to discourage its use among competitive athletes.

hGH and EPO are only among the first of the fruits of contemporary genetics that will pose difficult ethical problems as people seek to use them not to treat disease but for nontherapeutic purposes.

Challenges to Our Self-Understanding

There is a tendency in bioethics, as in other fields, to focus on the immediate, practical dilemmas posed by new developments. It may be that the most important challenges posed by the human genome project will not be the pragmatic concerns discussed thus far, but will have to do with the way we understand ourselves, our nature and significance, and our connections with our ancestors and descendants.

One manifestation of the genome initiative will be increased understanding of our genetic similarity with other species. Scientists have tried to estimate DNA homology between humans and other primates to clarify evolutionary relationships among primate species.[27] These early efforts to gauge genetic similarity seem to indicate that human DNA is strikingly similar to that of chimpanzees, at least as assessed by the relatively crude measures thus far available. Human DNA appears to be related to more distant species as well. Comparisons of synteny between mouse and human chromosomes show a substantial degree of correspondence, which suggests an evolutionary relationship, coincidentally giving researchers clues to where in the genome to search for a particular gene if its homolog has been found in mice (or humans).[28]

Evidence of our genetic relationship with other species is nothing more than additional confirmation of the theory of evolution. On the other hand, if more and more sophisticated tests of similarity continue to show how much like other species we are, genetically speaking, we may reevaluate not only our molecular but also our moral relationship with nonhuman forms of life.

Genetics as Explanation/Excuse?

One of the most common ethical and legal questions is whether and to what extent a person is responsible for something. We ask if the individual who committed a violent act was responsible for his behavior—morally blameworthy, legally culpable. We ask if the inventor was responsible for the new device (credit rather than blame being at stake here), the scientist for the discovery, the author for the idea, style, and specific words.

We also ask about responsibility for larger issues than discrete acts or inventions. We question whether the overweight person is responsible for his or her obesity, and by extension, for the illnesses to which obesity is a contributing factor, and perhaps for the expense of caring for those illnesses. We ask whether alcoholics are responsible for their addiction and for the consequences of acts committed while they were inebriated.

On an even broader level, we ask whether groups who do especially well (or especially poorly) in various social and economic realms experience the outcomes they do because of some intrinsic traits and abilities or because of lack of opportunity.

Genetics frequently provides an explanation or excuse for individual behaviors or traits, as well as for group differences. The genome project will enhance the tendency to give genetic explanations for individual and group differences in two ways. First, research will suggest genetic correlates for a wide range of human traits and behavior. Some may prove to be spurious, but others will withstand scrutiny. Whenever such a genetic correlate is suggested for some ethically, legally, or economically consequential outcome, there will be a temptation to explain it as fundamentally—that is exclusively or exhaustively—genetic, and hence outside the individual's responsibility or capacity to control. Researchers have recently reported a link between a dopamine receptor gene and alcoholism.[29] Others have warned against attributing too much significance to this finding.[30]

Second, in the initial rush of findings of genetic connections to important human traits and behavior, both scientists and the public may become too eager to embrace genetic explanations for a vast range of ethically significant phenomena. The phenomena in question may range from illness, including mental illness, addiction, occupational and environmental illness, to the educational and occupational attainments of different racial and ethnic groups.

History is rich with examples of scientific perspectives used inappropriately for political purposes. Frequently, genetic or earlier hereditarian theories have been the subject of such misuse. This first large-scale intelligence testing program in the U.S. was the Army alpha program, which was designed to screen incoming recruits according to their intellectual capacities. Certain groups did not fare well on the test, including immigrants from southern Europe. Because the results conveniently fit the beliefs of those overseeing the testing, they accepted the tests as valid measures of intrinsic ability rather than as the profoundly culture-bound instruments they were. They also overlooked the fact that many of these immigrants spoke English poorly or not at all; the test was in English.

In recent decades the debate has been over the heritability of intelligence and the reason for disparities in IQ test scores among racial and ethnic groups.[31] The political consequences of such debates are clear: if educational and economic inequalities are caused by social inequities, then society has an obligation to remedy those inequities; if, on the other hand, the inequalities in outcome are a function of (which in political debate can rapidly be translated into determined by) inherited differences, then we do not have to be so troubled, or worry about our own unjust actions.

The questions raised here also have important, in some instances profoundly important, legal ramifications that must be thoroughly explored. As George Annas has written:

> Although we are utterly unprepared to deal with issues of mandatory screening, confidentiality, privacy, and discrimination, we will likely tell ourselves that we have already dealt with them well. . . . (p. 20)[32]

The point here is not that we should ignore the influence of genetics on human affairs. Scientists should be the last people to abandon evidence in favor of sentimental, comforting illusion. Lucidity demands that we confront the truth as it is. Rather, we must learn not to overinterpret what we find. We must learn how to communicate effectively among ourselves and with the public about the limits of our knowledge.

Last, we must acknowledge the limited ethical and political significance of our genetic knowledge. When the founders of the United States wrote that all men (all people) are created equal, they did not mean this as a statement of biological fact, but as an ethical, legal, and political proclamation: before the collectivity of the state, all persons must be regarded as equal—each due equal respect, equal liberty, and equal protection, among other fundamental rights. The sciences of inequality, with genetics at the forefront, will force us to reinterpret what equal treatment and equal regard mean in an enormous range of contexts. But they need not threaten the ethical core of that commitment.

References

1. Jonsen, A. R., and Toulmin, S. (1988) *The Abuse of Casuistry: A History of Moral Reasoning.* University of California, Berkeley.
2. Wexler, N. S. (1989) The oracle of DNA. In *Molecular Genetics in Diseases of Brain, Nerve, and Muscle* (Rowland, L. P., Wood, D. S., Schon, E. S., and Di-Mauro, S., eds) Oxford University Press, New York.
3. Jenkins, J. B., and Conneally, P. M. (1989) The paradigm of Huntington disease. *Am. J. Hum. Genet.* 45, 169–175.
4. Bird, S. J. (1985) Presymptomatic testing for Huntington's disease. *JAMA* 253, 3286–3291.
5. Rommens, J. M., Ianuzzi, M. C., Kerem, B., Drumm, M. L., Melmer, G., Dean, M., Rozmahel, R., Cole, J. L., Kennedy, D., Hidaka, N., Zsiga, M., Buchwald, M., Riordan, J. R., Tsui, L.-C., and Collins, F. S. (1989) Identification of the cystic fibrosis gene: chromosome walking and jumping. *Science* 245, 1059–1065.
6. Wilfond, B. S., and Fost, N. (1990) The cystic fibrosis gene: medical and social implications for heterozygote detection. *JAMA* 263, 2777–2783.
7. Holtzman, N. A. (1978) Rare diseases, common problems: recognition and management. *Pediatrics* 62, 1056–1060.
8. Caskey, C. T., Kaback, M. M., and Beaudet, A. L. (1990) The American Society of Human Genetics statement on cystic fibrosis. *Am. J. Hum. Genet.* 46, 393.
9. Callahan, D. (1970) *Abortion: Law, Choice and Morality.* Macmillan, New York.
10. Luker, K. (1984) *Abortion and the Politics of Motherhood.* University of California, Berkeley.
11. Haldane, J. B. S. (1938) *Heredity and Politics.* Allen and Unwin, London.
12. Stokinger, H. E., and Mountain, J. T. (1963) Tests for hypersusceptibility to hemolytic chemicals. *Arch. Environ. Hlth.* 6, 57–64.
13. Severo, R. (1980) Screening of blacks by DuPont sharpens debate on genetic tests. *New York Times.* 4 February, p. 1.
14. Murray, T. H. (1983) Warning: screening workers for genetic risk. *Hastings Center Rep.* 13, (Feb.) 5–8.
15. United States Congress Office of Technology Assessment. (1983) *The role*

of genetic testing in the prevention of occupational disease. U.S. Government Printing Office, Washington, D.C.

16. Genetic Testing Committee to the Medical Section of the American Council of Life Insurance (1989) *The potential role of genetic testing in risk classification.* American Council of Life Insurance, Washington, D.C.

17. Anderson, W. F. (1984) Prospects for human gene therapy. *Science* 226, 401–409.

18. President's Commission for the Study of Ethical Problems in Medicine and Biomedical and Behavioral Research. (1982) *Splicing life: the social and ethical issues of genetic engineering with human beings.* U.S. Government Printing Office, Washington, D.C.

19. U.S. Congress Office of Technology Assessment. (1984) *Human gene therapy—a background paper.* Washington, D.C. OTA-BP-BA-32.

20. Human gene therapy subcommittee, NIH recombinant DNA advisory committee. (1989) Points to consider in the design and submission of protocols for the transfer of recombinant DNA into human subjects. National Institutes of Health, Bethesda, Maryland.

21. Walters, L. (1986) The ethics of human gene therapy. *Nature (London)* 320, 225–227.

22. Anderson, W. F. (1985) Human gene therapy: scientific and ethical considerations. *J. Med. Philos.* 10, 275–291.

23. Culliton, B. J. (1989) Gene test begins. *Science* 244, 913.

24. Marwick, C. Preliminary results may open door to gene therapy just a bit wider. *JAMA* 262, 1909.

25. Murray, T. H. (1987) The growing danger from gene-spliced hormones. *Discover* 8, 88–92.

26. Murray, T. H. (1983) The coercive power of drugs in sport. *Hastings Center Rep.* (Aug) 13, 24–30.

27. Koop, B. F., Goodman, M., Xu, P., Chan, K., and Slightom, J. L. (1986) Primate n-globin DNA sequences and man's place among the great apes. *Nature (London)* 319, 234–238.

28. McKusick, V. A. (1990) *Mendelian Inheritance in Man,* 8th ed. Johns Hopkins University Press, Baltimore.

29. Blum, K., Noble, E. P., Sheridan, P. J., Montgomery, A., Ritchie, T., Jagadeeswaran, P., Nogami, H., Briggs, A. H., and Cohn, J. B. (1990) Allelic association of human dopamine D_2 receptor gene in alcoholism. *JAMA* 263, 2055–2060.

30. Gordis, E., Tabakoff, B., Gold, D., and Berg, K. (1990) Finding the gene(s) for alcoholism. *JAMA* 263, 2094–2095.

31. Jensen, A. R. (1973) *Educability and Group Differences.* Harper and Row, New York.

32. Annas, G. (1989) Who's afraid of the human genome? *Hastings Center Rep.* 19, (July/August) 19–21.

Study Questions

1. What is the main point of Murray's article? Why do you think it might be correct? Why do you think it might be wrong? Explain and defend your answers.

2. What assumptions does Murray make about technology? about ethics? Do you think these assumptions are correct? Explain and defend your answers.

3. If society accepts Murray's main conclusions, what consequences might follow? Would these consequences be desirable? Explain and defend your answers.

4. What are the ethical dangers associated with genetic screening? Explain.

5. What are the ethical dangers associated with manipulation of human genotypes and phenotypes? Explain.

6. What is the point of Murray's examples of Huntington's disease and cystic fibrosis?

7. Does Murray believe that genetics can be an excuse for particular behavior? Explain.

4.9

A "Transgenic Dinner"? Social and Ethical Issues in Biotechnology and Agriculture

Laura Westra

1. Introduction

The December 1991 issue of *The Gene Exchange* (Vol. 2, No. 4, National Wildlife Federation, ed. M. Mellon, J.D., Director Biosafety Regulations), showed on its front page a "Transgenic Dinner," listing commonly used foods in their modified forms, comprising genetically engineered items as well as transgenics (see Fig. 1).

Biotechnology refers to genetic engineering, or recombinant DNA (R-DNA) technology; its goal is to transfer genetic traits between entirely different organisms (D. Pimentel, *et al.*, 1989). When the traits are transferred between different species (for instance, from fish, mammals, or viruses to plants), the modified organisms are termed "transgenic."

Scientific opinion is strongly divided on both risks and benefits of these novel practices, particularly because there has been no time, as yet, to amass enough evidence to enable the scientific community to predict consequences with some degree of accuracy, and to assess them. David Pimentel says:

> Nevertheless, some releases of genetically engineered organisms may have sobering ecological, social and economic effects. (D. Pimentel, *et al.*, 1989)

Originally published in *Journal of Social Philosophy*, Vol. 24 No. 3, Winter 1993: 215–232. Copyright © 1993 *Journal of Social Philosophy*.

Figure 1

A Dinner of Transgenic Foods*

Appetizers
Spiced Potatoes with Waxmoth genes
Juice of Tomatoes with Flounder gene

Entree
Blackened Catfish with Trout gene
Scalloped Potatoes with Chicken gene
Cornbread with Firefly gene

Dessert
Rice Pudding with Pea gene

Beverage
Milk from Bovine Growth Hormone (BGH)-
Supplemented Cows

*Federal permits for environmental release are pending or have been granted for all the transgenic plants and animals included on the menu. BGH is under consideration for approval as a veterinary drug.

This paper addresses some of these questions and effects from the standpoint of ethics, and argues for controls, caution and a general slowdown of the present agricultural/technological "race" on both moral and prudential grounds. Speaking of the absolute necessity of accepting limits in our lifestyle choices, with particular reference to chemical pollutants, the authors of *Beyond The Limits* say:

> The most intractable hazardous wastes are human-synthesized chemicals. Since they have never before existed on the planet, no organisms have evolved to break them down and render them helpless. (D. Meadows, *et al.*, 1992)

The dangers and the uncertainty endemic to these technologies must be faced from the moral standpoint, on both anthropocentric and non-anthropocentric grounds. In addition, to the possible harms and uncertainties introduced by present chemically based argicultural practices, I argue that there are additional dangers and uncertainties arising from these newer technologies, and that the present approach of both corporate bodies and government institutions is insufficient and inappropriate to protect the public.

The introduction of these mutations raises a number of serious questions, adding totally new dimensions to the already abundant literature existing on risk assessments and the hazards of the "technoscientific enterprise" (K. Shrader-Frechette, 1991; E. Agazzi, 1992). What is particularly disturbing is the fact that this "marvelous" new adventure, that is, biotechnology, is presently endorsed by the United States government, which finds itself at the forefront of a brand new field and is bent on encouraging the industry to forge ahead and market biotechnology aggressively worldwide. At a time when recession is prevalent in North America, and when public distrust and disenchantment with technology and science in the food industry is high, biotechnology promotes itself as the "better" alternative. Agricultural biotechnologies are generally researched and strongly promoted by multinational corporations, chemical and pharmaceutical conglomerates, primarily.

Chemical companies have acquired a very unsavory reputation. Accidents such as the one at Bhopal, and episodes such as the ALAR scare contribute to the public fear and distrust, as well as to a growing anger at the efforts of these companies to marginalize and trivialize the so-called "ignorance" and "irrationality" displayed by the people attempting to voice their concerns. What could be better then, than to represent the same chemical companies in a new guise, now bearing "gifts" of non-chemical, genetically engineered "solutions" to the agricultural problems that (to some extent) they themselves created? In other words, it is the "high tech" approach to agriculture, the high input ways developed in North Western Countries and introduced as the "better way" to less developed countries (hereafter known as LDCs) that have contributed significantly to soil erosion, increased use of pesticides with its accompaniment, pesticide resistant insects, and contaminated food. These conditions point to reduced productivity in the near future, and foster poor long-term prospects as well.

This benign "new" version of the same corporate bodies, when closely analyzed, also reveals a host of new and old moral problems. The revamped "green" look of these companies does not represent an effective change: the mild sheep reveal themselves as nothing but the same hungry wolves, albeit in (Green) sheep's clothing. Some of the moral problems raised by the new discoveries may be addressed from standard traditional moral theories. Questions of human rights to life, to information, knowledge, and consent, to freedom from harm, equal treatment and equal protection under the law, are all fruitful ways of addressing many of these difficulties (K. Shrader-Frechette, 1991). Yet some other issues may be obscured by attempting to deal with such

new quandaries through traditional principles designed long before
these difficulties existed.

Hence one of the recurrent themes in environmental thought is the
requirement for a changed mindset, one that is both holistic and non-
anthropocentric (M. Soulé, 1991; A. Naess, 1991; W. Jackson, 1991;
L. Westra, 1989). My own proposed "first principle of morality," the
"principle of integrity" (PI), suggests a categorical imperative of non-
interference with the integrity of life-support systems, superseding all
other (weak) individual rights, while supporting strong (life-protec-
tion) individual rights instead. From that standpoint, other problematic areas in biotechnologies can be isolated, and strong reasons can be
identified to differentiate the latter from other hazardous agricultural
practices. Further, my work distinguishes between Structural Integrity
(I_1) and Functional Integrity (I_2) in ecosystems. It is clear that, while the
latter may or may not be affected by the introduction of alien species,
and scientists may disagree on how much disruption their introduction
may cause, the former is indisputably affected.

These reasons are:

1. the assumption of moral inconsiderability of all other than humankind, hence the non-relevance of their physical integrity and
   the structural integrity of the ecosystem of which they are a part;
2. the tacit assumption that bioengineered species represent only a
   novelty in degree rather than in kind, hence, that their introduction into ecosystems is not especially problematic; and
3. the assumption that bioengineered organisms are essentially
   analogous to other agricultural technological "aids," hence, whatever the range of problems caused by the latter, it will also include those engendered by the former.

A brief discussion of these three areas of concern will indicate that
all three assumptions are false, thus, that biotechnological "advances"
cannot simply be added to the long list of questionable practices in
agriculture to which answers have not, as yet, been given (K. Shrader-
Frechette, 1991). The first point, or the possibility of moral considerability for the "physical integrity" of entities beyond humankind or, at
most, some of the higher mammals whose considerability has been
defended by many (T. Regan, 1983; P. Singer, 1990; S. Sapontzis, 1991;
R. Rodd, 1990; VandeVeer, 1987), represents a coherent position from
the standpoint of "integrity." Other than the above-mentioned work
by Rodd (1990), to my knowledge no one takes a position in defense
of the moral considerability of insects, fish, domestic fowl, and other

organisms primarily affected by biotechnology. From my own standpoint, aside from clear cases of self-defense (e.g., malaria-bearing mosquitoes, viruses, animal heartworms, and the like), *all* forms of life are owed respect, at least *prima facie* (L. Westra, 1989). I will return to this discussion in section three of this paper, after a brief outline of the profession and the products of the biotechnology industry, and today's "Dr. Frankensteins," in section two.

2. Today's Dr. Frankensteins and Non-Human Bioengineered Species: The Professionals and the Issues

Who are today's Dr. Frankensteins? For the most part, they are earnest young men and women, employed by large multinational conglomerates, Monsanto, Ciba-Geigy, Upjohn, and even USDA. Bioengineering is a profession like many others and the "products" of their professional activities are only "monstrous" from some specific perspectives. On the face of it, the final products consist of "new' engineered plants or domestic animals; but these often look as they always did, even though they are different in some significant way. A plant may have been bred with a virus so that the new creation is both "animal" and "vegetal" (an "aniplant" perhaps?), and it possesses traits the previous plant did not possess. These traits are desirable from the standpoint of economics and production: on the plus side, they increase yield, hence they promise to feed more people more efficiently. On the minus side, the new plant has now evolved into one with an inbred resistance to a specific herbicide. The result is that the bioengineered species—heralded as a step forward for environmental safety, and a step away from chemicals—represents *instead* a permanent, inescapable link to chemicals, as the new creation has a built-in tolerance for a specific herbicide.

Hence, the corporation gains twice: once, when it sells the biotechnology, and second, when it ensures thereby "permanent addiction" to its own patented herbicide. In contrast, the people and the environment, correspondingly, lose twice: once, because the proposed "safe" product ultimately is *not,* second, when other possibly safer, organic and sustainable choices are preempted instead. The October 1991 issue of *The Gene Exchange* (Vol. 2 [3]), lists 54 genetically engineered plants (USDA nos. 116–170), approved from February 5, 1991 to August 7, 1991 (four of these are still under review at this time). Nine of these exhibit "tolerance to herbicides" ranging from Glyphosate, chlorsulfuron, bromoxynil, glutosinate and 2, 4-D, either as their sole "engineered trait" or as one of a cluster of such traits. In most cases, the

USDA permission granting only "assumes" that approved tests have been conducted. Attempts have been made to regulate things in various locations. For instance, in 1989 "North Carolina enacted the first comprehensive state biotechnology law." Although it appeared to be a great step forward, and to provide for information to and input from the public, in essence its intent was "to protect and promote the industry." Some of its problems were: a) its ten-member review board only had one public interest representative; b) public participation was "at the discretion of the state regulators" and public hearings were permitted only when NCDA would determine that significant public interest and justification was present; and c) the Commissioner was the only one to determine if and when the public would be allowed to review CBI (confidential business information).

Where animals are involved, the new species is often "new" in only one significant aspect. BGH (or BST as it is sometimes known) is a hormone. Although scientists have not yet been able "to fully determine how BGH works at the cellular level, they know that the hormone stimulated bone and muscle growth, metabolic rate and food intake, and milk production. Monsanto, American Cyanamid, Eli Lilly, and Upjohn are producing the hormone, hoping to sell it to farmers who would inject their cows for increased milk productivity.

What, if any, are the differences between people working in the field of "monster" ("unnatural" or "man-made") creation today and in Dr. Frankenstein's time? The first and most significant difference is the relation between their respective work and their perceived "responsibility to their own species." Today's plant pathologists or other biotech specialists are motivated, *prima facie* at least, by the desire to "feed the hungry" through increased productivity. Anecdotally, often the response elicited by questions from a concerned environmentalist about ecological safety is "do you people want us to allow millions to starve?" Clearly, prudential considerations are the easiest to voice in that setting, as concerns for the "rights" or the integrity of animals and plants would be dismissed out of hand.

The implied dichotomy between sustainability and ecology does not exist, and I have addressed this question in detail elsewhere in regard to LDCs (L. Westra, K. Bowen, and B. Behe, 1991). In essence, the authors of *FOOD 2000* say that all attempts to separate environmental concern from agricultural progress can only give rise to "invidious distinctions between environment and development":

> These views are based on a misunderstanding of the issues. If a country
> wishes to pay attention to the economic costs (and benefits) of agricultural

production—and all countries do—it must deal with the environmental costs. (*FOOD 2000*, 1987)

Hence what is required is agricultural practices and regulations that are no longer "ecologically blind in their conception, funding and implementation" (*FOOD 2000*, 1987; cp. L. Westra, K. Bowen, and B. Behe, 1991). Nevertheless, even though ecological concern of this sort, let alone concern for animal ethics, is not viewed as a priority by biotech "creators," the human concern remains paramount: theirs is often an anthropocentric position of responsibility towards the human species, not a stance based on selfishness. The conflict faced by the original Dr. Frankenstein no longer exists. Rightly or wrongly his latter-day counterparts are convinced that what they do represents "progress" and is "for the betterment of mankind."

Unfortunately, this belief is shared by many of these professionals and is problematic at best. Thus, several areas of moral difficulty persist from the time of the original Frankenstein. Some of these are: a) the total lack of moral questioning of their own activities; b) the way the scientific enterprise is viewed; and most important, c) the presence of the same logical fallacy in their respective reasoning. The error of perceiving scientific questions as "purely technical" (a), rather than involving moral issues and values, persists today, as does the high and almost unassailable status of the "scientist" or "expert" (b). Enough has been written on both problems by such well-known thinkers as Kristin Shrader-Frechette, Mark Sagoff, Helen Longino and others that there is no need to repeat their arguments here. As far as the last item (c) is concerned, the logical confusion it indicates is still rampant today. Few acknowledge the vast gulf separating theory and practice, thus they confuse the right to free thought and belief with that of practice; hence, the right of free scientific investigation with that of action, or technological application (E. Agazzi, 1992). On the contrary, a strong case can be made for strong moral controls for many aspects of science as well (E. Agazzi, 1992).

At any rate, at least the right to activity (including technological productivity) is not privileged: it must be, as all activity, limited by the rights of others to safety and freedom from harm. Lately, a further dimension has been added to this obvious tenet. The definition of "moral patients" or the targets of such moral considerability is by no means as clearly defined as it might have been in Frankenstein's time. Hence, those who recommend caution and care are not improperly attempting to limit the soaring of the human (scientific) spirit: they are simply, minimally, recommending respect for others, and more will be said of this in the next section.

Biotechnology poses moral problems both at the theoretical and at the practical level. The theoretical problems need to be clearly worked out in a timely manner. Economic interests continue to expand the use of biotechnologies without allowing the public any input in the decisions to do so, or clearly understanding the moral and social implications and possible consequences. One is reminded of the nuclear industry, and its rush to license facilities without any clear understanding of how safe storage of radwaste material might be effected (K. Schrader-Frechette, 1983).

Nor is the problem only one of concepts, attitudes, and world-views: the urgent problem is one of establishing guidelines and rules to implement protection from harm. Equally important perhaps, is to ensure that guidelines are provided and extreme caution practiced in the case of continued uncertainty. Even if the present lack of clear conclusions about the consequences of the employment of these technologies in effect works "both ways," that is, it neither proves harm nor disproves it, the impossibility of using factual scientific evidence as a basis for policy remains. Biotechnologies have been heralded by their manufacturers as providing salvation from possible problems, yet they are far from problem-free. Dr. Margaret Mellon, J.D. (Director, Biotechnology Regulations, World-Wide Wildlife Federation), has listed no less than eight possible clusters of problems: 1. they may not be safely used in areas where living organisms (human and non-human) are malnourished or unhealthy; 2. the possible "combined impact of the rapid genetic improvement of the world's major commodity crops threatens to deprive millions in the Third World of their livelihoods," while advantaging multinationals and larger producers; 3. although these are presented "as ending the era of chemical agriculture," they will shackle the world to herbicides for the foreseeable future (as we saw earlier); 4. the promised increased yields may, paradoxically, increase the misery of millions of poor farmers; and finally, and perhaps worst of all: 5. "biotechnology diverts all of us from the better path of sustainable agriculture." Margaret Mellon says:

> The World Resources Institute . . . just released a study that concluded that if the costs of agriculture are fully accounted—that is, if the costs of water pollution and soil erosion were put into the equation—low input, alternative agricultural systems perform better than conventional agriculture, and the former, rather than biotechnology, is the path that will eventually feed people without destroying the environment. (M. Mellon, 1991; cp. D. Pimentel, *et al.*, 1989; D. Pimentel, *et al.*, 1992)

On the question of health, three more problems arise: 6. biotechnology poses environmental risks—traits can be transferred to crops' "weedy relatives," especially in LDCs; 7. once genes are put into food crops for pesticides, they cannot be washed off, and may pose risks to humans, as yet not fully comprehended (e.g., BT toxin); 8. risks exist from these technologies to wildlife, particularly fish.

Hence, it appears to be necessary to attempt a philosophical and ethical analysis of the issues involved, from a variety of standpoints, in the hope of capturing most of the problematic facets of these new technologies.

3. Ethical and Social Issues in Biotechnology and Agriculture

In the fact of so many serious problems, a thorough and comprehensive approach is needed: this approach must consider both traditional ethics positions and newer, more inclusive ones. The first and major consideration is that of the possibility of harm, hence, the evaluation of risk. Risk evaluation is not a univocal concept; as the recent analysis of Kristin Shrader-Frechette amply shows, it is neither a "social construct" nor a totally "value-free" objective "given" (1991). "Constructs don't kill people," and the possibility of value-free science (including risk evaluations) has been disproven not only in her work, but in that of many others (H. Longino, 1990; L. Laudan, 1984; M. Scriven, 1992; E. Agazzi, 1992).

If neither cultural, relativistic constructs nor technical absolute will yield clear-cut scientific answers to questions of risk-assessment and evaluation, then the views of lay people, that is, of "potential victims" (K. Shrader-Frechette, 1991), should be accorded far more respect than they are at present. Dismissing the public as "ignorant" or "irrational," and discounting their concerns as "not serious" but simply matters to be handled through public relations and advertising persuasion, so that the "experts," whether corporate or professional, can continue to pursue their economic interests, will then be seen as improper from the moral standpoint (M. Sagoff, 1988). Recent literature on these and related topics shows that even in North America the situation is, and has been for some time, far more complicated and harder to assess than the simplistic answers the public has been fed would indicate. Recent examples include a symposium sponsored by Cornell and Guelph Universities on "New Directions in Pesticide Use: Environment, Economics and Ethics" (1991) and the latest issue of the *Journal of Agricultural Ethics* (Vol. 4, No. 2, 1991) devoted entirely to opposing viewpoints on "agricultural biotechnology."

Further, much needs to be said about the most vulnerable, the poor, and the Third World, from the standpoint of traditional theories. A point worth considering is that the only scientific "certainty" in regard to possible predictions about the effects of toxic substances (particularly in regard to their interaction) is that no certainty is available. Scientists disagree on the appropriate methodology for testing as well as on population sample size and composition (K. Shrader-Frechette, 1991). More dangerous still, all exposure "standards" are based on the "average" person, normally taken to be male and adult, thus excluding critical and large segments of the population—e.g., women and children (L. Westra, *APA Newsletter*, 1991).

Under these circumstances, the use of biotechnologies in food, with neither labeling nor available public information (*The Gene Exchange*, 1992), might place all citizens in the position of uninformed, thus unconsenting subjects of experimentation. If medical doctors wish to experiment on human subjects for the "advancement of science," the "good of mankind," or any other motive, they are strictly limited and regulated according to both the Code of Nuremberg, and the Code of Helsinki. It seems appropriate that if scientific uncertainties make use of many of the technologies discussed, which are by all intents and purposes "experimental" (and all of us unwilling subjects of such experimentation), then tight codes and regulations are long overdue.

As far as the question of "risk" is concerned, K. Shrader-Frechette deals with technology transfer of United States pesticides to LDCs (1991). If biotechnologies are simply yet another form of regular technological agricultural aid, then all critiques of the latter will also be applicable to the former. Hence, the "joint" problems will be addressed before those unique to biotechnology are presented. Some of the factual points to emerge from the work mentioned are: "29% of all United States pesticide exports are products that are banned (20%), or not registered (9%) for use in the United States":

> This means that about 49,000 persons, many in developing nations, die annually from pesticides. One person is poisoned by pesticides every minute in developing countries. (K. Shrader-Frechette, 1991; R. Repetto, 1985)

Arguments advanced to discount such information as morally insignificant are (a) the "Social-Progress argument," (b) the "Countervailing Benefits argument," (c) the "Consent argument," and (d) the "Reasonable Possibility argument." Yet the claim that such hazardous "aid" constitutes a contribution to (a) the social progress of LDCs or provides them with (b) countervailing benefits is not sufficient if the price paid may be the blood of misinformed, non-consenting, innocent citizens. People have a right to not be seriously harmed, wherever they are, and there is something morally repellent about offering this argument in various ways (for instance, "we can't let people starve," etc.):

> The argument is that a bloody loaf of bread is sometimes better than no loaf at all, . . . and that food riddled with banned pesticides is better than no food at all. (K. Shrader-Frechette, 1991)

Such a position cannot be supported either on utilitarian grounds or from the standpoint of justice. From a practical standpoint, neither (a) nor (b) is a good argument: those who are primarily exposed to pesticides are—typically—not those who benefit the most. "Between 50 and 70 percent of pesticides used for underdeveloped countries are applied to crops destined for export" (K. Shrader-Frechette, 1991). Moreover, even the increased affluence in the country (an "indirect" benefit to the hungry) is most often used for "luxury consumer goods, urban industrialization, tourist facilities and office buildings," hence, neither directly nor indirectly benefiting the poor and hungry (K. Shrader-Frechette, 1991). Similarly, the "consent" argument (c) is based on factually incorrect premises. Most often, the poor in LDCs are *not* informed of the risks posed by the technologies they receive, nor is the "consent" of hungry, desperate people without other alternatives or choices "consent" in the proper sense of the term. Finally, it is not a "reasonable possibility" (d) that it is "impossible to prevent" the transfer of hazardous technologies (K. Shrader-Frechette, 1991).

These are some of the problems related to pesticide use in the Third World. Much of what is written in regard to that aspect of the problem can also be applied to problems at home. We in Western countries most of the time know very little, and *mutatis mutandis*, questions of unaccepted risk, lack of consent, lack of appropriate "countervailing benefits," and doubtful "progress" would apply here as well. Neither human rights (strong rights as in Alan Gewirth, 1982), nor our "good," nor justice of distribution of benefits and burdens, are considered— much less served—by the status quo.

Some believe that biotechnologies are simply another innovation in a long line of agricultural technologies. David Kline, addressing those who feel it is wrong to develop products that "farmers are forced to adopt" (in the sense that not to do so would be irrational from an economic viewpoint), says:

> . . . we do not have a very plausible moral principle in the case of those who oppose the production of bGH. The assertion that it is wrong to develop products that farmers are forced to adopt implies that it is wrong

to adopt tractors and fences, clearly an absurd implication. (D. Kline, 1991)

This is a clear case of the "slippery slope" fallacy. It is also a "faulty analogy," because it assumes what has not been proven, namely, that biotechnologies are essentially the same in all relevant respects as all other technologies. Further, Kline also assumes that "technologies are bodies of knowledge that produce results that are quite useful. Hence technologies are typically desirable" (D. Kline, 1991). The last point is by no means obviously true when stated in such a sweeping, unqualified manner (D. Meadows, *et al.*, 1992).

Thus it does not appear that Kline's position can be defended: as we will see, because biotechnologies raise problems over and above those brought about by other technologies, the two sorts of agricultural practices cannot be collapsed under one heading. Were that the case, enough could be said about and against them; the novel dimensions simply add to what appears to be overwhelming evidence in favor of regulation and restraint, rather than support for biotechnology.

Let us return now to the aspects that distinguish these from other practices. The *first* was the *assumption* of moral inconsiderability of all that is not human, and the non-relevance of their physical integrity (as well as that of the ecosystem they inhabit). One of the examples of this position comes from the quote from Kline, where he refers to "bGH." The way bovine growth hormone is symbolized is symptomatic of the moral difficulty alluded to here: why are *growth* and *hormone* emphasized, while the live animal—whose physical integrity is being tampered with—is not worthy of the same "honor"? This clearly ignores the abundant, vigorous literature that supports and defends animal concern (T. Regan, 1983; P. Singer, 1990; S. Sapontzis, 1989a; Sapontzis, 1991; VandeVeer, 1987). It is not necessary at this time to review all the authors who have taken a strong position on the question of the moral considerability of non-human animals, or to detail their arguments.

Such arguments can be and have been drawn from traditional theories such as Kantianism, Utilitarianism, or some form of Justice theory. Further, the assumption of non-considerability takes no account of either the philosophical or the scientific relevance of biodiversity and of the ecosystem approach in ecology and public policy. For instance, from my own approach, the *structural* integrity of ecosystems is also morally considerable (L. Westra, 1993). Thus, from the standpoint of traditional morality as recently extended to the non-human world as well as from the standpoint of recent thought (A. Leopold, 1969; A. Naess, 1990; L. Westra, 1989), the assumption of non-considerability of all beyond humankind cannot be made without a serious response to these arguments. Finally, the scientific emphasis on biodiversity and preservation of species is also not considered by the introduction of "exotics" into ecosystems (point 2. to be discussed). It is worth noting that even scientists, policy-makers, and legislators (M. Soulé, 1991; GLWQA, 1978; IJC, 1990), increasingly call for a non-anthropocentric approach, a new "mindset" and a reduction in the status of humans from "managers" and controllers, to simply members of the biota. These changes in worldview are advocated because of the failures of previous, traditional approaches to morality (particularly individualistic, contractarian theories), to re-direct moral thinking in an environmentally sound manner.

The *second* assumption ignores the possibility of long-term undesirable side-effects for all. It is not the *known* and manageable hazards that loom largest in the public mind, but the uncertainty arising out of practices and products that have no known "precedent" or history of previous use, or results on which to base present risk assessments (K. Shrader-Frechette, 1991).

As uncertainty is—and is perceived to be—the gravest problem in the use of all agricultural technologies, immediately a difference *in kind* emerges between biotechnology and previous practices. It is becoming increasingly evident that the introduction of "exotics" into an ecosystem is extremely problematic (*Science*, 1991). Many years later, even species that were introduced voluntarily for some human purpose, now create havoc. From the zebra mussels in the Great Lakes, to kudzu, or the California "superbugs," displaced organisms, which have been geographically removed from their previous habitat, create vast unforeseen problems in their new "homes." Hence, it is not plausible to simply assume that the introduction of completely new life-forms—organisms with *no previous habitat*, who are true "aliens" in their present environment—is going to have either a benign impact or none (D. Pimentel, *et al.*, 1989).

The structural integrity of an ecosystem is at stake (L. Westra, 1992). Hence, the effect of the introduction of bioengineered species is not only *practically* problematic, as it brings a high degree of uncertainty and the possibility of increased hazardous outcomes (on the basis of the "exotics" performance), but it is also theoretically problematic. It tampers with *both* the "macro" integrity of the ecosystem, and the *micro* integrity of at least two individual organisms. If one considers seriously the abundant literature reflecting research on the effects of the introduction of "exotic" species into ecosystems, it is clear that biotechnologies pose threats and produce uncertainties *quite* different from (and additional to) those arising from chemically based technologies and—pace Kline—*a fortiori* from such technologies as "fences and tractors."

One is led instead to ask the question Wes Jackson raises, "How long will it take the biotechnologists to come up with the equivalent of the ozone hole?" (1991). In fact, just as the doubtful promises of "better living through chemistry" were not kept, the "promises of biotechnology" appear equally unfounded. Jackson suggests that a Cartesian frame of mind, rational as "presuming to know in a vacuum," assumes the capacity of science or biotechnology to predict accurately, and thus may lead to catastrophic results *in addition* to those "achieved" by chemistry in the last 100 years (1991). A similar indictment of technology in general can be found in the work of Robert McGinn (1991): his discussion of "technological maximality" parallels that of Jackson's "progressive foundationalism" [*sic*], which he deems far worse than the "religious" variety.

The *third* assumption is that biotechnology introduces specific problems related to the (micro) integrity of the modified organisms. Neither the corporate bodies involved nor any government institutions and agencies are making any attempt to establish the sort of control that these technologies appear to require. This necessity encompasses three separate levels: 1. exhaustive testing beyond the laboratory (that is in the field) before commercial use is sanctioned; 2. open, public, and enforced review of *all* testing results prior to distribution; and 3. clear and complete labeling of all genetically engineered foods ("A Mutable Feast: Assuring Food Safety in the Era of Genetic Engineering," Environmental Defense Fund, Oct. 1991; *The Gene Exchange*, 1992).

These are national as well as global concerns. From the moral point of view, what is at issue is the obligation to test and to disclose. The reason why the moral obligation goes beyond the citizens' right to freedom from harm and hazards is the special threat transgenics pose to the right to know, and even the right to freedom of religious practice. Reflection will disclose that many may be potentially affected by new developments; from the physical standpoint, for instance, mothers concerned with the diet of minor children have a clear right to know the composition of what their children eat. Persons suffering from allergies also seem to have an equal right.

The new aspect introduced by transgenics is that of creating moral problems in connection with our freedom to know, to believe, and to practice: all those whose beliefs or religious orthodoxy include specific nutritional guidelines and prohibitions *also* have an indisputable right to know. These include vegetarians, vegans, Muslims, Jews ("Mutable Feast . . . ," 1991). Also included are Jehovah's Witnesses, who have already established their right to religious practice even in the case of life-threatening circumstances, from both a legal and a moral stand-point (M. Guarini, 1992). Both United States and Canadian laws reflect this belief, permitting the refusal of life-saving treatment on religious grounds. Since they deem the "ingestion" of human blood products through a transfusion an infliction of harm serious enough to warrant putting their life on the line, it seems reasonable that they would also have the right to refuse food that may contain objectionable compo-nents. Hence, their right *to know*, like Muslims and Jews, what food they are ingesting and what foreign substances, if any, such food might contain, should be protected.

Even beyond the scope of physical harm, implicit in so many other agricultural practices, biotechnologies introduce a non-physical threat, a totally new dimension; one that needs to be explored and weighed carefully before decisions are made to implement these products.

Some try to minimize or ignore these risks. Kline says:

> Within a capitalist society it is held that individuals are allowed to pursue whatever ends they choose as long as they do not *physically* and *directly* harm others. (My emphasis; Kline, 1991)

The case of the Jehovah's Witness's right to refuse treatment is a clear counter-example to the above statement: not all inadmissible harm is either "physical" or "direct." The tacit use of chemicals and other pesticides may infringe one's right to know and may pose a threat to one's health, but the additional possible non-physical harm is specific to biotechnology, and must be seriously considered by those who propose to address the moral questions arising from the develop-ments in this new field. To simply ignore these relevant and important aspects in order to subsume these technologies under previous catego-ries is to deliberately obscure the new questions they may raise.

When a well-known source of development support such as the World Bank supports two totally incompatible positions on the envi-ronment in direct conflict with one another, almost at the same time, one's concern about the desperate need for regulation and public information suddenly goes beyond philosophical considerations and argumentation and becomes a plea for activism. In December 1991, Lawrence Summers, vice-president and chief economist of the World Bank, issued some comments on a draft of the World Bank 1992 Global Economics Perspectives (GEP) Report which was to be presented at the United Nations Conference in Rio de Janeiro, June 1992. It started with the following words:

> Just between you and me, shouldn't the World Bank be encouraging more immigration of the dirty industries to the LDC's? I can think of three reasons. . . .

These were, in essence:

1. "measuring the costs of health-impairing pollution depends on the earnings lost due to increased morbidity and mortality," hence, "the economic logic behind dumping a load of toxic waste in the lowest-wage country is impeccable and we should face up to that";
2. "I've always thought that underpopulated countries in Africa are vastly *under*-polluted, compared to Los Angeles or Mexico City . . ."; and
3. "The concern over an agent that causes a . . . change in the odds of prostate cancer is obviously going to be much higher in a country where people (live long enough) to get prostate cancer, than in a country where mortality is 200 per 1,000 under age five."

Harper's magazine produced this leaked memo under the heading "Let Them Eat Toxics," almost at the same time when a sober memo from the Washington Bureau of the World Bank appeared in the national paper of Canada, *The Globe and Mail* (Victoria Day Issue, Toronto, May 19, 1992), under the title, "World Bank Urges Green Consciousness; Environmental Practices Must Be Given Higher Priority, Development Report Says." One wonders whether, in the context of the previous "memo," this official document represents serious and sincere moral concern, or whether it is yet another example of what Shrader-Frechette has termed a "public pacifier," simply a public relations effort to show that one is jumping on the environmental "band wagon." If sentiments such as those expressed in the first memo cited exist in so influential a figure at the Rio meetings, one can only be saddened by the failure and waste that will follow a conference on which we must rest all our hopes. There is evidence that "The biotechnology industry was the prime force responsible (through intensive lobbying) for the United States' failure to sign the Biodiversity Convention at the Earth Summit in Rio" (*The Gene Exchange*, Vol. 3, No. 2, July 1992).

4. Some Concluding Remarks

In the previous two sections biotechnology was discussed, both from the standpoint of its creation and its role (section 2), and from that of the moral difficulties it engenders (section 3). The last section also advanced the argument that biotechnology is different in kind from all other technological aids; thus, it is not only appropriate to assess it from the standpoint of the risks and hazards from which other technologies are judged, but also it is vital to analyze and recognize the differences, in order to consider clearly the novel, additional harms it may pose.

A brief overview of some of the more recent literature on the topic discloses many moral difficulties arising from chemical agricultural aids. Problems of risks and hazards comprising both morbidity and fatality are endemic to the use of pesticides everywhere, hence the protection of "strong" human rights and of the environment as a whole appear to require the adoption of guidelines to restrain uses and practices that might endanger all life. To this end, I have suggested elsewhere the adoption of 1. the Harm Principle, and 2. the Equity Principle (L. Westra, K. Bowen and B. Behe, 1991). Minimally, the same necessity for strong principles to guide both individual and policy choices should govern biotechnology. In addition, as the latter imposes additional risks and hazards, as well as new questions of consent, information, and even freedom of religious belief and practice, further considerations must be introduced.

Humans and all other members of the biota and ecosystems are at risk through biotechnology from the chemical industry beyond those things to which they have been previously exposed. A further consideration is the introduction of conflicts with individual and collective democratic rights, a problem some identified with technology (J. McDermott, 1990). Recent multinational trade agreements (United Nations Economic and Social Council, Draft Code of Conduct of Transnational Corporations, May 1990), do not even touch upon environmental issues. Concerns include payments, industrial relations, employment practices, but nothing on occupational health or wider environmental issues. But when trade is allowed to proceed untrammeled and is regulated only by business and economic considerations, all other aspects (and consequences) of their corporate activities will not be taken seriously (D. Meadows, et al., 1992).

In that case, the corporate bodies will be promoting, what they deem appropriate, rather than permitting public, democratic choices to prevail. To all intents and purposes, democracy even if nominally in place, cannot currently be brought to bear upon policies involving the use of biotechnologies. Decisions affecting the life, health, and freedom of choice and practice of millions, are taken out of their hands and given into those of non-elected bodies, whose choices are governed by their own internal agenda rather than by the public interest. These problems are even more acute in LDCs than they are in Western countries, but as I have tried to show, they are indeed our problems as well.

The consequences of these corporate choices and activities are too grave to permit them to continue to go on unchecked under the heading of free enterprise or the pursuit of profit. As we saw, biotechnology not only represents an attack on the right to life, to health, and to freedom from experimentation, but also on the right to freedom of conscience. Some deem the latter to be even superior to all others, and most Western democracies, for instance, are built upon the absolute respect for individual and group religious freedom. Most traditional moral theories support and defend these basic human rights; environmental doctrines will support respect beyond the species barrier and be especially critical of the introduction of inconclusively tested or untested organisms in the ecosystem. The *somewhat* similar precedent—the introduction of "exotics" into ecosystems—is an example of the unreliability of scientific prediction.

Finally, from the standpoint of the "principle of integrity," the use of biotechnologies is equally blameworthy, as it represents an unacceptable interference with both ecosystemic and organismic integrity at one and the same time.

Bibliography

Agazzi, E. 1992. *Il Bene, Il Male e La Scienza*, Le Dimensioni etiche dell'impresa scientifico-tecnologica, Rusconi, ed. Milano.

FOOD 2000, "Global Policies for Sustainable Agriculture," 1987, Zed Books, Ltd.

The Gene Exchange. 1991. Vol. 2 No. 4, National Wildlife Federation. Ed. M. Mellon, J.D. Director, Biosafety Regulations.

Gewirth, Alan. 1982. "Essays on Justification and Applications," *Human Rights.* Chicago: University of Chicago Press.

Guarini, Marcello. Personal communication, 1992.

Jackson, Wes. 1991. "Our Vision for the Agricultural Sciences Need Not Include Biotechnology," *Journal of Agricultural and Environmental Ethics.* Vol. 4, No. 2. Guelph: University of Guelph.

Kline, D. 1991. "We Have Not Yet Identified the Heart of the Moral Issues in Agricultural Biotechnology," *Journal of Agricultural and Environmental Ethics.* Vol. 4, No. 2. Guelph: University of Guelph.

Laudan, L. 1984. *Science and Values.* Berkeley: University of California Press.

Leopold, Aldo. 1969. *A Sand County Almanac.* Oxford: Oxford University Press.

Longino, Helen. 1990. *Science as Social Knowledge.* Princeton: Princeton University Press.

McDermott, J. 1990. "Technology, The Opiate of the Intellectuals," in *Technology and the Future.* Ed. 5th ed., Albert Teich. New York: St. Martin's Press: 100–125.

McGinn, R. 1991. *Science, Technology and Society.* Englewood Cliffs: Prentice-Hall.

Meadows, D. *et al.* 1992. *Beyond The Limits.* Vermont: Chelsea Green Press.

Mellon, M. 1991. "Agriculture, Biotechnology in Latin America: An Environmental Perspective," for the XVI International Congress of the Latin American Studies Association, VA, April 6.

Naess, A. 1989. *Ecology, Community and Lifestyle*, Tr. and Ed. David Rothenberg. Cambridge: Cambridge University Press.

Pimentel, David, *et al.* 1991. "Environmental and Economic Effects of Reducing Pesticide Use," *BioScience*, Vol. 41, No. 6., June: 402–409.

Pimentel, David, *et al.* 1989. "Benefits and Risks of Genetic Engineering in Agriculture," *BioScience*, Vol. 39, No. 9, October: 606–614.

Regan, T. 1983. *The Case for Animal Rights*. Berkeley: University of California Press.

Repetto, R. 1985. "Paying the Price: Pesticide Subsidies in Developing Countries," Research Report No. 2; Washington: World Resources Institute.

Rodd, R. 1990. *Biology, Ethics, and Animals*. Oxford: Clarendon Press.

Sapontzis, S. 1991. "We Should Not Manipulate the Genome of Domestic Hogs," *Journal of Agricultural and Environmental Ethics*, Vol. 4, No. 2, Guelph: University of Guelph: 177–185.

Sapontzis, S. 1987. *Morals, Reasons and Animals*. Philadelphia: Temple University Press.

Science, Vol. 255, February 1992: 1063.

Science 164, 1969: 666ff. M. Soulé on "Biodiversity."

Scriven, M. "1980. The Exact Role of Value Judgment in Science," *Introductory Readings in the Philosophy of Science*. Eds. E. D. Klemke, R. Hollinger, A. Kline. Boston: Prometheus Books.

Shrader-Frechette, K. 1991. *Risk and Rationality*. Berkeley: University of California Press.

Shrader-Frechette, K. 1983. *Nuclear Power and Public Policy: The Social and Ethical Problems of Fission Technology*. Boston: Kluwer Publishing.

Singer, P. 1990. *Animal Liberation*, 2nd ed. New York: Random House.

Soulé, M. 1991. "Biological Immigrants Under Fire," *Science*, December 6, Vol. 254: 1444–47.

VandeVeer, D. 1987(?). "Interspecific Justice," in *People, Penguins and Plastic Trees*. Wadsworth Publishing.

Westra, L. 1993. *The Principle of Integrity: An Environmental Proposal for Ethics*. Lanham, MD: Rowman & Littlefield Publishers, Inc.

Westra, L. 1991. "Towards 'Integrity' in the Great Lakes: A Feminist Perspective," *APA Newsletter on Feminism and Philosophy*, Special Issue, guest editors, K. Warren and N. Tuana.

Westra, L. 1989. "Respect, Dignity and 'Integrity': An Environmental Proposal for Ethics," *Epistemologia* XII: 91–124.

Westra, L.; K. Bowen; and B. Behe. 1991. "Agricultural Practices, Ecology and Ethics in the Third World," *Journal of Agricultural and Environmental Ethics*. Vol. 4, No. 1: 60–77.

I wish to express my thanks to many who read the first draft of this paper and contributed critiques, suggestions, and information—primarily Kira L. Bowen, David Pimentel and Kristin Shrader-Frechette.

Study Questions

1. What is the main point of Westra's article? Why do you think it might be correct? Why do you think it might be wrong? Explain and defend your answers.
2. What assumptions does Westra make about technology? about ethics? Do you think these assumptions are correct? Explain and defend your answers.
3. If society accepts Westra's main conclusions, what consequences might follow? Would these consequences be desirable? Explain and defend your answers.
4. Why do biotechnology and transgenics have serious implications for less developed countries? Explain.
5. What moral and value questions are raised by biotechnology (used for food production)?
6. How do economic values influence agricultural biotechnology and transgenics? What other values ought to play a role?
7. Does agricultural biotechnology threaten individual rights in any way? Why or why not?

4.10

Further Reading

Bella, David. 1989. *Catastrophic Possibilities of Space-Based Defense in Philosophy of Technology.* Norwell: Kluwer.

Beveridge, M. C. M., Ross, L. G., and Kelly, L. A. 1994. "Aquaculture and Biodiversity," *Ambio* 23, 8 (December): 497–502.

Bonnicksen, Andrea L. 1991. *In Vitro Fertilization: Building Policy from Laboratories to Legislatures.* New York: Columbia University Press.

Brown, Geoffrey. 1990. *The Information Game: Ethical Issues in a Microchip World.* Atlantic Highlands, N.J.: Humanities Press.

Burkhardt, Jeffrey. 1988. "Biotechnology, Ethics, and the Structure of Agriculture," *Agriculture and Human Values* 5 (Summer): 53–60.

Callahan, Daniel. 1986. "How Technology Is Reframing the Abortion Debate," *Hastings Center Report* 16 (Fall): 33–42.

Donchin, Anne. 1989. "The Growing Feminist Debate Over the New Reproductive Technologies," *Hypatia* 4, 3 (Fall): 136–49.

Draper, E. 1981. *Risky Business.* Cambridge: Cambridge University Press.

Dunlap, Riley E., and Kraft, Michael E. 1993. *Public Reactions to Nuclear Waste: Citizens' Views of Repository Siting.* Durham: Duke University Press.

Fielder, John H., and Birsch, Douglas. 1992. *The DC-10 Case: A Study in Applied Ethics, Technology, and Society.* Albany: SUNY Press.

Fimbel, Nancie. 1990. "Defining the Ethical Standards of the High-Technology Industry," *Journal of Business Ethics* (December): 929–48.

Freudenburg, William R. 1994. *Oil in Troubled Waters: Perceptions, Politics, and the Battle Over Offshore Drilling.* Albany: SUNY Press.

Freudenburg, William R., and Rosa, Eugene H. 1984. *Public Reaction to Nuclear Power: Are There Critical Masses?* Boulder, Colo.: Westview Press, for the American Association for the Advancement of Science, Washington, D.C.

453

Gardner, William. 1995. "Can Human Genetic Enhancement be Prohibited?" *Journal of Medical Philosophy* 20, 1 (Fall): 65–84.

Gendel, Steven M., et al. 1990. *Agricultural Bioethics: Implications of Agricultural Biotechnology.* Ames: Iowa State University Press.

Gewirth, Alan. 1982. *Human Rights Essays on Justification and Applications.* Chicago: Chicago University Press.

GLWQA. 1978. Great Lakes Water Quality Agreement.

Gould, Carol C. 1989. *The Information Web: Ethical and Social Implications of Computer Networking.* Boulder, Colo.: Westview Press.

Gray, Susan H. 1989. "Electronic Data Bases and Privacy: Policy for the 1990s," *Science, Technology, and Human Values* 14 (Summer): 242–57.

Hartwig, Michael J. 1995. "Parenting Ethics and Reproductive Technologies," *Journal of Social Philosophy* 26, 1 (Spring): 183–202.

Herrera, Christopher. 1993. "Some Ways That Technology and Terminology Distort the Euthanasia Issue," *Journal of Medical Humanities* 14, 1 (Spring): 23–31.

Heyd, David. 1993. "Artificial Reproductive Technologies: The Israeli Scene," *Bioethics* 7, 2–3 (April): 263–70.

Hollander, Rachelle D. 1990. *Moral Responsibility, Values, and Making Decisions about Biotechnology in Agricultural Bioethics: Implications of Agricultural Biotechnology.* Ames: Iowa State University Press.

Hull, Richard T. (ed.). 1990. *Ethical Issues in the New Reproductive Technologies.* Belmont, Calif.: Wadsworth.

IJC. 1990. Great Lakes–St. Lawrence Research Inventory 1990/91. International Joint Commission.

Jamieson, D. 1992. "Ethics, Public Policy and Global Warming," *Science, Technology, and Human Values* 17, 2 (Spring): 139–53.

Jennett, Bryan. 1994. "Medical Technology, Social and Health Care Issues." In Gillon, Raanan (ed.). *Principles of Health Care Ethics.* New York: Wiley.

Johnson, Branden B. 1994. *Explaining Uncertainty in Health Risk Assessment: Effects on Risk Perception and Trust.* Washington, D.C.: U.S. Environmental Protection Agency.

Juengst, Eric, and Siegel, Ronald. 1988. "Subtracting Injury from Insult: Ethical Issues in the Use of Pharmaceutical Implants," *Hastings Center Report* 18 (December): 41–46.

Kent, Theodore C. 1995. *Mapping the Human Genome: Reality, Morality, and Deity.* Lanham, Md.: University Press of America.

Kunreuther, Howard. 1988. "Nevada's Predicament: Public Perceptions of Risk Form the Proposed Nuclear Waste Repository," *Environment* 30, 8 (October): 16–20, 30–33.

Lamm, Richard D. 1989. "High Technology Health Care," *National Forum* 69 (Fall): 14–17.

Lauritzen, Paul. 1993. *Pursuing Parenthood: Ethical Issues in Assisted Reproduction.* Bloomington: Indiana University Press.

Longino, Helen E. 1992. "Knowledge, Bodies, and Values: Reproductive Technologies and Their Scientific Context," *Inquiry* 35, 3–4 (September–December): 323–40.

MacLean, Douglas, and Brown, Peter G. 1983. *Energy and the Future.* Totowa, N.J.: Rowman & Littlefield.

McGinn, Robert F. 1991. *Science, Technology, and Society.* Englewood Cliffs, N.J.: Prentice Hall.

Mitcham, Carl., and Siekevitz, Philip (eds.). 1989. *Ethical Issues Associated with Scientific and Technological Research for the Military.* New York: New York Academy of Sciences.

Moor, James H. 1985. "What Is Computer Ethics?" *Metaphilosophy* 16 (October): 266–75.

Morden, Michael. 1988. "Cyclosporine as an Ethical Catalyst: Recent Issues in Kidney Transplantation," *Public Affairs Quarterly* 2 (October): 31–45.

Morse, M. Steven. 1989. "Liabilities and Realities Faced in Biomedical Engineering," *National Forum* 69 (Fall): 46–47.

Mott, Peter D. 1990. "The Elderly and High Technology Medicine: A Case for Individualized, Autonomous Allocation," *Theoretical Medicine* 11, 2 (June): 95–102.

National Research Council. Committee to Review Risk Management in the DOE's Environmental Remediation Program. 1994. *Building Consensus Through Risk Assessment and Management of the Department of Energy's Environmental Remediation Program.* Washington, D.C.: National Academy Press.

———. Committee to Study the Impact of Information Technology on the Performance of Service. 1994. *Information Technology in the Service Society: A Twenty-First Century Lever.* Washington, D.C.: National Academy Press.

———. NRENAISSANCE Committee. 1994. *Realizing the Information Future: The Internet and Beyond.* Washington, D.C.: National Academy Press.

Ottensmeyer, Edward J. 1991. "Ethics, Public Policy, and Managing Advanced Technologies: The Case of Electronic Surveillance," *Journal of Business Ethics* (July): 519–26.

Pardeck, John T. 1989. "Microcomputer Technology in Clinical Practice: An Analysis of Ethical Issues," *Philosophy and Social Action* 15 (January–June): 43–50.

Parens, Erik. 1995. "The Goodness of Fragility: On the Prospect of Genetic Technologies Aimed at the Enhancement of Human Capacities," *Kennedy Institute Ethics Journal* 5, 2 (June): 141–53.

Perpich, Joseph G. (ed.). 1986. *Biotechnology in Society: Private Initiatives and Public Oversight*. New York: Pergamon Press.

Pimentel, D. et al. 1991. "Environmental and Economic Effects of Reducing Pesticide Use," *BioScience* 41, 6: 402–9.

Purdy, Laura M. 1987. "The Morality of New Reproductive Technologies," *Journal of Social Philosophy* 18 (Winter): 38–48.

Samuelson, Pamela. 1987. "Innovation and Competition: Conflicts Over Intellectual Property Rights in New Technologies," *Science, Technology, and Human Values* 12 (Winter): 6–21.

Sclove, Richard E. 1989. "From Alchemy to Atomic War: Frederick Soddy's 'Technological Assessment' of Atomic Energy, 1900–1915," *Science, Technology, and Human Values* 14 (Spring): 163–94.

Sherwin, Susan. 1995. "The Ethics of Babymaking," *Hastings Center Report* 25, 2 (March–April): 34–37.

Shrader-Frechette, Kristin. 1994a. "Hazardous Wastes and Toxic Substances." In Reich, W. (ed.), *Encyclopedia of Bioethics*. New York: Prentice-Hall.

———. 1994b. "High-Level Waste, Low-Level Logic," *Bulletin of the Atomic Scientists* 50, 6 (November/December): 40–45.

———. 1994c. "Risk and Ethics." In Lindell, B. (ed.), *Radiation and Society: Comprehending Radiation Risks*. Stockholm: Swedish Risk Academy.

———. 1993. *Burying Uncertainty: Risk and the Case Against Geological Disposal of Nuclear Waste*. Berkeley: University of California Press.

———. 1991a. "Ethical Dilemmas and Radioactive Waste," *Environmental Ethics* 13, 4 (Winter): 327–43.

———. 1991b. "Pesticide Policy and Ethics." In Blatz, C. (ed.), *Ethics and Agriculture*. Moscow: University of Idaho Press.

———. 1980. *Nuclear Power and Public Policy: Social and Ethical Problems with Fission Technology*. Boston: Kluwer.

Shrader-Frechette, K., and Wigley, D. 1995. "Consent, Equity, and Environmental Justice." In Westra, L., and Wenz, P. (eds.), *The Faces of Environmental Racism: Issues in Global Equity*. Lanham, Md.: Rowman and Littlefield.

Silver, Lee M. 1990. "New Reproductive Technologies in the Treatment of Human Infertility and Genetic Disease," *Theoretical Medicine* 11, 2 (June): 103–10.

Slovic, Paul. 1991. "Risk, Perception, Trust, and Nuclear Waste: Lessons from Yucca Mountain," *Environment* 33, 3 (April): 7–11, 28–30.

———. 1987. "Perception of Risk from Automobile Safety Defects," *Accident Analysis and Prevention* 19, 5 (October): 359–73.

Sorenson, John H. 1990. "Ethics and Technology in Medicine: An Introduction," *Theoretical Medicine* 11, 2 (June): 81–86.

Spinello, Richard A. 1995. *Ethical Aspects of Information Technology*. Englewood Cliffs, N.J.: Prentice-Hall.

Taylor, Carol. 1990. "Ethics in Health Care and Medical Technologies," *Theoretical Medicine* 11, 2 (June): 111–24.

Thoma, H. 1986. "Some Aspects of Medical Ethics from the Perspective of Bioengineering," *Theoretical Medicine* 7 (October): 305–17.

Westra, L., Bowen, K., and Behe, B. 1991. "Agricultural Practices, Ecology and Ethics in the Third World," *Journal of Agricultural Ethics* 4, 10: 60–77.

Index

About the Editors

Kristin Shrader-Frechette is Distinguished Research Professor, at the University of South Florida, in Environmental Sciences and Policy and in Philosophy. She has held Professorships at the University of Florida and the University of California. With undergraduate degrees in mathematics and physics and a Ph.D. in philosophy from the University of Notre Dame, Shrader-Frechette has held postdoctoral fellowships in ecology, economics, and hydrogeology. Since 1981, her research has been funded continuously by the U.S. National Science Foundation. She is author of 200 articles that have appeared in periodicals such as *Journal of Philosophy, Ethics, Environmental Ethics, Environmental Values, Philosophy of Science,* and *Science.* Shrader-Frechette also has published in biology journals such as *BioScience, Quarterly Review of Biology, Conservation Biology, Oikos,* and *Trends in Ecology and Evolution.* Her work has been translated into 11 different languages. Shrader-Frechette's 13 authored books include *Method in Ecology: Strategies for Conservation* (1993), *Burying Uncertainty: Risk and the Case Against Geological Disposal of Nuclear Waste* (1993), *Policy for Land: Law and Ethics* (1993), *Risk and Rationality* (1991), *Environmental Ethics* (1991), and *Nuclear Power and Public Policy* (1981, 1983). Editorial board member for 18 journals, Shrader-Frechette also is Associate Editor of *BioScience* and Editor-in-Chief of the Oxford University Press Monograph Series on "Environmental Ethics and Science Policy." She is President-Elect of the International Society for Environmental Ethics, President of the Risk Assessment and Policy Association, and Past President of the Society for Philosophy and Technology.

Laura Westra received the Ph.D. in philosophy from the University of Toronto. Currently she is associate professor of philosophy at the University of Windsor. She is a founding member of the International Society for Environmental Ethics (ISEE) and, at present, is ISEE secretary. Westra is author of two books, *An Environmental Proposal for Eth-*

ics: The Principle of Integrity (Rowman and Littlefield, 1994) and *Freedom in Plotinus* (Mellon, 1990) and coeditor of *Ethical and Scientific Perspectives on Integrity* (Kluwer, 1995), *Roots of Ecology in Ancient Greek Thought* (University Press of North Texas, 1996), and *Faces of Environmental Racism: Confronting Issues of Global Justice* (Rowman and Littlefield, 1995). She also has published numerous journal articles and chapters in books, most on environmental ethics and ancient, Hellenistic, and medieval philosophy.